Reliability Prediction for Microelectronics

Reliability Prediction for Microelectronics

Joseph B. Bernstein
Ariel University,
Israel

Alain A. Bensoussan
Toulouse,
France

Emmanuel Bender
Massachusetts Institute of Technology (MIT),
Cambridge,
USA

Registered Offices
John Wiley & Sons, Inc., 111 River Street, Hoboken, NJ 07030, USA
John Wiley & Sons Ltd, The Atrium, Southern Gate, Chichester, West Sussex, PO19 8SQ, UK

For details of our global editorial offices, customer services, and more information about Wiley products visit us at www.wiley.com.

Wiley also publishes its books in a variety of electronic formats and by print-on-demand. Some content that appears in standard print versions of this book may not be available in other formats.

Library of Congress Cataloging-in-Publication Data applied for:
ISBN: HB: 9781394210930, ePDF: 9781394210947, ePub: 9781394210954

Cover image: Wiley
Cover design by: © Richard Newstead/Getty Images

Set in 9.5/12.5pt STIXTwoText by Straive, Pondicherry, India
Printed and bound by CPI Group (UK) Ltd, Croydon, CR0 4YY

C9781394210930_070224

To our patient and dedicated wives:
Rina Batya
Revital

To my wife Corinne, my son Edwin,
and to the memory of my parents, Myriem and Isaac Bensoussan.

Contents

Author Biography

Joseph B. Bernstein
Professor, Ariel University, Ariel (Israel)

Biography

Professor Joseph B. Bernstein specializes in several areas of nano-scale micro-electronic device reliability and physics of failure research, including packaging, system reliability modeling, gate oxide integrity, radiation effects, Flash NAND and NOR memory, SRAM and DRAM, MEMS, and laser-programmable metal interconnect. He directs the Laboratory for Failure Analysis and Reliability of Electronic Systems, teaches VLSI design courses, and heads the VLSI program at Ariel University. His laboratory is a center of research activity dedicated to serving the needs of manufacturers of highly reliable electronic systems using commercially available off-the-shelf parts. Research areas include thermal, mechanical, and electrical interactions of failure mechanisms of ultra-thin gate dielectrics, nonvolatile memory, advanced metallization, and power devices. He also works extensively with the semiconductor industry on projects relating to failure analysis, defect avoidance, programmable interconnect used in field-programmable analog arrays, and repair in microelectronic circuits and packaging. Professor Bernstein was a Fulbright Senior Researcher/Lecturer at Tel Aviv University in the Department of Electrical Engineering, Physical Electronics. Professor Bernstein is a senior member of IEEE.

Alain A. Bensoussan
Thales Alenia Space France (1988–2019).
  Senior Engineer, Optics and Opto-electronics Parts Expert at Thales Alenia Space France, Toulouse (France) (2010–2019)
  Formerly, Technical Advisor for Microelectronic and Photonic Components Reliability at IRT Saint Exupery, Toulouse, France (2014–2017)

Biography

Dr. Alain Bensoussan is Doctor of Engineering (Dr.-Ing.) and docteur d'Etat from University Paul Sabatier (Toulouse, France) in applied physics. His field of expertise is on microelectronic parts reliability at Thales Alenia Space. He worked at Institut de Recherche (IRT) Saint Exupery (Aeronautic, Space, and Embedded Systems, AESE), Toulouse (France), as a technical adviser for microelectronic and photonic components reliability and was recognized at Thales Alenia Space as an expert on optics and opto-electronics parts. Dr. Alain Bensoussan's interests lie in several areas in microelectronics reliability and physics of failure applied research on GaAs and III-V compounds, monolithic microwave integrated circuits (MMIC), microwave hybrid modules, Si and GaN transistors, IC's and Deep-Sub-Micron technologies, MEMS and MOEMS, active and passive optoelectronic devices and modules. He has represented Thales Alenia Space in space organizations such as EUROSPACE and ESA for more than 15 years.

Emmanuel Bender

Postdoctoral Researcher at Ariel University with a Research Affiliate at the Massachusetts Institute of Technology (MIT), Cambridge, USA

Biography

Dr. Emmanuel Bender received the Ph.D. degree in electrical and electronics engineering from Ariel University, Ariel, Israel, in 2022. He specializes in statistical failure analysis of silicon VLSI technologies, including 16nm FinFETs. His work focuses on failure phenomena in packaged devices, including bias temperature instability, electromigration, hot carrier instability, and the self-heating effect. He applied the Multiple Temperature Operational Life (MTOL) testing method to generate reliability profiles on FPGA-programmed test structures in 45nm, 28nm, and 16nm technologies. He is currently working as a Postdoctoral Researcher with Ariel University and has a research affiliation with the Microsystems Technology Laboratories, MIT, with a primary focus on advanced packaging device failure analysis. Dr. Bender is a member of IEEE

E-mail address(es):
Joseph B. Bernstein: josephbe@ariel.ac.il
Alain A. Bensoussan: gmalain2451@gmail.com
Emmanuel Bender: bendere@mit.edu

Series Foreword

Wiley Series in Quality & Reliability Engineering
Dr. Andre V. Kleyner
Series Editor

TheWiley Series in Quality & Reliability Engineering aims to provide a solid educational foundation for both practitioners and researchers in the Q&R field and to expand the reader's knowledge base to include the latest developments in this field. The series will provide a lasting and positive contribution to the teaching and practice of engineering. The series coverage will contain, but is not exclusive to,

- Statistical methods
- Physics of failure
- Reliability modeling
- Functional safety
- Six-sigma methods
- Lead-free electronics
- Warranty analysis/management
- Risk and safety analysis

Wiley Series in Quality & Reliability Engineering

Reliability Prediction for Microelectronics
By Joseph B Bernstein, Alain A. Bensoussan, Emmanuel Bender
March 2024

Software Reliability Techniques for Real-World Applications
by Roger K. Youree
December 2022

System Reliability Assessment and Optimization: Methods and Applications
by Yan-Fu Li, Enrico Zio
April 2022

Design for Excellence in Electronics Manufacturing
Cheryl Tulkoff, Greg Caswell
April 2021

Design for Maintainability
by Louis J. Gullo (Editor), Jack Dixon (Editor)
March 2021

Reliability Culture: How Leaders can Create Organizations that Create Reliable Products
by Adam P. Bahret
February 2021

Lead-free Soldering Process Development and Reliability
by Jasbir Bath (Editor)
August 2020

Automotive System Safety: Critical Considerations for Engineering and Effective Management
Joseph D. Miller
February 2020

Prognostics and Health Management: A Practical Approach to Improving System Reliability Using Condition-Based Data
by Douglas Goodman, James P. Hofmeister, Ferenc Szidarovszky
April 2019

Improving Product Reliability and Software Quality: Strategies, Tools, Process and Implementation, 2nd Edition
Mark A. Levin, Ted T. Kalal, Jonathan Rodin
April 2019

Practical Applications of Bayesian Reliability
Yan Liu, Athula I. Abeyratne
April 2019

Dynamic System Reliability: Modeling and Analysis of Dynamic and Dependent Behaviors
Liudong Xing, Gregory Levitin, ChaonanWang
March 2019

Reliability Engineering and Services
Tongdan Jin
March 2019

Design for Safety
by Louis J. Gullo, Jack Dixon
February 2018

Thermodynamic Degradation Science: Physics of Failure, Accelerated Testing, Fatigue and Reliability
by Alec Feinberg
October 2016

Next Generation HALT and HASS: Robust Design of Electronics and Systems
by Kirk A. Gray, John J. Paschkewitz
May 2016

Reliability and Risk Models: Setting Reliability Requirements, 2nd Edition
by Michael Todinov
November 2015

Applied Reliability Engineering and Risk Analysis: Probabilistic Models and Statistical Inference
by Ilia B. Frenkel (Editor), Alex Karagrigoriou (Editor), Anatoly Lisnianski (Editor), Andre V. Kleyner (Editor)
October 2013

Design for Reliability
by Dev G. Raheja (Editor), Louis J. Gullo (Editor)
July 2012

Effective FMEAs: Achieving Safe, Reliable, and Economical Products and Processes Using Failure Modes and Effects Analysis
by Carl Carlson
April 2012

Failure Analysis: A Practical Guide for Manufacturers of Electronic Components and Systems
by Marius Bazu, Titu Bajenescu
April 2011

Reliability Technology: Principles and Practice of Failure Prevention in Electronic Systems
by Norman Pascoe
April 2011

Improving Product Reliability: Strategies and Implementation
by Mark A. Levin, Ted T. Kalal
March 2003

Test Engineering: A Concise Guide to Cost-Effective Design, Development and Manufacture
by Patrick O'Connor
April 2001

Integrated Circuit Failure Analysis: A Guide to Preparation Techniques
by Friedrich Beck
January 1998

Measurement and Calibration Requirements for Quality Assurance to ISO 9000
by Alan S. Morris
October 1997

Electronic Component Reliability: Fundamentals, Modelling, Evaluation, and Assurance
by Finn Jensen
November 1995

Preface

This book provides statistical analysis and physics of failure methods used in engineering and areas of Applied Physics of "Healthy" (PoH). The engineering and statistical analyses deal with the concept of remaining useful life (RUL) of electronics in the new deep sub-micron (DSM) and nano-scale technologies era and integrated electronics in operation. Many concepts developed in this book are of high interest to the benefit of various products and systems managers, manufacturers, as well as users in many commercial industries: aerospace, automotive, telecommunication, civil, energy (nuclear, wind power farms, solar energy, or even oil and gas).

A broad audience of practitioners, engineers, applied scientists, technical managers, experimental physicists, test equipment laboratory managers, college professors and students, and instructors of various continuing education programs were in mind to construct the overall structure of the book.

Engineering products must offer a worthwhile lifetime and operate safely during their service period under predefined conditions and environmental hazards. To achieve this goal, engineers must understand the ways in which the useful life-in-service of the product can be evaluated and incorporate this understanding into the design (by hardware or software). The tasks involved with reliability prediction are to analyze the design, test, manufacture, operate, and maintain the product, system, or structure at all stages of product development, manufacturing, and use, from the moment of design to the cessation of maintenance and repair to the end. Reliability standards are based on experience and consider random failure rate associated with an activation energy for a single failure mechanism. Today, we are using devices (smart phones, PC's) with IC's size nodes as low as a few atomic layers (<5 nm range). Reliability analysis concepts must consider multiple stress and multiple failure mechanisms activated simultaneously.

Scope

This physics of failure-based book is meant to teach reliability prediction for electronic system to offer a more accurate reliability estimation for electronic system by highlighting the problematic areas of conventional approaches and giving alternative suggestions to cover the shortcomings that lead to inaccurate estimations. It is the opinion of the authors that the major limitation in reliability prediction is the reliance on incorrect prediction mathematics that were improperly introduced early in the days of reliability physics methodology, started by Mil Handbook 217 and the like. These errors continue to propagate themselves to this day. Hence, the motivation for this book derives from the need we see that the practices of relying on incorrect statistical analysis and false combinations of physical phenomena lead to completely wrong approaches to reliability assessment and qualification.

We describe herein a competing failure mechanism approach based on an acceleration test matrix. We present a cell-based reliability characterization and a statistical comparison of constant failure rate approximations for various physical rate-based mechanisms. Our alternative suggestion should lead to correct reliability predictions and is justified mathematically rather than to assume a single "base" failure rate multiplied by vaguely justified "π" factors.

The problem at hand is not that conventional handbooks cannot give instructions for electronic system failure rate calculation or that they are not based on the physics of failure foundation; it is that they still apply fundamentally immature and incorrect assumptions. One example of such a basic false assumption is that there exists a "base" failure rate, λ_b, for some average condition and that small modifications, called "π" factors, can be multiplied to get a modified "true" failure rate. We will show here that this π-factor modification has no mathematical justification and that a proper sum-of-failure-rate approach would be much more consistent with today's understanding of reliability physics.

Our assumptions and suggestions need to be articulated more to build an electronic system reliability paradigm.

Introduction

"Zero failure" qualification reported by industries today is one of the criteria that blocks progress in the domain of reliability. One of the engineers' responsibilities is to do "conjecture" to find new ways to test not only the products but also the physical theories behind them. When many devices are tested and zero failures occur during the qualification test, there is no way to distinguish exactly which failure mechanism did NOT fail, since no failure was found. In that case, which is the only acceptable case by most industry standards, it is impossible to tell what acceleration factor can be assigned to that lack of failures, especially when we know from the beginning that multiple mechanisms compete for dominance at any operating condition.

The electronics industry, for example, takes this to the extreme in the JEDEC standards (formerly Joint Electron Device Engineering Council), where they propose a χ^2 statistic and allow for improperly adding degrees of freedom and a completely unjustified acceleration factor that can never be falsified, since there are no failures to be found. We will discuss this in more detail in what follows; however, we hope to show that there is no statistical validity to adding imaginary data that never occurred and, furthermore, to attributing acceleration factors that were never measured.

"Competing failure mechanisms" is one of the suggestions that could replace the "zero failure rate" paradigm. Once we accept it (now it is accepted even by industries and standard handbooks), we could apply an accelerating test matrix to provide accurate acceleration factors. Then not only the accurate lifetime of the system could be predicted but also flaws and weaknesses could be revealed.

Our primary purpose in this book is to challenge the "π" model of multiplying acceleration factors when these "adjustment" factors are reflected by multiple mechanisms that need to be separated. Secondly, we will reject the idea that a zero-failure test "results" can give you a predictable time to fail. Alternatively, we will illustrate a multiple-mechanism failure rate matrix approach that will accurately consider multiple failure mechanisms consistently and simultaneously, allowing for practical and accurate failure rate prediction and calculations.

1

Conventional Electronic System Reliability Prediction

The history of reliability engineering goes back to 1950s when electronics played a major role for the first time. At that time, there was great concern within the US military establishment for the reliability and maintainability of the current electronic systems. Many meetings and ad hoc groups were created to cope with the problems. Developing better parts, finding quantitative reliability for parts, and collecting field data on actual part failures to determine the root cause of problems were three major fields of research in those days.

When the complexity of electronic equipment began to increase significantly, and new demands were placed on system reliability, a permanent committee (AGREE) was established to identify the actions that could be taken to provide more reliable electronic equipment (1952). The reliability era began when the first Radio Corporation of America (RCA) report on reliability of electronic parts was released in 1956, the first time when reliability was defined as a probability. On the other hand, one of the first reliability handbooks titled Reliability Factors for Ground Electronic Equipment was published in 1956 by McGraw-Hill under the sponsorship of the Rome Air Development Center (RADC); while the McGraw-Hill handbook gave information on design considerations, human engineering, interference reduction, and a section on reliability mathematics, failure prediction was only mentioned as a topic under development.

Reliability prediction and assessment are traced to November 1956 with publication of the RCA release TR-1100, titled "Reliability Stress Analysis for Electronic Equipment," which presented models for computing rates of component failures. It was the first time that the concepts of activation energy and the Arrhenius relationship were used in modeling component failure rates. However, in 1960s, the first version of a military handbook for the reliability prediction of electronic equipment (MIL-HDBK-217) was published by the US Navy [1]. It covered a broad range of part types, and since then, it has been widely used for military and commercial electronics systems.

Reliability Prediction for Microelectronics, First Edition. Joseph B. Bernstein, Alain A. Bensoussan, and Emmanuel Bender.
© 2024 John Wiley & Sons Ltd. Published 2024 by John Wiley & Sons Ltd.

In July 1973, RCA proposed a new prediction model for microcircuits, based on previous work by the Boeing Aircraft Company. In the early 1970s, RADC further updated the military handbook and revision B was published in 1974. The advent of more complex microelectronic devices pushed the application of MIL-HDBK-2 17B beyond reason. This decade is known for development of new innovative models for reliability predictions. Then, RCA developed the physics-of-failure model, which was initially rejected because of the lack of availability of essential data.

To keep pace with the accelerating and ever-changing technology base, MIL-HDBK-217C was updated to MIL-HDBK-217D on January 15, 1982 and to MIL-HDBK-217E on October 27, 1986. In December 1991, MIL-HDBK-217F became a prescribed US military reliability prediction document. Two teams were responsible for providing guidelines for the last update. Both teams suggested:

1) that the constant failure rate (CFR) model could not be used;
2) that some of the individual wear-out failure mechanisms (like electromigration and time-dependent dielectric breakdown) could be modeled with a lognormal distribution;
3) that the Arrhenius-type formulation of the failure rate in terms of temperature should not be included in the package failure model; and
4) that stresses such as temperature change and humidity should be considered.

Both groups noticed that temperature cycling is more detrimental to component reliability than the steady state temperature at which the device is operating, so long as the temperature is below a critical value. This conclusion has been further supported by a National Institute of Standards and Technology (NIST), and an Army Fort Monmouth study which stated that the influence of steady-state temperature on microelectronic reliability under typical operating changes is inappropriately modeled by an Arrhenius relationship [2–4]. However, considering the ability to separate failure mechanisms by separate Arrhenius activation energies, it may be possible to return to the physics of failure (PoF) assumption that each mechanism will have a unique activation energy.

1.1 Electronic Reliability Prediction Methods

There are several different approaches to the reliability prediction of electronic systems and equipment. Each approach has unique advantages and disadvantages; several papers have been published on the comparison of reliability assessment approaches. However, there are two distinguishable approaches to reliability prediction, traditional/empirical, and PoF approach.

Traditional, empirical models are those that have been developed from historical reliability databases either from fielded applications or from laboratory tests [5].

Handbook prediction methods are appropriate only for predicting the reliability of electronic and electrical components and systems that exhibit CFRs. All handbook prediction methods contain one or more of the following types of prediction:

- Tables of operating and/or non-operating CFR values arranged by part type,
- Multiplicative factors for different environmental parameters to calculate the operating or non-operating CFR, and
- Multiplicative factors that are applied to a base operating CFR to obtain non-operating CFR [6].

MIL-HDBK-217 reliability prediction methodology which was developed under the activity of the RADC (now Rome Laboratory) and its last version released in February 1995 intended to "establish and maintain consistent and uniform methods for estimating the inherent reliability (i.e. the reliability of a mature design) of military electronic equipment and systems. The methodology provided a common basis for reliability predictions during acquisition programs for military electronic systems and equipment. It also established a common basis for comparing and evaluating reliability predictions or related competitive designs. The handbook was intended to be used as a tool to increase the reliability of the equipment being designed."

In 2001, the office of the US Secretary of Defense stated that ".... the Defense Standards Improvement Council (DSIC) decided several years ago to let MIL-HDBK-217 'die the death.' This is still the current OSD position, i.e. we will not support any updates/revisions to MIL-HDBK-217" [6].

Two basic methods for performing the prediction based on the data observation include the parts count and the parts stress analysis. The parts count reliability prediction method is used for the early design phases when not enough data is available, but the numbers of component parts are known. The information for parts count method includes generic part types (complexity for microelectronics), part quantity, part quality levels (when known or can be assumed), and environmental factors. Since equipment consists of the parts operating in more than one environment, the "parts count" equation is applied to each portion of the equipment in a distinct environment. The overall equipment failure rate is obtained by summing the failure rate for each component over its expected operating condition.

A part stress model is based on the effect of mechanical, electrical and environmental stress and duty cycles such as temperature, humidity, and vibration on the part failure rate. The part failure rate varies with applied stress and the strength–stress interaction determines the part failure rate. This method is used when most of the design is complete, and the detailed part stress is available. It is applicable during later design phases as well. Since more information is available at this stage, the result is more accurate than the parts count method.

The environmental factor gives the influence of environmental stress on the device. Different prediction methods have their own list of environmental factors suitable for their device conditions. For instance, the environmental factor of MIL-HDBK-217F covers almost all the environmental stresses suitable for military electronic devices except for ionizing radiation. The learning factor shows the maturity of the device; it suggests that the first productions are less reliable than the next generations [7, 8]. The parts stress model is applied at component level to obtain part failure rate (λ_p) estimation with stress analysis. A typical part failure rate can be estimated as:

$$\lambda_p = \lambda_b \cdot \pi_Q \cdot \pi_E \cdot \pi_A \cdot \pi_T \cdot \pi_V \tag{1.1}$$

where λ_b is the base failure rate obtained from statistical analysis of empirical data, the adjustment factors include: π_T (temperature factor), π_A (application factor), π_V (voltage stress factor), π_Q (quality factor), and π_E (environmental factor). The equipment failure rate (λ_{EQUIP}) can be further predicted through parts count method:

$$\lambda_{EQUIP} = \sum_{i=1}^{n} N_i \left(\lambda_g \cdot \pi_Q \right)_i \tag{1.2}$$

where λ_g is the generic failure rate for the ith generic part, π_Q is the quality factor of the ith generic part, N_i is the quantity of ith generic part and n is the number of different generic part categories in the equipment. To accommodate the advancement of technology, a reliability growth model was introduced in the handbook approach to reflect the state-of-the-art technology.

$$\lambda_p \propto exp[G_r(t_2 - t_1)] \tag{1.3}$$

where G_r is the growth rate, t_1 is the year of manufacture for which a failure rate is estimated, t_2 is the year of manufacture of parts on which the data were collected. It takes time to collect field data and obtain the growth rate G_r especially when the growth is fast. Furthermore, the validity of applying a reliability growth model without taking technology generation into consideration is not confirmed. It is said that the predictions based on the handbook approach usually lead to conservative failure rate estimation. However, the data is very vague as to the actual validity of the calculated failure rate and no correlation with reality has ever been shown.

Telcordia SR-332, BT-HRD-5, NTT, CNET, RDF 93 and 2000, SAE, Siemens SN29500, prediction of reliability, integrity and survivability of microsystems (PRISM) [9] and FIDES are all the instances of traditional prediction models which provide their own sources of data environment from ground military, civil equipment, automotive, Siemens products, Telecom, commercial-military and aeronautical and military, respectively. Most of these models have gained popularity over time because of their ease of use and uniqueness [10].

The stated purpose of Telcordia SR-332 is "to document the recommended methods for predicting device and unit hardware reliability and for predicting serial system hardware reliability." The methodology is based on empirical statistical modeling of commercial telecommunication systems whose physical design, manufacture, installation, and reliability assurance practices meet the appropriate Telcordia (or equivalent) generic and system-specific requirements. In general, Telcordia SR-332 adapts the equations in MIL-HDBK-217 to represent what telecommunications equipment experience in the field. Results are provided as a CFR, and the handbook provides the upper 90% confidence-level point estimate for the CFR [6].

The basis of the Telcordia math models for devices is the Black Box Technique. This parts count method defines a black box steady-state failure rate for different device types and parts count steady-state failure rate for units; the system-level failure rate is simply the sum of all failure rates of the units contained in a system.

The main concepts in MIL-HDBK-217 and Telcordia SR-332 are similar, but Telcordia SR-332 also can incorporate burn-in, field, and laboratory test data, using a Bayesian analysis [7]. All these methods, in the end, continue to rely on the "π" model for adjusting a base failure rate, making mathematical justification quite illusive.

PRISM was developed in the 1990s by the Reliability Analysis Center (RAC) under contract with the US Air Force. The latest version of the method, which is available as software, was released in July 2001. RAC Rates is the name of PRISM mathematical model for component failure rates; the component models are based on filed data driven from several sources. PRISM applies Bayesian methods to the empirical data to get the system-level predictions. This methodology considers the failures of components as well as those related to the system. However, the component models are the heart of the analysis. It provides different models for capacitors, diodes, integrated circuits, resistors, thyristors, transistors, and software. The total component failure rate is composed of:

1) operating conditions,
2) non-operating conditions,
3) temperature cycling,
4) solder joint,
5) electrical overstress (EOS).

Each mechanism is treated independently.

For components not having RAC Rates models, PRISM provides Non-electronic Parts Reliability and Electronic Parts Reliability Data books. A multitude of part types can be found in these data books with failure rates for various environments.

Unlike the other handbook based on the CFR models, the RAC Rate models do not have a separate factor for part quality level. Quality level is implicitly accounted for by a method known as process grading. Process grades address factors such as design, manufacturing, part procurement, and system management, which are intended to capture the extent to which measures have been taken to minimize the occurrence of system failures [6-8, 11].

RDF 2000, released in July 2000, is a French Telecom standard that was developed by the Union Technique de l'Electricite (UTE). It is the last version of CNET reliability prediction methodology developed by the Centre National d'Etudes des Telecommunications (CNET). "RDF 2000 provides a unique approach to failure rate predictions in that it does not provide a parts count prediction. Rather component failure is defined in terms of an empirical expression containing a base failure rate multiplied by factors influenced by mission profiles." These mission profiles contain information about operational cycling and thermal variations during various working phases [7].

FIDES prediction method attempts to predict the CFR experienced in the useful life portion of the classic bathtub curve. The approach models intrinsic failures such as item technology and distribution quality. It also considers extrinsic failures resulting from equipment specification, design, production, and integration, as well as selection of the procurement route. The methodology considers failures resulting from development and manufacturing, and the overstresses linked to the application such as electrical, mechanical, and thermal.

Some proponents claim that FIDES predictions are "close" to the observed failure rate; its predictions are somewhere in between PRISM predictions that are more optimistic and those of MIL-HDBK-217 which are more conservative [11, 12]. RAC PRISM and FIDES are two methodologies which take place in two stages; the key element is applying the Bayesian statistical techniques to the initial reliability prediction.

Due to the wide range of available traditional reliability predictions, several articles have been published to study and compare the prediction models and methodologies. As an instance of these efforts, IEEE Std 1413-1998 was developed to identify the key required elements for an understandable and credible reliability prediction and to provide its users with sufficient information to evaluate prediction methodologies and to effectively use their results.

A comparison of some of the more popular electronic reliability prediction tools is shown in Table 1.1 [10, 13–16].

Table 1.1 Electronic reliability prediction tool comparator.

Attribute		Electronic prediction tool				
		MIL-HDBK-217	IEC-TR-62380	TELCORDIA	PRISM	FIDES
Methodology	Version/date/source	F, Notice 2, February 1995	Edition 1 August 2004	SR-332, Issue 1, May 2001	1.5 July, 2003	Issue A October, 2021
	Distribution method	Handbook	Handbook	Handbook	Software	Handbook
	Updates anticipated	No	Yes	Yes	Yes	Yes
	Metric used	Failures per 10^6 operating hours	Failures per 10^6 operating hours	Failures per 10^9 operating hours	Failures per 10^6 calendar hours	Failures per 10^9 calendar hours
	Software available	Yes	Yes	Yes	Yes	Yes
	Environmental predefined choices	14	12	Limited to 5	37	7
	Additional environmental modifiers	None	None	None	Operating temperature, dormant temperature, amplitude and frequency of thermal cycles, relative humidity, vibrational level	Operating temperature, amplitude and frequency of thermal cycles, relative humidity, vibrational level, ambient pollution level, overstress exposure
	Part model type	Multiplicative	Multiplicative	Multiplicative	Additive	Additive
	Operating profile	No	Yes	No	Yes	Yes

(Continued)

Table 1.1 (Continued)

Attribute		Electronic prediction tool				
Methodology	MIL-HDBK-217	IEC-TR-62380	TELCORDIA	PRISM	FIDES
Thermal cycling	No	Yes	No	Yes	Yes
Thermal rise in part	Yes	Yes	No	Yes	Yes
Solder joints failures	No	Yes	No	Yes	Yes
Induced failures	No	No	No	Yes	Yes
Failure rate data base for other parts	Limited	Limited	No	Yes	Yes
Infant mortality	No	No	Yes	Yes	No
Dormant failure rate	No	No	No	Yes	Yes
Test data integration	No	No	Yes	Yes	No
Bayesian analysis	No	No	No	Yes	No

Source: Reproduced with permission from [16]/Quanterion Solutions Incorporated.

The most controversial aspects of traditional approaches could be classified as the concept of CFR, the use of Arrhenius relation, the difficulty in maintaining support data, problems of collecting good-quality field data, and the diversity of failure rates with the source data are other limitations of traditional approaches [4].

Despite the disadvantages and limitations of traditional-/empirical-based handbooks, they are still used by engineers; strong factors such as good performance centered around field reliability ease of use as well as and providing approximate field failure rates make them still popular. Crane survey shows that almost 80% of the respondents use Mil handbook, while PRISM and Telcordia have second and third places in the chart [17].

The first publications on reliability predictions for electronic equipment were all based on curve fitting a mathematical model to historical field failure data to determine the CFR of parts. None of them could include a root-cause analysis of the traditional/empirical approach. The PoF approach received a big boost when the US Army Material Command authorized a program to institute a transition from reliance exclusively on MIL-HDBL-217. The command's Army Material System Analysis Activity (Amsaa), Communications-Electronic Command (Cecom), the Laboratory Command (Lab-com) and the University of Maryland's Computer-Aided Life-Cycle Engineering (CALCE) group collaborate on developing a physics-of-failure handbook for reliability assurance. The methodology behind the handbook was assessing system reliability based on environmental and operating stresses, the material used, and the packaging selected. Two simulation tools called Computer-Aided Design of Microelectronic Packages (CADMP-2) and CALCE were developed to help with the assessment; CADMP-2 assesses the reliability of electronics at the package level while CALCE assesses the reliability of electronics at the printed wiring board level. Together, these two models provide a framework to support a physics-of-failure approach to reliability in electronic systems design.

PoF uses the knowledge of root-cause failure to design and do the reliability assessment, testing, screening, and stress margins to prevent product failures. The main task of PoF approach is to identify potential failure mechanisms, failure sites, and failure modes, the appropriate failure models, and their input parameters, determine the variability for each design parameter, and compute the effective reliability function. In summary, the objective of any physics-of-failure analysis is to determine or predict when a specific end-of-life failure mechanism will occur for an individual component in a specific application. A physics-of-failure prediction looks at each individual failure mechanism such as electromigration, solder joint cracking, die bond adhesion, etc., to estimate the probability of component wear-out within the useful life of the product. This analysis requires detailed knowledge of all material characteristics, geometries, and environmental conditions. The subject is constantly challenging, considering that new failure mechanisms are discovered and even the old ones are not completely explained. One of the most important advantages of the physics-of-failure approach is the accurate predictions of wear-out mechanisms and their cumulative effect on the time to fail. Moreover, since the acceleration test is one of the main aspects of finding the model parameters, it could also provide the necessary test criteria for the product. To sum up, modeling potential failure mechanisms, predicting the end-of-life, and using generic failure models effective for new materials and structures are the achievements of this approach [4].

The disadvantages of PoF approaches are related to their cost, complexity of combining the knowledge of materials, process, and failure mechanisms together, the difficulty of estimating the field reliability, and their inapplicability to the devices already in production as the result of its incapability of assessing the whole system [4, 18, 19]. The method needs access to the product materials, process, and data. Partly inspired and completed from Cushing et al. [15], Table 1.2 compares MIL-HDBK-217, FIDES approach and PoF.

Nowadays circuit designers have reliability simulations as an integral part of the design tools, like Cadence Ultrasim and Mentor Graphics Eldo. These simulators model the most significant physical failure mechanisms and help the designers meet the lifetime performance requirement. However, there are disadvantages which hinder designers to adopt these tools. First, the tools are not fully integrated into the design software because the full integration requires technical support from both the tool developers and the foundry. Second, they can't handle the large-scale design efficiently. The increasing complexity makes it impossible to exercise full-scale simulation considering the resources that simulation will consume. Chip-level reliability prediction only focuses on the chip's end-of-life, while the known wear-out mechanisms are dominant; however, these prediction tools do not predict the random, post-burn-in failure rate that would be seen in the field [5, 21].

By consideration of all the available reliability prediction tools and methods, they are all based on this original Mil-Handbook approach of assuming a base failure rate, λ_b, and multiplying that by what should be small correction factors, π. We will show in the following chapters that this fundamentally contradicts the presumption of a failure rate, λ, since failure rates add and do not multiply. This brings us to the established and known reliability assumption that linear rate mechanisms add and must be separated into a properly weighted sum-of-rates model.

Table 1.2 Comparison between MIL-HDBK-217 and physics of failure.

Issue	MIL-HDBK-217	FIDES	Physics of failure
Model development	The effectiveness of models in providing accurate design or manufacturing guidance is questionable since they have been developed based on assumed constant failure-rate data rather than root cause or time-to-failure data. As highlighted by Morris [20], the data available is often fragmented, requiring interpolation or extrapolation for the development of new models. Consequently, it is not appropriate to associate statistical confidence intervals with the overall model results.	FIDES and MIL-HDBK are both methodologies used for reliability prediction and analysis, but they have some key differences. FIDES offers advantages over MIL-HDBK in terms of utilizing real-world data, improving root cause analysis, and acknowledging the challenges of data completeness and quality. While statistical confidence intervals may not be directly associated with model results in FIDES, the emphasis on using real-world data helps mitigate some of the limitations observed in MIL-HDBK's model development approach. In addition, FIDES model is presented to be based on field failure data from few aerospace and military companies but is representing a small part of the worldwide original equipment manufacturers. The model assumes that each component has a constant failure rate, which is not relevant, in terms of part types failure.	Physics of Failure (PoF) models based on science and engineering first principles offer an alternative physical approach to reliability prediction and analysis compared to MIL-HDBK and FIDES models. PoF models rely on a deep understanding of the physical mechanisms and failure processes involved in a system, allowing for more accurate predictions and a clearer understanding of the underlying causes of failures. Unlike MIL-HDBK and FIDES, PoF models consider the fundamental physics and engineering principles governing the behavior of materials, components, and systems. This approach enables the modeling of complex interactions between different stress factors, such as temperature, mechanical stresses, humidity, and vibration, leading to a more comprehensive understanding of failure mechanisms. PoF models can support both deterministic and probabilistic applications. Deterministic PoF models focus on understanding and predicting specific failure modes, providing valuable insights into the system's weak

(Continued)

Table 1.2 (Continued)

Issue	MIL-HDBK-217	FIDES	Physics of failure
			points and guiding targeted mitigation strategies. On the other hand, probabilistic PoF models incorporate statistical variations and uncertainties in the inputs to provide a probabilistic assessment of reliability, considering the inherent variability in material properties, manufacturing processes, and operational conditions.
			By leveraging scientific principles and detailed knowledge of the system's behavior, PoF models can overcome some of the limitations of MIL-HDBK and FIDES approaches. They provide a solid foundation for reliability analysis, enabling engineers to make informed decisions based on the understanding of physical failure mechanisms rather than relying solely on empirical data or statistical models.
Device design modeling	The assumption of perfect designs in MIL-HDBK lacks substantiation as there is a notable absence of root-cause analysis for field failures. Additionally, MIL-HDBK-217 models fail to address wear-out issues adequately. Additionally, they do not adequately model the effects of derating rules, particularly concerning AC or RF	FIDES offers advantages over MIL-HDBK in terms of its emphasis on failure analysis, addressing wear-out issues, incorporating derating rules, and promoting adherence to industry standards. By considering these aspects, FIDES provides a more comprehensive and tailored approach to device design modeling, enhancing	Device Design Modeling involves the development of models that explicitly consider the impact of design, manufacturing, and operation on reliability by focusing on root-cause failure mechanisms. Among these models, thermal models play a crucial role and should be developed with utmost accuracy. It is essential to accurately determine and model

dynamic stresses and on-off power cycling conditions, where the consideration of such rules is insufficient.

reliability assessments and supporting improved design practices. Even though FIDES offers advantages in device design modeling, there are a few notable drawbacks to consider:

1) Limited Adoption: FIDES may have a narrower adoption compared to other methodologies, which can impact the availability of resources and community support.

2) Data Availability and Quality: The reliance on failure data in FIDES can be challenging if comprehensive and accurate data is not readily available. Incomplete or inaccurate data can affect reliability predictions.

3) Complexity and Learning Curve: Implementing FIDES requires familiarity with the methodology and statistical analysis, leading to a potential learning curve and additional training requirements.

4) Customization for Specific Applications: FIDES may not fully capture the specific requirements and failure mechanisms of every application or industry, requiring additional customization and adaptation efforts.

temperature rises in hot spots, taking into account the true static and dynamic stress conditions experienced by the device.

Additionally, it is important to model the biasing conditions, including voltage, current, and power excursions, as close as possible to the active area of the device. These biasing conditions can significantly impact the local variation in temperature, and accurately modeling them is crucial for a comprehensive understanding of the device's thermal behavior.

By incorporating both the accurate modeling of temperature variations and biasing conditions, designers can gain a deeper insight into the thermal performance of the device. This enables them to identify potential reliability issues, optimize the design for improved performance, and make informed decisions regarding the selection of materials, architectures, and operational parameters. Such a comprehensive approach to device design modeling enhances the overall reliability and longevity of electronic devices.

Device defect modeling

Device defect modeling faces limitations that hinder its effectiveness in various aspects. Firstly, these models do not explicitly consider the impact of manufacturing variation on reliability.

FIDES approach in Device Defect Modeling has shown promise in assessing effectiveness. It considers manufacturing variation, provides guidelines for defect definition and

Failure mechanism models can be used to: (i) relate manufacturing variation to reliability, (ii) determine what constitutes a defect and how to screen/inspect. This rely on electrical failure modes which can be

(Continued)

Table 1.2 (Continued)

Issue	MIL-HDBK-217	FIDES	Physics of failure
	Secondly, they lack clear guidelines for defining defects and conducting effective screening and inspection processes. Furthermore, the electrical and thermal aspects of design are not accurately modeled, particularly for static and dynamic conditions, which play a crucial role in uncovering infant failure, random defects and wear-out failures.	screening/inspection, and accurately models electrical and thermal aspects. However, areas of improvement include ensuring data completeness and quality, continuous updates and adaptation to evolving technologies, and industry-specific considerations. By addressing these aspects, FIDES can further enhance its effectiveness in Device Defect Modeling, leading to improved reliability and reduced defects in devices. The FIDES methodology does not account for the variability of technical factors such as supplier selection, product performance related to process variability. It only provides a point estimate prediction that is subjectively adjusted for various factors. The process factor of the FIDES methodology is based on audit results and is calculated using an empiric formula subject to change with respect to the maturity of the product. FIDES divides the product development life-cycle into seven phases and assigns a weight to each, termed contribution phase. The user of the methodology can change the distribution of weights. The auditor can also determine whether some questions are irrelevant and omit them.	modeled in term of variation with time as the signature of a seed failure mechanism under activation. PoF is then a concrete tool to identify and survey hidden device defects. Such information can lead to precisely defined screening efficiency method.

	MIL-HDBK-217	FIDES	
			Potential areas for improvement in FIDES Device Defect Modeling include ensuring data completeness and quality, continuous updates and adaptation to evolving technologies, and addressing industry-specific considerations.
Device screening	MIL-HDBK-217 promotes screening without recognition of potential failure mechanisms highlights potential limitations in its approach, which may lead to inadequate screening practices and compromised reliability assessments.	By considering a wide range of failure modes and mechanisms, FIDES aims to enhance screening methodologies and improve overall reliability predictions. The approach strives to maintain a comprehensive understanding of failure mechanisms, enabling accurate identification and mitigation of potential risks. FIDES also emphasizes the need to keep up with evolving technologies and industry standards to overcome limitations and maintain its effectiveness in promoting robust screening practices for enhanced reliability.	In order to effectively perform screening for device reliability, it is crucial to consider multiple factors and adopt a comprehensive approach. This involves utilizing failure mechanism models that relate manufacturing variation to reliability and provide insights into defect identification and screening.
			It is essential to accurately model the electrical and thermal stress conditions driving specific failure modes. This entails a thorough understanding of the device's electrical behavior and the variation in temperature. For instance, in the case of hot carrier effects, knowledge of their dependency on low-temperature stress conditions can help define appropriate voltage and current conditions for accelerated aging. Similarly, for scenarios involving RF swing in breakdown conditions or impact ionization, understanding the electrical stress conditions is crucial for accurate simulation and screening.
Device screening: lack of failure mechanism understanding	MIL-HDBK-217 primarily relies on empirical data and standardized formulas to estimate reliability. However, it does not emphasize the need for a thorough understanding of the underlying failure mechanisms. This omission limits the ability to accurately predict failure modes and develop targeted screening strategies.	FIDES recognizes the importance of understanding failure mechanisms and incorporates this knowledge into its approach. By considering failure mechanisms, FIDES aims to enhance the screening process and improve the overall reliability assessment of devices. This emphasis on failure mechanism understanding allows for a more thorough analysis of potential	Screening should also consider the impact of temperature fluctuations experienced by the device in real-world operating conditions. Incorporating the actual temperature

(Continued)

Table 1.2 (Continued)

Issue	MIL-HDBK-217	FIDES	Physics of failure
		failure modes and the implementation of appropriate screening practices. FIDES strives to ensure that failure mechanisms are adequately considered to mitigate risks effectively and improve device reliability.	variations into the screening process allows for more realistic reliability assessments. Moreover, screening efforts should account for the influences of mechanical stresses, humidity, vibration, and other critical stress factors that can significantly affect device reliability.
Device screening: insufficient consideration of stress factors	MIL-HDBK-217's screening recommendations often focus on generic stress factors, such as temperature and voltage, without delving into the specific failure mechanisms that might be affected. This approach overlooks the importance of considering stress factors that are directly relevant to the failure modes in question, resulting in ineffective screening practices.	The FIDES approach acknowledges the significance of precise modeling and understanding of the diverse stress factors that devices may experience throughout their operational lifecycle. By incorporating a comprehensive analysis of stress factors such as electrical, thermal, mechanical, and environmental stresses, FIDES aims to provide a more thorough and accurate assessment of device reliability. This consideration of stress factors enables FIDES to identify potential failure modes and implement appropriate screening practices that effectively address the specific stress factors relevant to each device. The user of the FIDES methodology can only include the stresses that FIDES lists for the particular component family. FIDES does not include for example, relative humidity as one of the stresses for capacitors, even though literature shows that humidity is a significant factor impacting the reliability of some capacitors.	As technology evolves, new failure modes and mechanisms emerge, requiring a flexible and adaptable screening approach. We need to implement a proactive approach to mitigate the potential for overlooking hidden failure modes by adopting a more comprehensive screening methodology. Standard screening procedures may not be sufficient to capture all possible failure modes, particularly those that are hidden or not explicitly recognized. PoF addresses the dynamic nature of technology and the continuous emergence of new failure modes and mechanisms. Unlike traditional approaches that rely on empirical data and statistical models, PoF focuses on understanding the fundamental physical and mechanistic aspects of failure. The dynamic nature of technology necessitates a flexible and adaptable screening approach. PoF recognizes this need and provides a framework that allows for the incorporation of new failure modes and mechanisms as they arise. By understanding the underlying physics and mechanisms behind failures, PoF enables engineers to

| Device screening: neglect of application-specific considerations | Different applications have unique operating conditions and environmental factors that can significantly influence failure mechanisms. However, MIL-HDBK-217 does not sufficiently account for these application-specific considerations when promoting screening approaches. As a result, the screening efforts may not effectively target the relevant failure mechanisms for a given application. | FIDES is similar to MIL-HDBK-217 and may still have limitations in accounting for all application-specific considerations and environmental factors that can influence failure mechanisms. While FIDES aims to incorporate these factors into its approach, it may not comprehensively capture every possible influence on failure mechanisms in all scenarios. Therefore, despite its efforts to address application-specific considerations, there may still be instances where certain factors are not adequately accounted for in the screening process. Continuous improvement and refinement of the FIDES approach are necessary to enhance its ability to capture a wide range of application-specific influences on failure mechanisms. The methodology used in FIDES does not account for the part and assembly materials, geometry, and architecture, and will result in inaccurate and misleading predictions. | proactively identify and mitigate potential failure modes, even those that are hidden or not explicitly recognized.

The comprehensive screening methodology offered by PoF goes beyond the limitations of standard procedures. It considers the interplay between various stress factors, such as thermal, electrical, and mechanical stresses, and their impact on the reliability of a device. This holistic approach helps uncover potential failure mechanisms that may not be captured by traditional methods, ensuring a more accurate assessment of reliability.

Additionally, PoF emphasizes the importance of conducting root cause analysis and understanding the underlying physics governing failure mechanisms. By delving deeper into the fundamental causes of failures, PoF provides valuable insights into how failures occur, allowing for the development of targeted screening strategies and design improvements.

Overall, the PoF approach is considered the best approach because it offers a proactive, adaptable, and comprehensive screening methodology that considers the evolving nature of technology and the need to mitigate the potential for overlooking hidden failure modes. By understanding the underlying physics and focusing on root cause analysis, PoF enables more accurate reliability |

(Continued)

Table 1.2 (Continued)

Issue	MIL-HDBK-217	FIDES	Physics of failure
Device screening: potential for overlooking hidden failure modes	By not explicitly recognizing potential failure mechanisms, MIL-HDBK-217 may inadvertently lead to the oversight of hidden failure modes that are not captured by the standard screening procedures. This can result in unexpected failures in the field, impacting the reliability of the system.	FIDES, unlike MIL-HDBK-217, aims to address the potential for overlooking hidden failure modes by emphasizing a more comprehensive approach to screening. FIDES recognizes that standard screening procedures may not capture all possible failure modes, especially those that are hidden or not explicitly recognized. As a result, FIDES strives to incorporate advanced methodologies and techniques that allow for the identification and mitigation of these hidden failure modes.	assessments and supports the development of robust and reliable electronic systems.
Device screening: limitations in addressing new technologies	MIL-HDBK-217 was initially developed for older technologies and may not adequately address the failure mechanisms and screening needs of emerging technologies. Without recognizing and adapting to the unique failure mechanisms associated with new technologies, the effectiveness of screening efforts may be compromised. Developing and maintaining current design reliability models for devices is an impossible task.	However, it is important to note that while FIDES makes efforts to address hidden failure modes, it may still have limitations in fully capturing all possible scenarios. There is always a potential risk of overlooking certain hidden failure modes due to the complexity and variability of different systems and environments. Therefore, continuous improvement and refinement of the FIDES approach, along with industry feedback and collaboration, are essential to enhance its ability to uncover and mitigate hidden failure modes effectively. FIDES contains a static list of basic	

		failure rates of components but does not provide any method of updating those failure rates using the user's reliability data or experience. It neither provides the data used to calculate the basic failure rates nor informs the user of the method for translating failure data to the tabulated basic failure rates. FIDES does not state the data collection timeline (for which generation of parts the models are meant, if any). Since electronic materials, designs, and processes change very rapidly, most of the models, even if they were correct, which they are not, would be outdated.	While Arrhenius models provide a straightforward framework, PoF considers also the non-Arrhenius relation to accelerating parameters.
Use of Arrhenius Model	The use of the Arrhenius Model in MIL-HDBK-217 presents both advantages and limitations. While it provides a simple framework with a focus on steady-state temperature, it may not readily accommodate temperature changes and might oversimplify the treatment of different failure mechanisms.	In the FIDES approach, the fluctuation of temperature is considered by allowing for explicit temperature change inputs in the modeling process. This means that instead of relying solely on steady-state temperature, FIDES considers the actual temperature variations experienced by the device in real-world operating conditions. In that statement, FIDES acknowledges that devices are subject to dynamic thermal environments where temperature can fluctuate over time. These fluctuations can be caused by factors such as changes in ambient temperature, operational conditions, power cycling, or other environmental factors. To accurately assess the device's reliability, FIDES incorporates these temperature variations explicitly in its modeling process. It takes into	PoF acknowledges the importance of experimental data and empirical observations to validate and refine reliability models. Through a combination of laboratory testing, field data analysis, and physical understanding, PoF provides a robust framework to assess reliability that goes beyond simplistic models like Arrhenius.
Use of Arrhenius Model – primary emphasis on steady-state temperature	Highlighting the significance of steady-state temperature set the primary stress factor to be reduced for improving reliability. This approach simplifies the design process by focusing on a single parameter that can have a significant impact on device performance and longevity.		By considering the specific failure mechanisms associated with a particular system or component, PoF allows for a more accurate and realistic assessment of reliability. This approach enables engineers to capture the complex dependencies between temperature, stress, and failure modes,
Use of Arrhenius Model – limitation in accepting	One limitation of MIL-HDBK-217 is that its models do not readily accept explicit temperature change inputs.		

(Continued)

Table 1.2 (Continued)

Issue	MIL-HDBK-217	FIDES	Physics of failure
explicit temperature change inputs	This restriction hampers the ability to accurately account for transient temperature variations, which can occur in real-world operating conditions. Ignoring such temperature fluctuations may lead to less accurate reliability predictions.	account the specific temperature profiles or temperature cycles that the device is expected to encounter during its operational lifecycle. This enables a more accurate assessment of the device's performance and longevity, as it reflects the realistic temperature fluctuations that occur during operation. By explicitly accounting for temperature changes, FIDES provides a more comprehensive and accurate representation of the device's behavior under varying thermal conditions.	leading to improved predictions and more effective design decisions.

PoF recognizes that different failure mechanisms can contribute to the overall reliability of a system. Instead of treating all failure mechanisms as a single entity, PoF allows for a mechanism-specific approach. It considers the unique characteristics, behaviors, and interactions of each failure mechanism, enabling a more accurate analysis of their impact on system reliability. PoF helps in identifying how different mechanisms may interact or influence each other, leading to a more comprehensive assessment of reliability. |
| Use of Arrhenius Model – lumping acceleration models from different failure mechanisms | MIL-HDBK-217 employs a methodology that combines different acceleration models from various failure mechanisms into a single framework. This approach, although convenient from a modeling perspective, may overlook the distinct characteristics and dependencies of individual failure mechanisms. As a result, the reliability estimations may lack the necessary accuracy and precision required for specific failure modes. | While FIDES acknowledges the importance of addressing distinct characteristics and dependencies of individual failure mechanisms, one limitation is that it does not explicitly consider the simultaneous effects of multiple failure mechanisms. Unlike MIL-HDBK-217, which combines different acceleration models into a single framework, FIDES focuses on individual mechanisms separately. While this approach allows for a more detailed analysis of each mechanism, it may overlook the complex interactions and dependencies that can occur when multiple failure | By conducting laboratory testing and analyzing field data, PoF can verify and refine the models that describe the behavior of different failure mechanisms. This empirical approach provides evidence for the interdependencies between failure mechanisms and helps in improving the accuracy of reliability predictions.

PoF offers a flexible framework that can adapt to different types of failure mechanisms and their interdependencies: see chapter 5 of this handbook for the MTOL approach. |
| Use of Arrhenius Model – limited consideration of other stress factors | While MIL-HDBK-217 primarily emphasizes temperature, it may not adequately address the influence of | | |

Operating temperature	other stress factors that can significantly impact reliability, such as mechanical stresses, humidity, vibration, or electrical stresses. This limited consideration of multiple stress factors can lead to suboptimal reliability improvements if other critical stressors are not appropriately accounted for. The consideration in MIL-HDBK-217 is confined to steady-state temperature alone. However, the accuracy of assessing the impact of steady-state temperature is questionable as it lacks a foundation in root-cause analysis and time-to-failure data.	mechanisms act concurrently. As a result, the reliability estimations provided by FIDES may not fully capture the combined effects of multiple stress factors, potentially leading to a less comprehensive understanding of overall device reliability. Unlike MIL-HDBK-217, the FIDES approach recognizes the limitations of solely considering steady-state temperature in reliability assessments. FIDES aims to address this limitation by providing explicit support for the impact of operational temperature cycling on device reliability. Additionally, FIDES considers the combined effect of contaminants, such as ionic or hydrogen, and biasing conditions, which MIL-HDBK-217 does not explicitly address.	PoF is well-suited to address the constraints related to operating temperature, operational temperature cycling, and their combined effect with contaminants. Different failure mechanisms may exhibit varying sensitivities to temperature changes. By incorporating accurate temperature models and data, PoF can assess the impact of temperature variations on the reliability of a system.
Operational temperature cycling and combined to contaminants	MIL-HDBK-217 does not provide explicit support for considering the impact of operational temperature cycling on reliability, nor does it account for the combined effect of contaminants (such as ionic or hydrogen) and biasing conditions. Additionally, there is no provision for superposing the effects of temperature cycling and vibration in the reliability analysis within the handbook.	Furthermore, FIDES acknowledges the need to account for the superposition of temperature cycling and vibration effects in reliability analysis. By incorporating these considerations, FIDES offers a more comprehensive approach to reliability assessment,	Repeated temperature fluctuations can induce additional stress on the components and accelerate certain failure mechanisms, such as fatigue or thermal fatigue. Contaminants, such as ionic or hydrogen species, can interact with the materials and components, leading to accelerated degradation or failure. By considering the interactions between contaminants and

(Continued)

Table 1.2 (Continued)

Issue	MIL-HDBK-217	FIDES	Physics of failure
		considering multiple factors that can influence failure mechanisms and performance degradation. These differences in approach make FIDES a more robust methodology for capturing the complex interactions and effects of various stressors on device reliability. However, one drawback of the FIDES approach is that it may require more detailed and specific data compared to MIL-HDBK-217. The accurate modeling and analysis of operational temperature cycling, combined effects of contaminants and biasing conditions, and the superposition of temperature cycling, and vibration rely heavily on comprehensive and reliable data. Obtaining such data may pose challenges in certain situations, particularly for unique or specialized applications where limited or fragmented data is available. Therefore, while FIDES offers a more comprehensive approach, the availability and quality of data play a	temperature, PoF enables engineers to assess the synergistic effects and predict the impact on system reliability. Maximum Rating and Derating: PoF takes into account the maximum rating definitions and derating guidelines provided by component manufacturers. It recognizes the importance of operating components within their specified limits to ensure reliability and prevent premature failures.

	crucial role in its successful implementation and reliability assessments.	
Input data required	Does not model critical failure contributors, such as materials, architectures, and realistic operation stresses. Minimal data in, minimal data out. It is important to acknowledge that MIL-HDBK-217 was developed as a general reliability prediction handbook, aiming to provide broad guidance for a wide range of systems. While its simplicity may be advantageous in certain scenarios, it also introduces limitations in terms of accurately capturing the complexities associated with critical failure contributors.	For FIDES, there is a greater emphasis on capturing critical failure contributors, such as materials, architectures, and realistic operation stresses. The FIDES approach strives to incorporate more comprehensive and detailed input data, allowing for a more accurate representation of the system's reliability. By considering these critical factors, FIDES aims to provide more robust and tailored reliability predictions, addressing the limitations of MIL-HDBK-217 in capturing the complexities associated with failure contributors.	PoF considers the influence of input data required by promoting a comprehensive understanding of materials, architectures, and operating conditions. It emphasizes the importance of accessing relevant information from material datasheets, research publications, design documentation, environmental testing, failure analysis, and collaborative networks.

PoF takes into account the influence of design factors such as materials and architectures on the time to failure and failure sites. It gives means to consider the specific failure modes and mechanisms relevant to the device or assembly and predicts their occurrence based on the design characteristics. By understanding how different design choices affect reliability, designers can make informed decisions to improve the performance and longevity of the system. |
| Output data | A quote from a proponent representative sheds light on the matter, stating that the handbook is not designed to predict field reliability and, overall, it does not perform well in providing accurate predictions in an absolute sense [20]. | In terms of output data, FIDES aims to provide reliable predictions of field reliability. Unlike MIL-HDBK-217, FIDES do not suffer from the same limitation of not being designed to predict field reliability accurately. It focuses on providing more precise and meaningful output data, enabling engineers and decision-makers to | PoF considers the influence of output data required by providing designers with valuable insights into the impact of materials, architectures, loading conditions, and associated variations. It predicts the time to failure and failure sites for key failure modes and mechanisms, allowing designers to prioritize their efforts and make informed decisions to enhance reliability. Whether |

(Continued)

Table 1.2 (Continued)

Issue	MIL-HDBK-217	FIDES	Physics of failure
		make informed decisions regarding the system's reliability performance. This distinction highlights the commitment of FIDES to offer improved and more reliable predictions compared to the shortcomings of MIL-HDBK-217. However, it is important to note that the availability of comprehensive and accurate input data for FIDES can sometimes be a challenge. The limitation of requiring detailed information on materials, architectures, and realistic operation stresses can potentially impact the effectiveness and applicability of the FIDES approach, especially in cases where detailed data is limited or unavailable. Therefore, careful consideration should be given to data availability and quality when implementing FIDES for reliability predictions.	through deterministic or probabilistic treatment, PoF enables a comprehensive understanding of reliability factors and supports effective design improvements. PoF models enable the prediction of the time to failure for key failure modes and mechanisms. By considering the stress factors, material properties, and environmental conditions, PoF models can estimate the expected lifespan of the device or assembly. It provides insights into the locations or sites where failures are likely to occur within a device or assembly. It considers factors such as stress concentrations, weak points in the design, and potential failure mechanisms to identify the areas prone to failure.
DoD/industry acceptance	Mandated by government, 30-year record of discontent. Not part of the US Air Force Avionics Integrity Program (AVIP). No longer supported by senior US Army leaders.	FIDES has been introduced in 2004 by a consortium of French industrials (led by DGA (French DoD) and MBDA, including Thales, Eurocopter, Airbus, Nexter), with an update in 2009. See FIDES model coverage at: https://www.fides-reliability.org/en/node/555.	Adopting PoF as a best practice in industry offers a more accurate, comprehensive, and scientifically grounded approach to reliability assessment. While there may be initial investments in terms of expertise and resources, the long-term cost savings, prevention of failures, and optimization of

Conclusion	The FIDES methodology models the failures whose origins are intrinsic (item technology or manufacturing and distribution quality) and extrinsic (equipment specification and design, selection of the procurement route, equipment production and integration) to the items studied.	design decisions make PoF a compelling choice over MIL and FIDES.
	The FIDES methodology covers non-operating phases, whether standby periods between utilizations or actual storage.	PoF's dedication to continuous peer review, collaboration with industry leaders, and integration of Artificial Intelligence technologies positions it as a superior approach. These factors contribute to the refinement and advancement of PoF models, ensuring their relevance and effectiveness in addressing complex reliability challenges in the era of AI.
The models lack formal peer review as they have not been submitted to engineering societies or technical journals. Additionally, future tri-service coordination is a matter of concern.	Unlike MIL-HDBK, FIDES do not undergo formal peer review through submission to engineering societies or technical journals. This lack of external scrutiny and expert input may limit the ability to identify and address potential shortcomings or biases in the reliability models.	MIL and FIDES, with their reliance on empirical models or simplified approaches, may not provide the same level of accuracy and cost-effectiveness. They may overlook hidden failure modes or fail to adequately account for the complexities of modern technologies, potentially leading to unexpected failures and increased costs down the line.
	In summary, FIDES offers the advantage of utilizing real-world data and promoting collaboration, while MIL-HDBK benefits from its long-standing use and simplicity. The choice between the two approaches depends on factors such as the availability of data, industry requirements, and the desired level of accuracy and customization in reliability assessments.	PoF is actively engaging with leading electronics companies, agencies, and industries worldwide to develop new software and documentation.
		The involvement of AI in PoF can contribute to automated data collection, analysis, and decision-making processes. AI algorithms can process vast amounts of data and extract valuable insights, enabling faster and more efficient reliability assessments. This

(Continued)

Table 1.2 (Continued)

Issue	MIL-HDBK-217	FIDES	Physics of failure
			combination of PoF and AI technologies has the potential to revolutionize the field of reliability engineering by providing advanced tools and methodologies.
Relative cost of analysis	Cost is high compared with value added. Can misguide efforts to design reliable electronic equipment.	The cost associated with implementing the FIDES methodology can be perceived as high when compared to the value it adds. It is important to carefully consider the cost-benefit ratio to ensure that the investments made in applying FIDES are justified. However, it is crucial to note that while the cost may be significant, it is intended to ensure more accurate reliability assessments and minimize the potential for misguided efforts in designing reliable electronic equipment. By investing in the appropriate resources and training, organizations can maximize the benefits of the FIDES approach and make informed decisions that lead to improved product reliability and customer satisfaction.	PoF offers flexibility in data acquisition, allowing for tailored approaches that align with specific project requirements and budgets. This flexibility in data acquisition translates to adaptable cost structures, making PoF accessible to a wider range of industries and organizations. By designing for reliability and minimizing defects throughout the operational life of a product, PoF enables organizations to achieve higher reliability, lower failure rates, and reduced costs, thereby enhancing the overall quality and performance of their products.

1.2 Electronic Reliability in Manufacturing, Production, and Operations

"Health Monitoring" covers the monitoring of the state of health of a system, continuously or intermittently, from direct or indirect observations, to anticipate and take the best decisions, whatever their nature. One can model a degradation process in various ways. In the absence of any observation of the phenomenon, reliability models, such as the Weibull law, can be obtained from statistical failure data on similar products. To estimate the probability of failure after a given duration of operation, these models can be supplemented by acceleration models and accelerated life tests.

The issue of reliability estimation turns out to be different when a product is subject to a degradation process that can be quantified over time and on which a threshold limit of acceptability can be set. It is then possible to follow the evolution of such a process during trials, to assess the predicted reliability of the product, or during the operational life. To know his state of health from day to day, estimates of the remaining useful life (RUL) can be generated and decisions about using the component can be processed accordingly. The conditions of use and the environment conditions can accelerate the degradation process based on the specific physics of the failure mechanism being activated.

The identification of a direct or indirect observable (indicator) of the degradation and the knowledge of the influential stress factors is the domain of the component expertise. The measurement of degradation can be carried out under different stress conditions or require a return to standard conditions. By observing the variations in degradation per unit of time, the phenomenon can be modeled by a Gamma process, if the degradation is always increasing (monotonic deterioration), or by a Wiener process if the degradation is decreasing temporarily by phenomena of recovery or even healing. The degradation phenomenon can be stationary or non-stationary and be influenced by environmental conditions.

In the absence of direct observation, certain degradation phenomena can be modeled by a hidden Markov process. The algorithm and the Hidden Markov models were first described in a series of articles by Baum [22] and his peers at the IDA Center for Communications Research, Princeton in the late 1960s and early 1970s. As written in Wikipedia, this describes a Markov process whose states, representative of levels of progressive degradation, are partially observable through probabilistic indicators (a vibrational spectrum or an oil color will, for example, provide information on the state of degradation of a mechanism). The Viterbi algorithm is then used to find the most probable sequence of transitions from a sequence of observations and the Forward-Backward, also called de Baum-Welch, to estimate the parameters of the model.

Other phenomena can be modeled by piecewise deterministic Markovian processes (PDMP). Developed by Davis [23], this type of hybrid dynamic reliability process makes it possible to associate random characteristics with continuous components in interaction (environmental variables that will influence and be influenced by the system, for example).

The use of dynamic Bayesian networks can also be considered to model degradation processes. These represent the evolution of random variables as a function of a discrete sequence.

1.2.1 Failure Foundation

Why and how high- or low-quality components can fail in comparable time scale? Of course, reducing temperature stress conditions can improve long-term mission operation but not only temperature. Indeed, applying lower bias voltage, current, and power are other options. How to manage Time-to-Failure for warranty and limits with predictable outcomes?

The failure event pattern is detailed in the following Figure 1.1a, the general failure pattern overview presents the alternatives a component or a structure may fail by way of a permanent or reversible mechanism, either from a sudden catastrophic malfunction or by a continuous degradation trend and in addition with or without partial recovery.

What are the root causes of component failure? What characterizes component failure? This is related to the definition of trackers as detailed.

The foundation of a failure is described in behind basic sequences:

- To identify the *actors* or the *stressors* (Figure 1.1b) (or in other terms, which stress parameters are the origin of the failures),
- The *tracers* or the *observers* (Figure 1.1c) (types of measurands whose evolution is a quantification correlated to the level of failure),
- The *trackers* or the *signatures* (Figure 1.1d) (a selected electrical or mechanical parameter which quantifies the type and level of failure).

Stressor levels (the actors) can be measured directly or indirectly, such as junction temperature from package temperature and thermal resistance, for example.

1.2.2 Reliability Foundational Models (Markovian, Gamma, Lévy, Wiener Processes)

A paper published in 2020 by Cabarbaye et al. [24] provides an example of how the degradation analysis method from experiments can be implemented. To characterize a degradation, two families of models can be considered: continuous models and state-space (or multi-state) models. The first relate to a physical quantity, such as a crack propagation, a current degradation, or a temperature rise, for example,

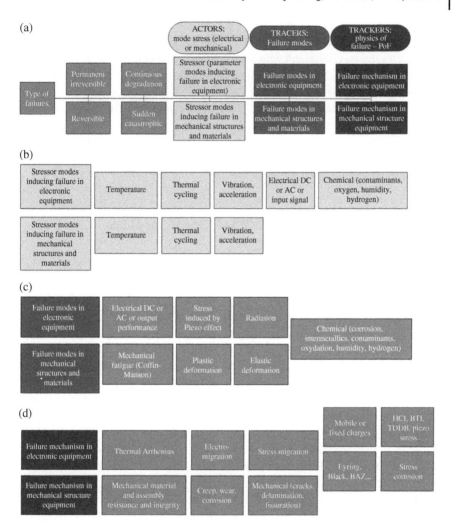

Figure 1.1 (a) General failure pattern overview. (b) How are failures induced by stress conditions (*tracer modes*). (c) Identify failure mode *tracers*. (d) PoF based on *trackers* or *signatures*.

whose evolution trajectory we seek according to a physics model. The second considers a series of pronounced states of degradation and transition laws between these states. They bring together members of the family of Markovian models such as PDMP.

The standard Accelerated Life Testing model (ALT), which includes most of the acceleration models used in reliability (Arrhenius, Eyring, Black, …), can then be used again on the assumption that only the scale factor of the degradation law is modified and not its form. The acceleration factor (AF) is then like the one considered for random failures but with different parameter values. If a threshold limit of acceptability can be set, the degradation model is transformed into a reliability model.

Continuous degradation models are generally based on the family of incremental Lévy processes [from the French mathematician (1886–1971)] independent and stationary (each increase depends only on the time interval), including the Gamma process, the Wiener process, and the compound Poisson process (degradation linked to an accumulation of shocks).

1.2.3 Correlation Versus Causation and Representativeness of Trackers

In reliability modeling, it is important to differentiate between correlation and causation: A correlation between variables, however, does not automatically mean that the change in one variable is the cause of the change in the values of the other variable. Causation is a cause-effect process and indicates that one event is the result of the direct influence of a variable stress parameter. In that sense, tracers are both types of correlated and causation parameters. While a tracker should be the "best parameter" showing causation to a failure mechanism. When several failure mechanisms are concurrent, several trackers can be identified and the trick consists in selecting, if possible, the most representative parameter of the existing multiple failure mechanisms.

Consequently, a physical model may be deduced from a tracker which could be biased so as not to accurately reflect the observed degradation of one (or several) failure mechanism(s). This adds "fuzzy" information which contributes to discrediting a failure mode model to predict the occurrence of failure. So, it looks like our quest to predict failures of a particular constituent being in a sub-system, an electronic card, or a component installed on it (of which it has its history of manufacturing, assembly process, and environmental applied stress conditions) may not yield the desired result. Hence, it is important to ask:

What is the representativeness of tracers and trackers?

The description of the allegory of the jigsaw puzzle given hereafter is a simplistic view of the situation.

Puzzle pieces can be similar in strength, stiffness, size, shape, and variability of colors, but at the same time, dissimilar with some drawbacks: indeed, this is because each puzzle piece is different in form, fit, and function (F3), and furthermore a single piece does not tell you the full sense of the picture.

Tracker#02

Tracker#0x

Tracker#ref01

Tracker#02

Tracker#ref02

Should a single valid Tracker be used to
describe true component reliability figure?

There is something erroneous to describe the reliability of a product using one single representative element. Moreover, reliability is also influenced by variability, inherent in all manufacturing processes. To build a usable and accurate reliability model, you must start by observing the tracers, select those that are closest to a faithful description of reality, and validate a series that is the most representative in reliability signature.

We must pay attention to know what to measure and how to control this range of variation. The conditions of usage like bias voltage or input signal must be defined with respect to the maximum rating limits allowed for a product (note they are also distributed with some variability). Variation also exists in the environments that engineered products must withstand. Temperature, mechanical stress, vibration spectra, and many other varying factors must be considered.

The measurement conditions of trackers must be set as generic and reproducible as possible. In steady-state life test experiments, it is recommended to measure the trackers under typical bias at room temperature conditions. But in real life, it is not always possible to measure the trackers in such stable conditions and use of sensors can be an option to track the change in power consumption or local temperature provided the causation is established and validated.

What is the definition of a failure? How to quantify it? Is it better to describe failure in absolute or in relative value? How can a series of tracker models accurately represent the entire figure of a product? The trackers are buoys to know where the potential wells of poor reliability are and how an accumulation of trackers can inform about the total generic reliability level.

1.2.4 Functional Safety Standard ISO 26262

Saying that, the automotive Functional Safety Standard ISO 26262 introduced the concept of the Probabilistic Metric for random Hardware Failures (PMHF) associating it with the failure rate of the safety critical system or subsystem. In an SAE publication, Kleyner and Knoell [25] presented this concept and discussed how to calculate both the failure rate and the average probability of failure per hour in terms of definitions, sources of the data, applications, and advantages and disadvantages.

Testing is an essential part of any engineering development program. There is one dilemma and conflict inherent in reliability testing as part of an integrated test program, however. To obtain information about reliability in a cost-effective way, that is also quick, it is necessary to generate failures. Only then can safety margins be ascertained. On the other hand, failures interfere with functional and environmental testing. In addition, lot-to-lot and device-to-device variability affects the probability of failure as well and another major source of variability is the range of production processes involved in converting designs into hardware. These considerations are exposed in the book "Practical Reliability Engineering" published by O'Connor and Kleyner [26] and assumes *"the old design concept of 'test-analyze-and-fix' (TAAF), in which reliance is placed on the test program to show up reliability problems, no longer has a place in modern design and manufacturing due to shorter design cycles, relentless cost reduction, warranty cost concerns, and many other considerations. Therefore, the reliability should be 'designed-in' to the product using the best available science-based methods. This process, called Design for Reliability (DfR) begins from the first stages of product development and should be well integrated through all its phases."*

Testing to demonstrate reliability when failures occur during tests can be analyzed using the life data analysis methods. The downside of the test-to-failure method is longer test times compared to equivalent success-run tests. Degradation analysis introduced is one way of demonstrating reliability within a relatively short period of time. Talking about parts reliability is not only considering the intrinsic reliability component capability but also defining the least stressful conditions of use taking into consideration extreme values external parameters can achieve during the mission. In other words, "a hammer blow can always destroy any good device." Consequently, it is equally important to select parts and assembly processes, and to have a good knowledge of the environment conditions (biasing/temperature). Knowing this, we enter the world of "Physics of Failures" (PoF) or "Physics of Health" (PoH).

The cause of a product failure is because of a disruption in the weakest link (a fragile part, a degraded aged part, or due to a sudden external overstress event). Either the breakage is foreseeable due to a parametric performance degradation with time, or it is catastrophic with now forewarning. In the wear-out case, statistical model variation with time is applicable (semi-Markovian or non-Markovian processes) and we can refer to wear-out failure mechanisms. In the case of a sudden break or catastrophic failure, it is by nature modeled as a random failure mechanism indicated only by a time to fail. Both may exit and must be combined to assess what we call the RUL of a product/device.

DfR is a set of approaches, methods, and best practices that are supposed to be used at the design stage of the product to minimize the risk that it might not meet the reliability requirements, objectives, and expectations. Statistical methods provide the means for analyzing, understanding, and knowing variation. These statistical representations can be only simplified models and we can't be sure they correctly describe the PoF or PoH. They help visualize mathematically some behavior, but the true life may be quite different.

As mentioned in chapter 12 of [26], *reliability testing should be considered part of an integrated test program, which should include:*

1) *Functional testing, to confirm that the design meets the basic performance requirements.*
2) *Environmental testing, to ensure that the design can operate under the expected range of environments.*
3) *Statistical tests, as described in Chapter 11, to optimize the design of the product and the production processes.*
4) *Reliability testing, to ensure (as far as is practicable) that the product will operate without failure during its expected life.*
5) *Safety testing, when appropriate.*

1.2.5 Additional Considerations

A component, however good as it may be, degrades under the effect of the applied stresses. A component is tested, manipulated, assembled, and embedded on an electronic card simply due to the manufacture of that electronic card. Then the accumulated effects of all the stressors evolve over time during the operational life of the equipment. How do the statistics related to the quality of the component batch, to the quality of the reporting and test processes, act and interact (signature statistics and environment statistics)? And how do these statistics change over time? Do they all evolve at the same rate? What impact do they have on the failure rate observed and described by the bathtub curve?

Until now, we measured the signature degrading over time when measured under reproducible conditions at room temperature and typical bias conditions (non-stress). During high-temperature endurance life tests, the components are biased under high-temperature conditions, which can induce stress levels different from the nominal operational ones. At high temperatures, the electrical characteristics of the components behave differently and the derating rates on the current or the voltage defined when cold are not the same as those when hot. They may, therefore, not be representative of the operational behavior of the components. The models deduced from these facts are therefore questionable. It is thus necessary to be extremely careful with the interpretation of the facts which can be distorted by this multi-statistical bias and high-temperature impact.

We must also mention another important concept which is the level of confidence of the trackers. What are the systematic drifts attached to them? And how to minimize them?

There are several definitions of systematic drifts and care must first be taken in defining what is called bias. We call bias the difference between the overall objective reliability measure (as a value deduced from the consideration of all the trackers) compared to the reliability deduced on a single tracker. The greater this difference, the more the measured reliability deviates from the overall objective reliability. When the bias is zero, there is no difference between the overall reliability and the reliability measured by a single tracker.

1.3 Reliability Criteria

Reliability is defined as the probability that a system (component) will function over some time period t. If T is defined as a continuous random variable, the time to failure of the system (component), then reliability can be expressed as:

$$R(t) = Pr\{T \geq t\} \tag{1.4}$$

where $R(t) \geq 0$, $R(0) = 1$ and $lim_{t \to \infty} R(t) = 0$

For a given value of t, $R(t)$ is the probability that the time to failure is greater than or equal to t, and $F(t)$ could be defined as:

$$F(t) = 1 - R(T) = Pr\{T < t\}, F(0) = 0, \ lim_{t \to \infty} F(t) = 1 \tag{1.5}$$

Then $F(t)$ is the probability that a failure occurs before t.

$R(t)$ is referred to as the probability function and $F(t)$ as the cumulative distribution function (CDF) of the failure distribution. A third function is defined by:

$$f(t) = \frac{dF(t)}{dt} = -\frac{dR(t)}{dt} \tag{1.6}$$

is called the probability density function (PDF) and has two properties:

$$f(t) \geq 0, \quad \text{and} \quad \int_0^\infty f(t)dt = 1 \tag{1.7}$$

Both the reliability function and the CDF represent areas under the curve defined by $f(t)$. Therefore, since the area beneath the entire curve is equal to one, both the reliability and the failure probability will be defined so that:

$$0 \leq R(t) \leq 1, \ \text{and} \ 0 \leq F(t) \leq 1 \tag{1.8}$$

The mean time to failure (MTTF) is defined by:

$$MTTF = E(T) = \int_0^\infty t \cdot f(t)dt \qquad (1.9)$$

which is the mean or the expected value of the probability distribution defined by $f(t)$.

Failure rate or Hazard rate function provides an instantaneous rate of failure:

$$Pr\{t \le T \le t + \Delta t\} = R(t) - R(t + \Delta t) \qquad (1.10)$$

The conditional probability of a failure in the time interval from t to $t + \Delta t$ given that the system has survived to time t is:

$$Pr\left\{t \le T \le t + \Delta t \middle| T \ge t\right\} = \frac{R(t) - R(t + \Delta t)}{R(t)} \qquad (1.11)$$

and then:

$\dfrac{R(t) - R(t + \Delta t)}{R(t)\Delta t}$ is the probability of failure per unit of time (failure rate). Set:

$$\lambda(t) = \lim_{\Delta t \to \infty} \frac{-[R(t + \Delta t) - R(t)]}{\Delta t} \cdot \frac{1}{R(t)} = \frac{-dR(t)}{dt} \cdot \frac{1}{R(t)} = \frac{f(t)}{R(t)}$$

$$(1.12)$$

Then $\lambda(t)$ is known as the instantaneous hazard rate or failure rate function. The failure rate function $\lambda(t)$ provides an alternative way of describing a failure distribution. Failure rates in some cases may be characterized as increasing (instantaneous failure rate – *IFR*), decreasing (*DfR*), or constant (*CFR*) when $\lambda(t)$ is an increasing, decreasing, or constant function. Table 1.3 shows the relationship among the measures.

1.3.1 The Failure Rate Curve for Electronic Systems

An important form of the hazard rate is a bathtub curve. Systems having this hazard rate function experience decreasing failure rates early in their life cycle (infant mortality), followed by a nearly CFR (useful life), followed by an increasing failure rate (wear-out); this curve may be obtained as a composite of several failure distributions or as a function of piecewise linear and CFRs.

The reliability of electronic devices is represented by the failure rate curve (Figure 1.2) (called the "bathtub curve"). The curve can be divided into the three following regions: (i) initial failures, which occur within a relatively short time after a device starts to be used, (ii) random failures, which occur over a long period of time, and (iii) wear-out failures, which increase as the device nears the end of its life.

Table 1.3 The relationship among the measures.

	$F(t)$	$R(t)$	$f(t)$	$\lambda(t)$	
$F(t)$	$F(t)$	$1 - R(t)$	$\int_0^t f(x)dx$	$1 - exp-\left(-\int_0^t \lambda(x)dx\right)$	Cumulative distribution function (CDF) or probability of failure
$R(t)$	$1 - F(t)$	$R(t)$	$\int_t^\infty f(x)dx$	$exp\left(-\int_0^t \lambda(x)dx\right)$	Reliability (R)
$f(t)$	$\dfrac{dF(t)}{dt}$	$\dfrac{dR(t)}{dt}$	$f(t)$	$\lambda(t) \cdot exp\left(-\int_0^t \lambda(x)dx\right)$	Probability density function (PDF)
$\lambda(t)$	$\dfrac{\frac{dF(t)}{dt}}{1-F(t)}$	$\dfrac{d(\ln R(t))}{dt}$	$\dfrac{f(t)}{\int_t^\infty f(x)dx}$	$\lambda(t) = \dfrac{f(t)}{R(t)}$	Instantaneous failure rate (IFR) or hazard function

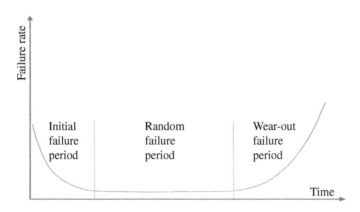

Figure 1.2 Failure rate curve.

"Initial failures" are considered to occur when a latent defect is formed, for example, during the device production process and then becomes manifest under the stress of operation. For example, a defect can be formed by having tiny particles in a chip in the production process, resulting in a device failure later. The failure rate tends to decrease with time because only devices having latent defects will fail, and these devices are gradually removed. At the start of device operation, there may be electronic devices with latent defects included in the set.

These will fail under operating stress, and as they fail, they are removed from the set based on the definition of the failure rate. The initial failure period can therefore be defined as the period during which devices having latent defects fail. The failure rate can be defined as a decreasing function since the number of devices having latent defects decreases as they are removed.

Most of the microelectronic device initial defects are caused by defects built into devices mainly in the wafer process. The most common causes of these defects are dust adhering to wafers in the wafer process and crystal defects in the gate oxide film or the silicon substrate, etc. The defective devices surviving in the sorting process (during the manufacturing process) may be shipped as passing products. These types of devices which are inherently defective from the start often fail when stress (voltage, temperature, etc.) is applied for a relatively short period. Finally, almost all of these types of devices fail over a short period of time and the failure rate joins the region called random failure part. The process of applying stresses for a short period of time (before shipping the product) to eliminate the defective devices is called screening ("burn-in").

"Random failures" occur once devices having latent defects have already failed and been removed. In this period, the remaining high-quality devices operate stably. The failures that occur during this period can usually be attributed to randomly occurring excessive stress, such as power surges, and software errors. Moreover, memory software errors and other phenomena caused by α rays and other high-energy particles are sometimes classified as randomly occurring failure mechanisms. There are also phenomena such as electrostatic discharge (ESD) breakdown, overvoltage (surge) breakdown (EOS), and latch-up which occur at random according to the conditions of use. However, these phenomena are all produced by the application of excessive stress over the device's absolute maximum ratings. So, these are classified as breakdowns instead of failures, and are not included in the random failure rate.

"Wear-out failures" occur due to the aging of devices from wear and fatigue. The failure rate tends to increase rapidly in this period. Semiconductor devices are therefore designed so that wear-out failures will not occur during their guaranteed lifetime. Accordingly, for the production of highly reliable electronic devices, it is important to reduce the initial failure rate to ensure the long life, or durability against wear-out failures. Wear-out failures are failures rooted in the durability of the materials comprising semiconductor devices and the transistors, wiring, oxide films, and other elements, and are an index for determining the device life (useful years). In the wear-out failure region, the failure rate increases with time until ultimately all devices fail or suffer characteristic defects.

Wear-out failures are failures rooted in the durability of the materials comprising semiconductor devices and the transistors, wiring, oxide films and other elements, and are an index for determining the device life (useful years). In the wear-out failure region, the failure rate increases with time until ultimately all devices fail or suffer characteristic defects.

1.3.2 Basic Lifetime Distribution Models

There are a handful of parametric models that have successfully served as population models for failure times arising from a wide range of products and failure mechanisms. Sometimes there are probabilistic arguments based on the physics of the failure mode that tend to justify the choice of model. Other times the model is used solely because of its empirical success in fitting actual failure data. Some of the models used in the electronic area are:

1) Exponential
2) Weibull
3) Extreme value
4) Lognormal

The exponential model, with only one unknown parameter, is the simplest of all life distribution models. The failure rate reduces to the constant for any time. The exponential distribution is the only distribution to have a CFR. The key factors are:

The probability density function (PDF),

$$f(t) = \lambda e^{(-\lambda t)}$$

The Cumulative Distribution Function (CDF),

$$F(t) = 1 - e^{(-\lambda t)}$$

$$\text{Since } R(t) = 1 - F(t), \quad R(t) = e^{(-\lambda t)}$$

(1.13)

The Instantaneous Failure Rate (IFR) function is given by

$$h(t) = \frac{f(t)}{R(t)} = \lambda$$

Although $1/\lambda$ is the average time of failure, it is not the same as the time when half the population would have failed. This median time to failure, or T_{50}, occurs in the interval for which the cumulative frequency distribution registers 50%. For a continuous population, the mean parameter is expressed by $Mean = \int_0^\infty x \cdot f(x) dx$. The variance of a random variable X is defined as the expected value of $(X - \mu)^2$, that is, $V(x) = E([X - \mu]^2)$.

So, for the exponential distribution, we get:

$$Mean = \frac{1}{\lambda}, Median = \frac{\ln 2}{\lambda}, Variance = \frac{1}{\lambda^2}$$

(1.14)

The CFR property, the exponential distribution is an excellent model for the long flat "intrinsic failure" portion of the Bathtub Curve. Since most components and systems spend most of their lifetimes in this portion of the Bathtub Curve, this justifies frequent use of the exponential distribution (when early failures or wear-out is not a concern). Its usefulness is to approximate a curve by piecewise straight-line segments. We can approximate any failure rate curve by week-by-week or month-by-month constant rates that are the average of the actual changing rate during the respective time durations. Then it would be possible to approximate any model by piecewise exponential distribution segments patched together. Some natural phenomena have a CFR (or occurrence rate), for example, the arrival rate of cosmic ray alpha particles or Geiger countertics. The exponential model works well for inter-arrival times (while the Poisson distribution describes the total number of events in each period). When these events trigger failures, the exponential life distribution model will naturally apply.

Weibull, on the other hand, is a very flexible life distribution model with two parameters.

The goal is to find a CDF that has a wide variety of failure rate shapes. With the constant $h(t) = \lambda$, we get just one possibility. Allowing any polynomial form of the type $\lambda(t) = \lambda \cdot \gamma \cdot (\lambda \cdot t)^{\gamma - 1}$ for a failure rate function achieves this objective. To derive $f(t)$, it is easier to start with the CDF $F(t)$. Setting

$$F(t) = 1 - exp\left(-\int_0^t \lambda(x)dx\right), \text{ or } F(t) = 1 - exp - (\lambda \cdot t)^\gamma \tag{1.15}$$

We get the probability density function (PDF),

$$f(t) = \frac{dF(t)}{dt} = \frac{\gamma}{t} \cdot (\lambda \cdot t)^\gamma \cdot exp - (\lambda \cdot t)^\gamma$$

Since $\quad R(t) = 1 - F(t), \quad R(t) = exp - (\lambda \cdot t)^\gamma \tag{1.16}$

The Instantaneous Failure Rate (IFR) function is given by

$$h(t) = \frac{f(t)}{R(t)} = \frac{\gamma}{t} \cdot (\lambda \cdot t)^\gamma$$

For $t > 0$, other key formulas are given by:

$$Mean = \frac{1}{\lambda} \cdot \Gamma\left(1 + \frac{1}{\gamma}\right), Median = \frac{1}{\lambda}(ln\, 2)^{\frac{1}{\gamma}}$$

$$Variance = \frac{1}{\lambda^2}\Gamma\left(1 + \frac{2}{\gamma}\right) - \left[\frac{1}{\lambda}\Gamma\left(1 + \frac{1}{\gamma}\right)\right]^2 \tag{1.17}$$

with α the scale parameter (the Characteristic Life), γ (gamma) the Shape Parameter, and Γ the Gamma function with $\Gamma(N) = (N - 1)!$ for integer N.

When $\gamma = 1$, the Weibull reduces to the Exponential Model. Depending on the value of the shape parameter, γ the Weibull model can empirically fit a wide range of data histogram shapes.

Weibull is widely used because of its flexible shape and ability to model a wide range of failure rates (Weibull has been used successfully in many applications as a purely empirical model); besides, Weibull model can be derived theoretically as a form of Extreme Value Distribution, governing the time to occurrence of the "weakest link" of many competing failure processes. This may explain why it has been so successful in applications such as capacitors, ball bearings, and relay and material strength failures. Another special case of the Weibull occurs when the shape parameter is 2. The distribution is called the Rayleigh Distribution and it turns out to be the theoretical probability model for the magnitude of radial error when the x and y coordinate errors are independent normal with 0 mean and the same standard deviation.

Extreme value distributions are the limiting distributions for the minimum or the maximum of a very large collection of random observations from the same arbitrary distribution. Gumbel [27] showed that for any well-behaved initial distribution (i.e. $F(x)$ is continuous and has an inverse), only a few models are needed, depending on whether you are interested in the maximum or the minimum, and if the observations are bounded above or below. In the context of reliability modeling, extreme value distributions for the minimum are frequently encountered. For example, if a system consists of n identical components in series, and the system fails when the first of these components fails, then system failure times are the minimum of n random component failure times. Extreme value theory says that independent of the choice of component model, the system model will approach a Weibull as n becomes large. The same reasoning can also be applied at a component level if the component failure occurs when the first of many similar competing failure processes reaches a critical level. The distribution often referred to as the Extreme Value Distribution (Type I) is the limiting distribution of the minimum of many unbounded identically distributed random variables. The PDF and CDF are given by:

$$f(x) = \frac{1}{\beta}e^{\frac{x-\mu}{\beta}}e^{-e^{\frac{x-\beta}{\beta}}} \text{ where} - \infty < x < \infty, \ \beta > 0, \tag{1.18}$$

$$F(x) = 1 - e^{e^{\frac{x-\mu}{\beta}}} \text{ where} - \infty < x < \infty, \beta > 0 \tag{1.19}$$

If the x values are bounded below (as is the case with times of failure), then the limiting distribution is the Weibull.

Extreme Value Distribution model could be used in any modeling application for which the variable of interest is the minimum of many random factors, all of which can take positive or negative values, try the extreme value distribution as a likely candidate model. For lifetime distribution modeling, since failure times are bounded below by zero, the Weibull distribution is a better choice. Weibull distribution and the extreme value distribution have a useful mathematical relationship. If $t_1, t_2, \ldots, t_n$ are a sample of random times of fail from a Weibull distribution, then $ln(t_1), ln(t_2), \ldots, ln(t_n)$ are random observations from the extreme value distribution. In other words, the natural log of a Weibull random time is an extreme value random observation. If a Weibull distribution has shape parameter γ and characteristics life α, then the extreme value distribution value distribution (after taking natural logarithms) has $\mu = ln\ \alpha,\ \beta = 1/\gamma$.

Because of this relationship, computer programs and graph papers designed for the extreme value distribution can be used to analyze Weibull data. The situation exactly parallels using normal distribution programs to analyze lognormal data, after first taking natural logarithms of the data points.

The lognormal life distribution, like the Weibull, is a very flexible model that can empirically fit many types of failure data. The two-parameter form has parameters = the shape parameter and T_{50} = the median (a scale parameter). If time to failure, t_f, has a lognormal distribution, then the (natural) logarithm of time to failure has a normal distribution with mean $\mu = ln(T_{50})$ and standard deviation σ. This makes lognormal data convenient to work with; just take natural logarithms of all the failure times and censoring times and analyze the resulting normal data. Later on, convert back to real time and lognormal parameters using σ as the lognormal shape and $T_{50} = e^{\mu}$ as the (median) scale parameter.

$$f(t) = \frac{1}{\sigma t \sqrt{2\pi}} e^{-\left(\frac{1}{2\sigma^2}\right)(ln\,t\,-\,ln\,T_{50})^2} \tag{1.20}$$

$$F(t) = \int_0^T \frac{1}{\sigma \cdot t \sqrt{2\pi}} \cdot e^{-\left(\frac{1}{2\sigma^2}\right)\cdot(ln\,t\,-\,ln\,T_{50})^2} dt = \Phi\left(\frac{ln\,t\,-\,ln\,T_{50}}{\sigma}\right) \tag{1.21}$$

$\Phi(z)$ denoting the standard normal CDF and $Mean = T_{50} \cdot exp\left(\frac{\sigma^2}{2}\right)$.

The lognormal PDF and failure rate shapes are flexible enough to make the lognormal a very useful empirical model. In addition, the relationship to the normal (just take natural logarithms of all the data and time points and "normal" data) makes it easy to work with mathematically, with many good software analysis programs available to treat normal data. The lognormal model can theoretically match many failure degradation processes common to electronic failure mechanisms including corrosion, diffusion, migration, crack growth, electromigration, and, in general, failures resulting from chemical reactions or processes. Applying the Central Limit Theorem to small additive errors in the log domain and justifying a normal model is equivalent to justifying the lognormal model in real time when a process moves toward failure based on the cumulative effect of many small "multiplicative" shocks. More precisely, if at any instant in time, a degradation process undergoes a small increase in the total amount of degradation that is proportional to the current total amount of degradation, then it is reasonable to expect the time to failure (i.e. reaching a critical amount of degradation) to follow a lognormal distribution [28].

1.4 Reliability Testing

Reliability testing is a series of laboratory tests carried out under known stress conditions to evaluate the life span of a device or system. Reliability tests are performed to ensure that semiconductor devices maintain their performance and functions throughout their life. These reliability tests aim to simulate and accelerate the stresses that the semiconductor device may encounter during all phases of its life, including mounting, aging, field installation, and operation. The typical stress conditions are defined in the testing procedure described later in this document. Reliability tests are performed at various stages of design and manufacture. The purpose and contents differ in each stage. When testing semiconductor devices, (i) the subject and purpose of each test, (ii) the test conditions, and (iii) the judgment based on test results must be considered.

1.4.1 Reliability Test Methods

Reliability tests are performed under known stress conditions. A number of standards have been established including Japan Electronics and Information Technology Industries Association (JEITA) standards, US military (MIL) standards, International Electrotechnical Commission (IEC) standards, and Japan Industrial Standards (JIS). The testing procedures and conditions differ slightly from one another, but their purpose is the same. The test period including the number of cycles and repetitions, and test conditions are determined in the same manner as the number of samples, by the quality level of design and the quality level required by the customer. However, mounting conditions, operating periods, as well as the acceleration of tests are considered to select the most effective and cost-efficient conditions. Reliability tests must be reproducible. It is preferable to select a standardized testing method. For this reason, tests are carried out according to international or national test standards.

The definition of a device failure is important in planning reliability tests. It is necessary to clarify the characteristics of the device being tested and set failure criteria. These failure criteria must be used in determining whether any variations in the device characteristics before and after the test are in an allowable range. The reliability of a device is usually determined by its design and manufacturing process. Therefore, a reliability test is conducted by taking samples from a population within the range that the factors of design and manufacturing process are the same.

The sampling standard for semiconductor devices is determined by both the product's reliability goal and the customer's reliability requirements to achieve the required level of "lot" tolerance percent defective (LTPD). The reliability tests for materials and processes are performed with the number of samples determined independently. For newly developed processes and packages, however, there are cases in which the existing processes or packages cannot be accelerated within the maximum product rating or in which new failures cannot be detected within a short period of time. In these cases, it is important to perform reliability testing based on the failure mechanism by means of Test Element Groups (TEGs). To ensure reliability during product design, we conduct testing according to the failure mechanism in the same process as the products, clarify the acceleration performance for temperature, electric field, and other factors, and reflect this in the design rules used in designing the product.

The standards and specifications related to the reliability of semiconductor devices can be classified as shown in Table 1.5. Reliability methods are often subject of standardization in several organizations. In those cases, the standards were usually derived from a common source, e.g. MIL-STD-883, and only minor differences exist between different standards.

The MIL standards have been used as industry standards for many years, and also today many companies still refer to the MIL standards. With the advent of new technologies, other standards have become more important. MIL-STD-883 establishes test methods and other procedures for integrated circuits. It provides general requirements, mechanical, environmental, durability and electrical test methods, and quality assurance methods and procedures for procurement by the US military. The test methods described in MIL-STD-883 include environmental test methods, mechanical test methods, digital electrical test methods, linear electrical test methods, and test procedures. MIL-STD-750 specifies mechanical, environmental, and electrical test methods and conditions for discrete semiconductor devices. MIL-STD-202 specifies mechanical and environmental test methods for electronic and electrical components. MIL-HDBK-217 is a US military handbook for reliability prediction of electronic equipment. This handbook introduces methods for calculating the predicted failure rate for each category of parts ion electronic equipment, including e.g. mechanical, components, lamps, and advanced semiconductor devices such as microprocessors.

Table 1.4 shows the subject, purposes, and contents of some reliability tests carried out at some laboratories.

The standards and specifications related to the reliability of semiconductor devices can be classified as shown in Table 1.5. Reliability methods are often subject of standardization in several organizations. In those cases, the standards were

Table 1.4 Examples of reliability testing conducted when new products are developed.

Phase	Purpose	Context	Contents
Development of semiconductor products	To verify that the design reliability goals and the customer's reliability requirements are satisfied	Quality approval for the product developed	The following and other tests are carried out as required 1) Standard tests 2) Accelerated tests 3) Marginal tests 4) Structural analysis
Development or change of materials and processes	To verify that the materials and processes used for the product under development satisfy the design reliability goals and the customer's reliability requirements To understand the quality factors and limits that are affected by materials and processes	Quality approval for wafer process/package developed or changed	Products are used to perform acceleration tests and other analyses as required with attention paid to the characteristics and changes of materials and processes
Pilot run before mass production	To verify that the production quality is at the specified level	Quality approval for mass production	This category covers reliability tests for examining the initial fluctuation of parameters that require special attention, as well as fluctuations and stability in the initial stage of mass production

Table 1.5 Classification of test standards.

Class	Example
International standard	IEC
National standards	USA (ANSI), CNES, DGA
Governmental standards	MIL, ESA
Industrial standards	EIA, JEDEC; JEITA, SIA, SAE
Standard by expert groups	ESDA, FIDES
Standards by semiconductor users	AEC Q100

usually derived from a common source, e.g. MIL-STD-883, and only minor differences exist between different standards.

The MIL standards have been used as industry standards for many years, and also today many companies still refer to the MIL standards. With the advent of new technologies, other standards have become more important. MIL-STD-883 establishes test methods and other procedures for integrated circuits. It provides general requirements; mechanical, environmental, durability, and electrical test methods; and quality assurance methods and procedures for procurement by the US military. The test methods described in MIL-STD-883 include environmental test methods, mechanical test methods, digital electrical test methods, linear electrical test methods, and test procedures. MIL-STD-750 specifies mechanical, environmental, and electrical test methods and conditions for discrete semiconductor devices. MIL-STD-202 specifies mechanical and environmental test methods for electronic and electrical components. MIL-HDBK-217 is a US military handbook for reliability prediction of electronic equipment. This handbook introduces methods for calculating the predicted failure rate for each category of parts ion electronic equipment, including e.g. mechanical, components, lamps, and advanced semiconductor devices such as microprocessors.

The IEC was founded in 1908 and is the leading global organization that prepares and publishes international standards (ISs) for electrical, electronic, and related technologies. IEC standards serve as a basis for national standardization and as references when drafting international tenders and contracts. IEC publishes IS, Technical specifications (TS), technical reports (TR), and guides. Examples of reliability-related IEC standards are IEC 60068 (which deals with environmental testing, this standard contains fundamental information on environmental testing procedures and severities of tests. It is primarily intended for electrotechnical products), IEC 60749 series (deals with mechanical and climatic test methods; these standards are applicable to semiconductor devices including discrete and integrated circuits), IEC 62047 (about micro-electromechanical devices), IEC 62373/74 on wafer level reliability, IEC/TR 62380 about reliability data handbook Universal model for reliability prediction of electronics components, printed circuit boards (PCBs) and equipment.

JEDEC Solid State Technology Association (formerly Joint Electron Device Engineering Council) is the semiconductor engineering standardization division of the Electronic Industries Alliance (EIA), a US-based trade association that represents all areas of the electronics industry. JEDEC was originally created in 1960 to cover the standardization of discrete semiconductor devices and later expanded in 1970 to include integrated circuits.

JEDEC spans a wide range of standards related to reliability. These standards are among the most referenced by semiconductor companies in the United States and Europe. The JEDEC committees for reliability standardization are JC-14

Committee on quality and reliability of solid-state products, JC-14.1 Subcommittee about reliability test methods for packaged devices, JC-14.2 Subcommittee on wafer level reliability, JC-14.3 Subcommittee about silicon devices reliability qualification and monitoring., JC-14.4 Subcommittee on quality processes and methods, JC-14.6 Subcommittee about failure analysis, JC-14.7 Subcommittee on Gallium Arsenide reliability and quality standards. JEDEC publishes standards, publications, guidelines, standard outlines, specifications, and ANSI and EIA standards.

JEITA was formed in 2000 from a merger of The Electronic Industries Association of Japan (EIAJ) and Japan Electronic Industry Development Association (JEIDA). Its objective is to promote the healthy manufacturing, international trade and consumption of electronics products. JEITA's main activities are supporting new technological fields, developing international cooperation, and promoting standardization.

Principal product areas covered by JEITA are digital home appliances, computers, industrial electronic equipment, electronic components and electronic devices (e.g. discrete semiconductor devices and ICs). JEITA standards related to semiconductor reliability are ED4701 which containing environmental and endurance test methods, ED4702 about mechanical tests, ED4704 on failure mode driven tests (wafer level reliability [WLR]), EDR4704 about guidelines for accelerated testing, EDR4705 on SER, EDR4706 about FLASH reliability JEITA standards are available online.

The Automotive Electronics Council and the ESD association are the most important examples of user groups and expert groups, respectively. Table 1.6 shows the reliability test standards [29, 30].

The purpose of semiconductor device reliability testing is primarily to ensure that shipped devices, after assembly and adjustment by the customer, exhibit the desired lifetime, functionality, and performance in the hands of the end user. Nevertheless, there are constraints of time and money. Because semiconductor devices require a long lifetime and low failure rate, testing devices under actual usage conditions would require a great amount of test time and excessively large sample sizes. The testing time is generally shortened by accelerating voltage, temperature, and humidity. In addition, statistical sampling is used, considering the similarities between process and design, so as to optimize the number of test samples. Temperature, temperature and humidity, voltage, temperature difference, and current are the accelerated stresses which can be applied to devices.

Reliability tests are performed under known stress conditions. The testing procedures and conditions differ slightly from one another although their purpose is the same. The test period including the number of cycles and repetitions, and test conditions are determined in the same manner as the number of samples, by the quality level of design and the quality level required by the customer. However,

Table 1.6 Reliability test standard.

Japan Electronics and Information Technology Industries Association (JEITIA) Standards	
EIAJ ED-4701/001	Environmental and endurance test method for Semiconductor Devices (General)
EIAJ ED-4701/100	Environmental and endurance test method for Semiconductor Devices (Lifetime Test I)
EIAJ ED-4701/200	Environmental and endurance test method for Semiconductor Devices (Lifetime Test II)
EIAJ ED-4701/300	Environmental and endurance test method for Semiconductor Devices (Strength Test I)
EIAJ ED-4701/400	Environmental and endurance test method for Semiconductor Devices (Strength Test I)
EIAJ ED-4701/50	Environmental and endurance test method for Semiconductor Devices (Other Tests)
US Military (MIL) Standards	
MIL-STD-202	Test method for Electronic and Electrical Parts
MIL-STD-883	Test method and Procedures for Microelectronics
International Electromechanical Commission (IEC) Standards	
IEC 60749	Semiconductor Devices – Mechanical and climatic test methods
IEC 60068-1	Environmental testing Part 1: General and guidance
IEC 60068-2	Environmental testing Part 2
Join Electron Devices Engineering (JEDEC) Standards	
JESD 22	Series Test Methods
JESD 78	IC Latch-Up Test
JEP 122	Failure Mechanisms and Models for Semiconductor Devices
European Cooperation for Space Standardization (ECSS) [26]	
ECSS-Q-ST-60-02C	Space product assurance – ASIC, FPGA and IP Core product assurance
ECSS-Q-30-08A	Components reliability data sources and their use
ECSS-Q-ST-30-11C	Derating – EEE components
Japanese Industrial Standards (JIS)	
JIS C 00xx	Environmental Testing Methods (Electrical and Electronics) Series
CENELEC Electronic Components Committee (CECC)	
CECC 90000	General Specification Monolithic Integrated Circuits
CECC 90100	General Specification Digital Monolithic Integrated Circuit

mounting conditions, operating periods, as well as the acceleration of tests are considered to select the most effective and cost-efficient conditions.

The definition of a device failure is important in planning reliability tests. It is necessary to clarify the characteristics of the device being tested and set failure criteria. These failure criteria must be used in determining whether any variations in the device characteristics before and after the test are in an allowable range. The reliability of a device is usually determined by its design and manufacturing process. Therefore, a reliability test is conducted by taking samples from a population within the range that the factors of the design and manufacturing process are the same. The sampling standard for semiconductor devices is determined by both the product's reliability goal and the customer's reliability requirements to achieve the required level of LTPD. The reliability tests for materials and processes are performed with the number of samples determined independently. For newly developed processes and packages, however, there are cases in which the existing processes or packages cannot be accelerated within the maximum product rating or in which new failures cannot be detected within a short period of time. In these cases, reliability testing based on the failure mechanism by means of TEGs is performed. To ensure reliability during product design, different tests are performed, according to the failure mechanism in the same process as the products, to clarify the acceleration performance for temperature, electric field, and other factors, and reflect this in the design rules used in designing the product [32].

1.4.2 Accelerated Testing

As mentioned by JEDEC, accelerated testing is a powerful tool that can be effectively used in two very different ways: in a qualitative or a quantitative manner. Qualitative accelerated testing is used primarily to identify failures and failure modes while quantitative accelerated testing is used to make predictions about a product's life characteristics (e.g. MTTF, B10 life, etc.) under normal use conditions. In accelerated testing, the quantitative knowledge builds upon the qualitative knowledge. Accelerated testing to be used in a quantitative manner for reliability prediction requires a physics-of-failure approach, i.e. a comprehensive understanding and application of the specific failure mechanism involved and the relevant activating stresses.

In general, accelerated lifetime tests are conducted under more severe stress conditions than actual conditions of use (fundamental conditions). They are methods of physically and chemically inducing the failure mechanisms to assess, in a short time, device lifetime, and failure rates under actual conditions of use.

By increasing the degree of stressors (e.g. temperature and voltage), like by increasing the frequency of the applied stress, using tighter failure criteria, or using test vehicles/structures specific to the failure modes, the degree of acceleration is increased.

The stress methods applied in accelerated lifetime tests are constant stress, and step stress methods. The constant stress method is a lifetime test where stress, such as temperature or voltage, is held constant and the degree of deterioration of properties and time to failure lifetime distribution are evaluated. In the step stress method, contrary to the constant stress method, the time is kept constant, and the stress is increased in steps and the level of stress causing failure is observed.

Typical examples of tests using the constant stress method, the step stress method, and the cyclic stress method, a variation of the constant stress method, are shown in Table 1.7.

An accurate quantitative accelerated test requires defining the anticipated failure mechanisms in terms of the materials used in the product to be tested. It is important to determine the environmental stresses to which the product will be exposed when operating and when not operating or stored, choosing a test or combination of tests based on the failure mechanisms that are anticipated to limit the life of the product. Acceleration models that should be considered include:

- Arrhenius Temperature Acceleration for temperature and chemical aging effects
- Inverse Power Law for any given stress
- Miner's Rule for linear accumulated fatigue damage
- Coffin-Manson non-linear mechanical fatigue damage
- Peck's Model for temperature and humidity combined effects
- Eyring/Black/Kenney models for temperature and voltage acceleration.

Table 1.8 describes each of these models, their relevant parameters, and frequent applications of each.

For a known or suspected failure mechanism, all stimuli affecting the mechanism based on anticipated application conditions and material capabilities are identified. Typical stimuli may include temperature, electric field, humidity, thermomechanical stresses, vibration, and corrosive environments. The choice of accelerated test conditions is based on material properties and application requirements; the time interval should be reasonable for different purposes as well. Accelerated stressing must be weighed against generating fails that are not pertinent to the experiment, including those due to stress equipment or material problems, or "false failures" caused by product overstress conditions that will never occur during actual product use. Once specific conditions are defined, a matrix of

Table 1.7 Typical example of accelerated lifetime tests.

Applied stress method	Purpose	Accelerated test	Main stressor	Failure mechanism
Constant stress method	Investigation of the effects of constant stress on a device	High-temperature storage test	Temperature	Junction degradation, impurities deposit, ohmic contact, inter-metallic chemical compounds
		Operating lifetime test	Temperature Voltage Current	Surface contamination, junction degradation, mobile ions, equilibrium molecular dynamic (EMD)
		High-temperature high humidity storage	Temperature Humidity	Corrosion surface contamination, pinhole
		High-temperature high humidity bias	Temperature Humidity Voltage	Corrosion surface contamination, junction degradation, mobile ions
Cyclic stress method	Investigation of the effects of repeated stress	Temperature cycle	Temperature difference Duty cycle	Cracks, thermal fatigue, broken wires and metallization
		Power cycle	Temperature difference Duty cycle	Insufficient adhesive strength of ohmic contact
		Temperature humidity cycle	Temperature difference Humidity difference	Corrosion, pinhole, surface contamination
Step stress method	Investigation of the stress limit that a device can withstand	Operation stress	Temperature Voltage Current	Surface contamination, junction degradation, mobile ions, EMD
		High-temperature reverse bias	Temperature Voltage	Surface contamination, junction degradation, mobile ions, TDDB

Table 1.8 Frequently used acceleration models, their parameters, and applications.

Model name	Description/ parameters	Application examples	Model equation
Arrhenius Acceleration Model	Life as a function of temperature or chemical aging	Electrical insulation and dielectrics, solid state and semiconductors, intermetallic diffusion, battery cells, lubricants and greases, plastics, incandescent lamp filaments	$Life = A_0 \cdot e^{-\frac{E_a}{k_B T}}$ where $Life$ = median Life of a population A_0 = scale factor determined by experiment e = base natural logarithms E_a = activation energy (unique for each failure mechanism) k_B = Boltzmann's constant = 8.62×10^{-5} eV/K T = temperature (in K)
Inverse Power Law	Life as a function of any given stress	Electrical insulation and dielectrics (voltage endurance), ball and roller bearings, incandescent lamp filaments, flash lamps	$\dfrac{Life\ at\ normal\ stress}{Life\ at\ accelerated\ stress}$ $= \left(\dfrac{Accelerated\ stress}{Normal\ stress}\right)^N$ where $N = Acceleration\ factor$
Miner's Rule	Cumulative linear fatigue damage as a function of flexing	Metal fatigue (valid only up to the yield strength of the material)	$CD = \sum\limits_{i=1}^{k} \dfrac{C_{Si}}{N_i} \leq 1$ where: CD = Cumulative damage C_{Si} = Number of cycles applied @ stress Si N_i = number of cycles to failure under stress Si (determined from a S-N diagram for that specific material) K = number of loads applied

(Continued)

Table 1.8 (Continued)

Model name	Description/ parameters	Application examples	Model equation
Coffin-Manson	Fatigue life of metals (ductile materials) due to thermal cycling and/or thermal shock	Solder joints and other connections	$Life = \dfrac{A}{(\Delta T)^{B}}$ where $Life$ = Cycles to failure A = scale factor determined by experiment B = scale factor determined by experiment ΔT = Temperature change
Peck's	Life as a combined function of temperature and humidity	Epoxy packaging	$\tau = A_0(RH)^{-2.7}e^{\left(\frac{0.79}{k_B T}\right)}$ where τ = median life (time-to-failure) A_0 = scale factor determined by experiment RH = relative humidity
Peck's Power Law	Time to failure as a function of relative humidity, voltage and temperature	Corrosion	
Eyring/ Black/ Kenney	Life as a function of temperature and voltage (or current density – Black)	Capacitors, electromigration in aluminum conductors	$\tau = \dfrac{A}{T}e^{\left(\frac{B}{k_B T}\right)}$ where τ = median life (time-to-failure) A = scale factor determined by experiment B = scale factor determined by experiment
Eyring	Time to failure as a function of current, electric field and temperature	Hot carrier injection, surface inversion, mechanical stress	$TF = B \cdot I_{sub}^{-N} \cdot e^{-\frac{E_a}{k_B T}}$ where TF = time-to-failure B = scale factor determined by experiment I_{sub} = peak substrate current during stressing N = 2 to 4 E_a = −0.1 to −0.2 eV (apparent activation energy is negative) $TF = B \cdot (T_0 - T)^{-N} \cdot e^{-\frac{E_a}{k_B T}}$

Table 1.8 (Continued)

Model name	Description/ parameters	Application examples	Model equation
Thermo-mechanical stress	Time to failure as a function of change in temperature	Stress generated by differing thermal expansion rates	where TF = time-to-failure B = scale factor determined by experiment T_0 = stress-free temperature for metal (approximate metal deposition temperature for aluminum) $N = 2$–3 $E_a = 0.5$–0.6 eV for grain boundary diffusion, ≈ 1 eV for intra-grain diffusion

accelerated tests embodying these stimuli must be developed that allows separate modeling of each individual stimulus. A minimum of two accelerated test cells is required for each identified stimulus; three or more are preferred. Several iterations may be required in cases where some stimuli produce only secondary effects. For instance, high-temperature storage and thermal cycle stressing may yield similar failures, where high-temperature storage alone is the root stimulus.

Test sample quantity, the definition of failure criteria (i.e. the definition of a failure depends on the failure mechanism under evaluation and the test vehicle being used), and choosing an appropriate distribution (fitting the observed data to the proper failure distribution), together determine the AFs [29,30,32].

2

The Fundamentals of Failure

The binomial distribution is well suited for describing the probability of an event with two outcomes. This mathematical tool can be applied to fundamental random motion of particles or energy in a closed system. To develop the necessary insights into the physics of failure, it is important first to understand a few universal physical concepts. Since nearly all failure events are random, in principle, it is possible to predict the likelihood of these events based on a knowledge of the statistical behavior of an ensemble of microscopic and submicroscopic events and the energies associated with each.

In thermodynamic terms, the only reason anything happens is that it is energetically favorable to the universe. The whole system will always seek a minimum energy configuration, and if the result is a mechanical failure, that is what we as consumers and manufacturers must live with. The whole physics of failure approach to predicting when things fail is based on the belief that the energies of a system can be determined, and a lifetime can be predicted, based on knowledge of all the applied stresses and the material properties.

The process of breaking down a system into a collection of atoms and electrons for the purpose of understanding the fundamental properties is known as statistical mechanics. Most of the mechanical, electrical, and thermal properties can be understood from first principles using a few simple yet powerful models. This chapter is not going to present an exhaustive review but will go over a few of the most basic properties that are responsible for most of the predictable failures in mechanical and electronic parts.

Here we will first develop an understanding of diffusion based on the principle of a random walk followed by a discussion of chemical potential and thermal activation energy, which will physically justify the Arrhenius relation. To further our fundamental insights, a discussion of surface chemistry and basic oxidation processes will be presented.

Reliability Prediction for Microelectronics, First Edition. Joseph B. Bernstein, Alain A. Bensoussan, and Emmanuel Bender.
© 2024 John Wiley & Sons Ltd. Published 2024 by John Wiley & Sons Ltd.

The concepts presented in this chapter are designed to give general ideas on how to solve reliability problems by the physics of failure approach. All of physics based on what I learned from Professor David Adler, of blessed memory, from MIT is that when all the probabilities of any set of configurations are calculated, the combinations are all equally likely. Any configuration of microstates is random and equally probable based on its redundancy and energy level. From this fundamental principle of thermodynamics, we can derive the necessary equations and understandings of the second law of thermodynamics that all systems that produce useful outcomes will increase the entropy of the universe. This increased entropy will only lead to disorder and eventually a failure of that system which performs work.

2.1 The Random Walk

The principle of a random walk has been used to describe nearly all fundamental phenomena that occur in nature. It forms the basis for diffusion, as will be seen in this chapter, and is applied to mechanics as well as any type of random event. The formalism that will be presented here is extremely simplistic and one-dimensional, but the methods extend logically to two, three, and more dimensions to model any number of fundamental physical effects, including mechanics that lead to failures.

Picture a board game that is played where a sequence of equally marked squares are lined up and a playing piece starts in the middle at a square marked *GO*(0) and a coin is flipped by the player as shown in Figure 2.1. If the coin comes up a head, the piece moves one place to the right; a tail moves the piece one place to the left. In our game, we will first assume that the heads and tails are equally likely. This game is easy to visualize since after exactly one flip, the piece is equally likely to be one place ahead or one place behind the starting point, but there is zero probability the piece will be on *GO*. The mean position is still 0 and the standard deviation is exactly 1.

After two turns, the piece is twice as likely to be back on square zero as to be two boxes to the right or the left. The probability distribution for the piece residing on box −2, 0, and 2 (the negative sign indicates to the "left") are 1/4, 1/2, and 1/4, respectively, as seen in Figure 2.2. This result is intuitive just by counting the

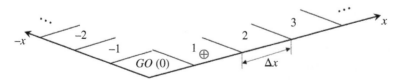

Figure 2.1 A game board representation of a particle moving randomly.

Figure 2.2 Probability distribution of the game piece after two rolls of the dice.

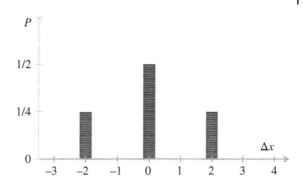

possible places to which the game piece can move after exactly two trials. The mean position is still zero and the standard deviation for two flips is $\pm\sqrt{2}$. This is the definition of a one-dimensional random walk process.

If this game were to be played over again many times, the mean position will always be zero. The symmetry of the game is due to the equal probability of landing a head ($p = 1/2$) and a tail ($q = 1/2$). The distribution function for N flips is binomial with a mean of zero and a standard deviation of $\pm\sqrt{N}$. If we define the spread in x, $\pm\sigma_x = \sqrt{N}$, we see that the spatial confinement squared (variance) is proportional to the number of times the game has been played in each time interval. This gives us what is known as a dispersion relation, where the variance in space, σ_x^2 is proportional to the number of times any motion is made, N, assuming also that exactly one move is taken at every interval.

It is also clear from this game, that if a weighted coin were used, where the probability of flipping a head would be slightly greater than flipping a tail ($p > q$), then the distribution would no longer be symmetric and the mean would drift to the right. The net position of the piece would be determined by the mean along the x direction,

$$\mu_x = (p - q) \cdot N \cdot \Delta x \tag{2.1}$$

and the standard deviation,

$$\sigma_x = \sqrt{4 \cdot N \cdot p \cdot q} \cdot \Delta x \tag{2.2}$$

for any probability, p, and for any number of flips, N, because of the binomial nature of only two randomly occurring possible outcomes.

The probability of being at any point is exactly determined by the binomial series. Thus, these values describe exactly the mean and standard deviation of an ensemble of matter diffusing while under a condition that favors one direction over another. This is shown graphically in Figure 2.3.

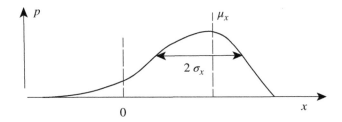

Figure 2.3 Probability distribution of a one-dimensional random walk having a slightly preferred direction over the other.

Suppose further that a second condition was made on the motion of the playing piece, it is only moved if a second condition was met with a probability p_c. This conditional probability could be the result of an energetic barrier to the diffusive process. This is a more common condition in the real world since very often defects are not free to move within a solid. The consequence of the secondary probability (a temperature-dependent term, for example) would be a significant modification of the distribution. The dispersion would be diminished by the factor of this conditional probability. The number of moves taken, N, out of a total number of possible moves, M, is also binomially distributed with a mean number of flips and, thus, we can make the substitution,

$$N = p_c \cdot M$$

The net motion and standard deviation will, thus, be modified from Eqs. (2.1) and (2.2) so that

$$\mu'_x = (p - q) \cdot p_c \cdot M \cdot \Delta x \tag{2.3}$$

where Gaussian approximations hold for the binomial distributions:

$$\mu'_x \approx \sqrt{8M \cdot pq \cdot p_c q_c} \cdot \Delta x \tag{2.4}$$

The result of this conditional probability, which may be dependent on the temperature or a factor other than concentration or time, is a reduction of the net motion for a given total number of flips. The addition of energetic constraints will tend to diminish diffusion as compared to a system with fewer constraints.

2.1.1 Approximate Solution

The actual solution to this problem becomes extremely complicated in the general case because the motion is a random event that is dependent on a second random event. When the probabilities approach 1 or zero, a Poisson distribution must be used, resulting in:

$$\sigma = \mu$$

and joining the probabilities is not a simple task. Yet, we can take the board game analogy one step further. If the time steps are at a regular time interval, Δt, which is the inverse of a frequency, ν,

$$M = \frac{t}{\Delta t} = \nu \cdot t$$

where the frequency is some net average over a long time. Now, the net motion in the direction, x, is directly related to the time, t, which is substituted into Eq. (2.3) to yield:

$$\mu''_x = (p - q)p_c \cdot \nu t \cdot \Delta x \tag{2.5}$$

where μ''_x represents the net motion in time, t. The standard deviations in space with respect to time give the combined probabilities of motion in one direction versus another, p, and the probability that a motion has occurred at all, p_c. Consequently, the standard deviation is:

$$\mu''_x \approx \sqrt{8 \cdot pq \cdot p_c q_c \nu t}\, \Delta x \tag{2.6}$$

2.1.2 Constant Velocity

In the limit where the coin has two heads, or where we say that $p = 1$ and $q = 0$, and the probability of motion at every flip is unity, the probability of landing on a numbered square is exactly unity if a motion occurs with probability A. This is one limiting case of dispersive transport. The net velocity, v, of particles is constant:

$$v_x = \frac{d\mu_x}{dt} = \nu \cdot \Delta x \cdot p_c \tag{2.7}$$

The dispersion of the particle is due only to the conditional probability, p_c, and the standard deviation about the mean is:

$$\sigma_x(t) = \sqrt{4 \cdot p_c q_c \cdot \nu t} \cdot \Delta x \tag{2.8}$$

so long as the distribution is approximated by a normal distribution. A more complicated Poisson distribution will have to be assumed when p approaches 1 or 0. Nonetheless, the dispersion in x is proportional to the square root of time.

The standard deviation is a measure of the net dispersion in space for a given amount of time after the particles begin to move. In the case that the conditional probability of a hop, p_c is very small,

$$p_c \approx 0$$

The standard deviation due to the dispersion is approximated by the Poisson distribution, so:

$$\sigma_x \approx \mu_x$$

and the total dispersion in x reduces to:

$$\sigma_x(t) = \sqrt{4 \cdot pq \cdot \nu t} \cdot (p_c \cdot \Delta x) \tag{2.9}$$

This expression relates the dispersion in space, σ_x, to time, t, and is known as the dispersion relation because it relates the spatial motion with time.

The binomial nature of a random walk process is universally applicable to all sorts of reliability models as well as to statistical mechanics. This is the basic formulation for Markov Processes where there are two possible states and there is a certain probability associated with remaining in either state or changing between states. These are all diffusive processes which are purely statistical in nature. Consequently, it is important not to extrapolate the normal distribution approximation beyond the limits of statistical validity. This is also why real failure events rarely actually fit into a normal distribution and other plotting schemes must be utilized. Again, we need to understand that all our calculations derive from the random occurrences that occur only because they can while invoking the fundamental principle that all possible outcomes are equally likely.

Randomness is the driving force behind all fundamental properties of physics, especially failure mechanisms. Failure events are always statistically distributed. When studying extremely large ensembles of atoms and defects, the randomness leads to exact calculations of the physical distribution of these particles because the statistical deviation is based on the square root of the number of possible outcomes. When the number of events is small compared to the number of atoms making up a system, it becomes impossible to predict with any great precision how long a product will last or when it will fail.

The random walk model helps explain conceptually how diffusion takes place purely because of the randomness of an ensemble. Simply because an object, a particle, or a defect has a probability of being located at a particular place at a particular time, we say that this particle will, in fact, be located there with the determined probability. This is the fundamental theorem of statistical mechanics, which states explicitly that the occupancy of all available states is equally likely in a closed system in equilibrium. It is the belief in this theorem that reliability predictions are made based on knowledge of the inherent physics involved with the failure.

2.2 Diffusion

Diffusion is the most fundamental physical failure mechanism that one can study. This describes nearly all the slowly moving activities that lead to the accumulation of defects or impurities that can lead to failure over a long period of time. We will first develop the formalism for particle diffusion and then derive the equation for thermal diffusion. Both phenomena can occur independently of each other, and both are thermally activated processes. So, temperature gradients can affect particular (particle) diffusion. We mean any distinguishable defect, contaminant, or particle of any type residing in a large mass of basically homogeneous material.

2.2.1 Particle Diffusion

We begin the diffusion formalism with many particles in a one-dimensional system such as was described earlier. Each particle is assumed not to interact with another. So, the motion of any one is completely independent of the others. It can be thought of as many game pieces sharing the same board, again referring to Figure 2.1. We therefore allow any number of particles to be in any box at any time. We will refer to a box with size, Δx, at space number 1 that is in contact with two other equally sized and spaced boxes on either side numbered 0 and 2. We may thus define the flux of particles across the boundaries between the boxes,

$$J_{01} = D(N_1 - N_0)/\Delta x \tag{2.10}$$

where J is the flux from box 0 to box 1, N is the concentration of particles in each box, and D is a constant of diffusivity.

Furthermore, we can assume that the total number of particles in the system is conserved. Thus, any particle may enter box 1 from box 0 or box 2, which has three possible effects; it leaves from where it came, it leaves out the opposite side, or it remains in the box. We can define these motions as fluxes, J, which allows us to express the conservation of particles (mass) in the following way,

$$J_{01} = J_{12} + \frac{\partial}{\partial t}N_1 \tag{2.11}$$

where the flux and particular concentration are functions of time (Figure 2.4).

The concentration of particles in the one dimension, x, is simply the number of particles divided by the spacing. Thus, we can now talk generally about concentration, C, at a given point in space, x, and time, t,

$$C(x_i, t) = \frac{N_i}{\Delta x} \tag{2.12}$$

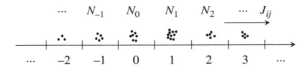

Figure 2.4 A series of equally spaced slots for N_i particles per slot.

which is time dependent. Here we introduced the time and space dependence on concentration. This allows us to define the flux as proportional to a diffusivity, D, by generalization of 2.10:

$$J(x,t) = -D\frac{\partial}{\partial x}C(x,t) \tag{2.13}$$

and the instantaneous gradient of flux at any point is the change in concentration in time as follows:

$$\frac{\partial}{\partial x}J(x,t) = -\frac{\partial}{\partial x}C(x,t) \tag{2.14}$$

which leads to a single differential equation relating the concentration in time and space for any initial condition of C, then after a few substitutions, the diffusion equation is derived:

$$D\frac{\partial^2}{\partial x^2}C(x,t) = \frac{\partial}{\partial t}C(x,t) \tag{2.15}$$

where D is the diffusion constant, also known as the diffusivity.

The diffusivity is simply the rate constant relating the step distance in space, Δx, to the step time, Δt. As can be seen from inspection, and by analogy to the random walk with equal probabilities for motion in two directions, the diffusivity relates net motion in x with time steps, t:

$$D = \frac{\Delta x^2}{\Delta t} = \nu \cdot \Delta x^2 \cdot p_c \tag{2.16}$$

where p_c is some relative probability and ν is some effective frequency. This expression is the same as what we defined before as the dispersion relation, where D describes the dispersion in space with respect to time. Therefore, the dispersion of matter due to diffusion in space is related to the square root of time by the diffusivity, D:

$$L_D(t) = \sqrt{D \cdot t} \tag{2.17}$$

where L_D is the characteristic diffusion length along the x-axis, which is equivalent to the standard deviation.

The exact solution to the diffusion equation depends on the boundary conditions and the initial profile of the diffusion species. In the case of particle diffusion, mass is generally conserved, unless the material is allowed to leave the system. Defects in a solid, however, can be created at a constant rate depending on the stress in the material and the temperature. At the same time, these defects and other impurity atoms will diffuse through the solid at a rate determined by the material properties.

For massive transport, including diffusion of atoms and electrons in a solid, the diffusivity is usually a single energy barrier event. That is to say that it is thermally activated. A thermally activated process is one where the probability of an event taking place is exponentially dependent on the temperature by the following relation:

$$D(T) = D_0 \cdot e^{-\frac{E_A}{k_B T}} \tag{2.18}$$

where D_0 is the infinite temperature diffusivity, E_A is the thermal activation energy of the diffusing species within the medium, k is Boltzmann's constant, and T is temperature in Kelvin. This concept will be developed in the next section.

The diffusion equation is also valid in three dimensions, although the solutions are much more intricate and less intuitive. More generally, the equation may be expressed as:

$$D \cdot \nabla^2 C(x, y, z, t) = \frac{\partial}{\partial t} C(x, y, z, t) \tag{2.19}$$

and can be solved for any set of boundaries and initial conditions. The basic form of this relation that distinguishes it as the diffusion equation is that the second derivative in space is directly proportional to the first derivative in time.

This equation only solves the random walk problem with exactly equal probabilities of diffusion in any direction. There is no drift or forcing term in this expression. So, it does not solve the general problem described earlier, which includes random walks with an inequity in likelihood of moving in one preferred direction over another.

2.3 Solutions for the Diffusion Equation

A general solution may be found for this equation by method of separation of variables. This technique will always find a set of solutions for a Fourier series. A solution to the diffusion equation will always be found for the concentration as a function of x and t may be guessed using the principle of separation of variables, to have the form:

$$C(x, t) = X(t) \cdot T(t) \tag{2.20}$$

where $X(x) = e^{\frac{x}{\lambda}}$ and $T(t) = e^{\frac{t}{\tau}}$.

For a characteristic length constant, x, and a time constant, t. Since X and T are independent of each other, the diffusion equation can be expressed as

$$D\frac{X''}{X} = \frac{T'}{T} \tag{2.21}$$

which is solved parametrically by equating the two constants in space and time:

$$\frac{D}{\lambda^2} = \frac{1}{\tau} \tag{2.22}$$

or, by rearranging, the following is received:

$$\lambda = \sqrt{D \cdot \tau} \tag{2.23}$$

where λ is the diffusion length occurring over a time scale, τ. By analogy to the random walk problem for $p = q = 1/2$, the standard deviation in x is proportional to the square root of the elapsed time, t, which is precisely the same as the diffusion length, L_D.

Two solutions to the one-dimensional diffusion equation approximate the form of transient diffusion in a system of infinite extent for very short times. The boundary conditions are that the concentration approaches zero as $x \to \infty$, and the concentration is finite at $x = 0$. There are only two general forms of the solution to this equation: a Gaussian and an error function, *erf* (alternately, the complementary error function, *erfc*).

2.3.1 Normal Distribution

The Gaussian solution solves for the boundary condition that the total dose or integral of the concentration is constant,

$$K = \int\limits_{-\infty}^{\infty} C(x,t)dx \tag{2.24}$$

for all time, t. The exact solution for one-dimensional diffusion from the origin $(x = 0)$ is

$$C(x,t) = \frac{K}{\sqrt{2\pi \cdot D \cdot t}} \cdot e^{\frac{x^2}{2Dt}} \tag{2.25}$$

where D is the diffusivity of the species.

The important characteristic of a Gaussian diffusion distribution is that the prefactor to the exponential decreases at the square root of time increases. This solution is exact in one dimension if the boundaries are very far away compared to the

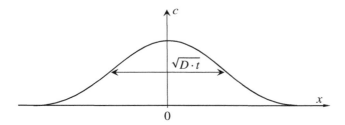

Figure 2.5 A normally distributed concentration for a given time.

diffusion length, L_D. In two or more dimensions, the solution becomes much more complicated and beyond the simple treatment presented here, but the main idea remains the same and the concentration at the origin of the diffusive species decreases with the inverse square root of time (Figure 2.5).

For one-dimensional diffusion from a finite boundary into a semi-infinite half-space, the Gaussian solution also holds true except the concentration factor, K, must be divided by 2 to account for the half-space integral from $x \rightarrow \infty$. If the total dose of the diffusive species is present at an initial time, $t = 0$, the concentration as a function of time follows the Gaussian solution 2.25 until the diffusion length approaches the boundary of the sample or the temperature decreases, lowering the diffusivity.

2.3.2 Error Function Solution

The second exact solution to the semi-infinite one-dimensional diffusion equation has the form of an error function (*erf*). The error function is often defined in different ways, but its shape is always equal to the integral of a Gaussian. The error function can be defined here as:

$$erf = 2 \int_0^\infty \frac{K}{\sqrt{2\pi \cdot D \cdot t}} \cdot e^{\frac{x^2}{2Dt}} \cdot dx \qquad (2.26)$$

where x is the relative position in the one dimension. This definition of the error function may differ from other tables, but it is normalized here to equal 0 at $x = 0$, and 1 as $x \rightarrow \infty$. When using tables for exact values of *erf*, it is important to be sure that the solution is normalized correctly.

One minor problem exists with the solution in Equation 2.26, the boundary conditions do not match those required for a constant finite concentration at $x = 0$. Thus, we have to take the complimentary error function, *erfc*, which will be defined here as:

$$erfc = 1 - erf \qquad (2.27)$$

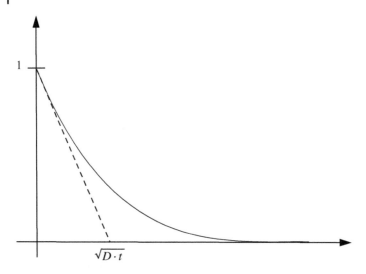

Figure 2.6 A complementary error function solution to the diffusion equation.

which goes from 1 at $x = 0$, to 0 as $x \to \infty$. This form is the normalized solution for a constant surface concentration for all time:

$$C(0, t) = 1 \tag{2.28}$$

and the characteristic diffusion depth is still described by:

$$L_D = \sqrt{D \cdot t} \tag{2.29}$$

for any $t > 0$ (Figure 2.6).

These transient solutions are limited to the time frame of the diffusion occurring in an infinite extent. As the time approaches a characteristic amount depending on the thickness of the sample, the diffusion approaches a steady state and these two solutions are no longer valid.

2.3.3 Finite Thickness

Assume that a material has a length, L, and the diffusion begins at one end, $x = 0$. The characteristic time for diffusion to proceed across this sample is found simply by rearranging to find:

$$t_D = \frac{L^2}{D} \tag{2.30}$$

where t_D is the characteristic diffusion time through a fixed length, L. For times much shorter than t_D, the material can be considered infinite in extent. However, when $t > t_D$, the solution approaches that of a steady-state diffusion.

The steady state is defined as when the concentration profile becomes independent of time,

$$\frac{\partial}{\partial x} C \to 0 \qquad (2.31)$$

which changes the diffusion equation, (2.19) to that of a Laplacian,

$$D \cdot \nabla^2 C = 0 \qquad (2.32)$$

which is easily solved in one dimension. The solution to this equation will always be a straight line with a constant slope.

The boundary conditions will determine the slope and the total concentration, but the general form will remain the same. If there is a source and a sink on either side of the material, then this equation represents a constant flux across the length as assumed before in Eq. (2.10). Hence, we have come full circle and have shown that diffusion is a passive process which occurs only because of randomness on the atomic scale. If the concentration of diffusing species is statistically very large, an exact profile can be calculated given the appropriate boundary conditions. Again, we see that these natural processes proceed in accordance with the fundamental principle that all possible outcomes are equally likely.

2.3.4   Thermal Diffusion

Now that the fundamentals of diffusion have been described for particles, the same treatment can be extended to thermal diffusion in solids, or in nonconvective liquids or gases. Heat diffuses because of the same basic mechanism as particles. Molecules or atoms at a certain kinetic energy can randomly impart their energy to adjacent molecules. The energetic flow through the medium will impose a temperature change and is regarded as heat flow through that medium.

Every material has a thermal conductivity associated with the propensity or the resistance to transfer thermal energy. This is a material property which is often dependent on the temperature itself but will be assumed to be constant while we develop the formalism. We begin by taking a cross-section of a material of width, Δx, which is uniformly at a temperature, T_1 (one-dimensional), on one surface and at a second temperature, T_2, at the other surface. The thermal flux density, Φ, (heat per unit area) is proportional to the negative gradient in temperature:

$$\Phi(x, t) = -\sigma_{th} \cdot \frac{\partial T}{\partial x} \qquad (2.33)$$

where σ_{th} is the thermal conductivity.

In like manner to the particle diffusion, we know that the change in thermal flux of a system is related to its temperature by the mass density, ρ_m, and constant volume heat capacity, C_v by conservation of energy, in one dimension (Figure 2.7).

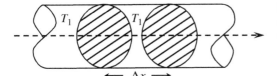

Figure 2.7 A representation of one-dimensional thermal conduction through a volume.

$$-\frac{\partial}{\partial x}\Phi(x,t) = \rho_m \cdot C_v \cdot \frac{\partial T}{\partial t} \tag{2.34}$$

This states that the difference in flux entering a body and leaving that body results in a net temperature rise in time. These are Fick's laws of diffusion and allow us to derive the diffusion equation for heat in solids or gasses or any material with a density, ρ_m.

By differentiating Φ in Eq. (2.31), with respect to x, to get a second derivative and substituting into Eq. (2.32), we get:

$$D_{th}\frac{\partial^2}{\partial x^2}T(x,t) = \frac{\partial}{\partial t}T(x,t) \tag{2.35}$$

which is the same diffusion equation as (2.15) where:

$$D_{th} = \frac{\sigma_{th}}{\rho_m \cdot C_v} \tag{2.36}$$

which is the thermal diffusion coefficient, also called "diffusivity".

Unlike massive diffusion, the constant for thermal diffusion is not necessarily independent of temperature, but temperature dependence on D_{th} makes Eq. (2.33) a very difficult equation to solve in closed form. However, engineering approximations can often be made to make the solution more tractable. Otherwise, there are several numerical simulation tools that solve the diffusion equation self-consistently for temperature dependent diffusivity, D_{th}.

Of course, heat diffuses equally in three dimensions as well, so the generalized diffusion equation becomes like (2.19)

$$D_{th} \cdot \nabla T = \frac{\partial}{\partial t}T \tag{2.37}$$

where the thermal diffusivity is the same in three as it is in one dimension. The solutions to this relation are in general more complicated than in one-dimension, but the relative shape of the profile is the same. These diffusion equations are important when changes are occurring with time. After a long time has elapsed and the temperatures have equilibrated, the second term goes to zero and we end up with steady-state conduction based on the temperature differences at the boundaries.

Thermal diffusion is almost always isotropic within a uniform material. Unlike particular (particle) diffusion which can be affected by temperature gradients, or charged particles, which can be forced by an electric field, thermal diffusion always occurs equally in all three dimensions within a homogeneous material. This is a passive phenomenon since there is little that can force a material to be thermally diffusive in a preferred direction within a homogeneous material. Interfaces and layered materials, on the other hand, can be engineered to directionally enhance and inhibit thermal diffusion in various directions.

2.4 Drift

Drift is a reversible, linear, net flow of particles (ions or electrons) under the influence of an electric or other force-inducing fields. A force that is exerted on a system of particles will tend to induce motion in one preferred direction. In the case of ions in an electric field, this is described by Ohm's law:

$$J_s = \sigma_s \cdot E \tag{2.38}$$

which states that the current density of a particular species, s, is proportional to the electric field, E. The current density is directed along the lines of the electric field. This drift component is equivalent to the increased probability of diffusion in one preferred direction.

The way to generalize the diffusion equation to include a propensity to diffuse in one direction, analogous to the lopsided random walk, is to include a velocity term in the diffusion equation. This term will be a vector quantity, such as $\overline{v_z}$.

$$D\nabla^2 C + \overline{v_z} \cdot \frac{\partial}{\partial z} C = \frac{\partial}{\partial t} C \tag{2.39}$$

where z is the axis of net drift. This is the combined drift-diffusion equation that describes any random motion with a driving term that will result in some net motion of particles. Of course, any number of additional driving terms may be added to this expression and solved rigorously to fully describe the distribution of particles: be they impurity atoms; defects; electrons in a semiconductor; or any diffusive species in a gas, liquid, or solid.

In the case of N_s, species available for drift, allowing each having a charge, q (for example, with ionic species or electrons), the net current density can also be described by the equation:

$$J_s = q \cdot N_s \cdot \overline{v_z} \tag{2.40}$$

where the actual distribution of particle velocities is a random number with an average value of $\overline{v}$.

The actual motion is due to random accelerations and collisions in the general direction of the forcing field. Drift is a linear motion, equivalent to having the probability of motion in one direction be certain, with $p = 1$ (and $q = 0$), so long as a transition occurs with probability (p_c) as described in Eq. (2.5). This average velocity is assumed to be proportional to the electric field in the case of simple (linear) drift:

$$\overline{v_z} = \mu_s \cdot \overline{E_z} \tag{2.41}$$

where E_z is the electric field in the z-direction and μ_s is the linear species mobility, affected by the field.

Of course, we should keep in mind that this is usually only a first-order approximation since the influence of the field on velocity is usually much more complicated, but it can always be linearized to first order. Thus, the drift component of particle motion is linearly affected by the electric field, while the drift component is anisotropic (equally likely in any dimension). Hence, one can predict these two components separately and combine them linearly to obtain a combined drift and diffusion relation.

Nearly the entire field of semiconductor device physics and nonequilibrium thermodynamics is involved with solving Eq. (2.37) in various configurations. The solution to this is well beyond the scope of this introductory treatment, but it is instructional to know the derivation of the equation. In terms of failure physics, these are the most important relations that control the degradation of materials and components under normal usage.

2.5 Statistical Mechanics

We have discussed briefly in the previous section the energy associated with transport and heat transfer. The energetics of a system and the interaction of these systems determine the probabilities of one event happening over another. Diffusion, as we have shown, is the consequence of an ensemble of random events which leads to net motion of atoms, defects, heat, or anything else. What drives diffusion is simply the fact that it can happen, and a diffusion profile is precisely described by the probabilities associated with the moves. The solutions to the diffusion equation are probability functions and we assume that the ensemble of diffusive species is sufficiently large that the concentration profile is predicted by the probability function. The predictability is a direct consequence of this fundamental principle that all possible outcomes are equally likely.

Statistical mechanics offers the ability to predict the precise distribution in time and space of an ensemble of particles, atoms, impurities, or anything else based purely on the assumption of randomness. This is the principle that justifies the idea that failures and accumulations of defects can be determined. This idea is what drives a great deal of effort in using the physics of failure approach to analyzing the reliability of a product or a component. The limitation on predicting behaviors of materials in mechanical and electronic design is only in our human ability to keep track of all the energies associated with all the parts and interfaces. It is, therefore, instructional to include some fundamental principles of statistical mechanics to help elucidate the ways one might go about predicting failure mechanisms.

By calculating the energies of a system of particles, it is possible to determine the most likely configuration and the variance from that configuration simply by determining the distribution of energies associated with each. In the following sections, we will cover some of the basic consequences of these fundamental statistical mechanics principles and how they apply universally to many aspects of physics. The following concepts are used to describe most of the physical mechanisms that can be used to model failures in mechanical and electronic systems.

2.5.1 Energy

Energy is a scalar quantity that has two forms: potential and kinetic. Whenever energy is used to describe the state of a particle or a system of particles, there is always a relative energy that is implied. In the case of gravity, for example, we can visualize relative potential energy as being proportional to the height of an object over the place it may fall. If the object is dropped, its potential energy will convert to kinetic energy and a new equilibrium will be established.

In the case of a massive object being dropped from a height, h, it always has a potential energy that is relative to another point in space that is lower, until it reaches the ground. When, it is said to be at its lowest possible energy, literally called the ground state. Once in the ground state, an object will not be able to transfer any more of its energy into another system unless someone takes the object and moves it back to a higher place.

Potential energy is the general quantity of energy that can be transferred to its surrounding environment through some mechanism; like being dropped. Kinetic energy is thus defined as the energy contained in motion of particles that are transferred from one potential energy state to another. The quantity of kinetic energy imposed on another body is referred to as 'work' and has the same units as energy. Work is the result of kinetic energy that is transferred but cannot be recovered. Work is, thus, the nonconserved energy that is lost by the system but added to the rest of the universe, thus increasing total entropy of the surroundings.

The total energy of a system is the sum of the kinetic and potential energies. Kinetic energy may be converted back into potential energy or, otherwise, escape in the form of heat, sound, or some other form of work. As an example, consider a weight being dropped from a height in a vacuum (no heat due to friction of air). As the object falls, its potential energy due to gravity is converted to kinetic energy. When the object hits the ground, two things can happen: it can bounce back up (conservative), or it can crash into the ground and stop (nonconservative). More likely than not, some combination of both will occur. Some energy will be converted to sound and heat and some of the energy will go into plastic deformation of the mass itself or the surface (a crack or a deformation of some sort). The remaining elastic energy will be redirected back to the object as a recoil, causing it to bounce.

We say that the total energy in the system has potential energy, P, kinetic energy, K, and lost energy due to work, W. The sum of these energies is constant throughout the experiment. All three values are a function of time and location along the downward trajectory of the object (whether the object is a rubber ball or a lead weight):

$$H = P + K + W \tag{2.42}$$

where H is the total energy of the system. This expression is known as a Hamiltonian and can be used to describe any system and the energy associated with it. A Hamiltonian will always have the form of potential energy, kinetic energy and source or sink terms. The units of all these terms are the same, and they are units of energy. If the loss term, W, is zero for all time, the system is completely conservative, and it will result in oscillatory behavior.

For example, take a ball in a completely conservative system. There is no friction and all the energy upon impact is directly transferred back to the ball. The resulting motion is completely oscillatory. The position of the ball can be determined at any time simply by determining the energies associated with the closed system. In this idealized system with no loss or increase in entropy, the potential energy, P, at any height off the ground, z, is:

$$P = mgz \tag{2.43}$$

where m is the mass of the ball and g is the gravitational constant (Figure 2.8).

The kinetic energy, as a function of z at any time $t > 0$, is:

$$K = \frac{1}{2}m\left(\frac{dz}{dt}\right)^2 \tag{2.44}$$

where the derivative of z with respect to time is the instantaneous velocity.

So, if we know the initial height from which the ball was dropped, h, then the total energy at any time is the same as the energy at that time when $K = 0$.

$$H = mgz + \frac{1}{2}m\left(\frac{dz}{dt}\right)^2 = mgh \tag{2.45}$$

Figure 2.8 A ball drop from height, *h*.

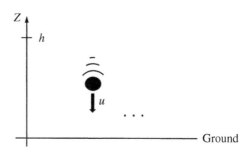

This can be expressed as a single relation for all time after dividing by *m*,

$$\left(\frac{dz}{dt}\right)^2 = 2g(h - z) \tag{2.46}$$

This can be solved *z* for $0 < z < h$,

$$z(t) = h - \frac{1}{2}g\,t^2 \tag{2.47}$$

where $0 \leq t \leq \sqrt{2h/g}$.

Of course, in this example, the ball will bounce up again along the same trajectory, substituting $2\sqrt{2h/g} - t$ for $t > \sqrt{2h/g}$, as seen in Figure 2.9. The period of oscillation is $2\sqrt{2h/g}$; the time to drop and return to its original position. The difference in final height from the initial height is an exact measure of all the work done to the floor, air, and the bail itself as a result of the impact. By keeping track of the energies of the states of a system, it is possible to predict the physical behavior or measure the net energy that was lost to the atmosphere.

Figure 2.9 Position versus time of the bail dropping and bouncing without loss off the surface.

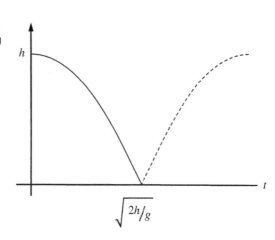

This energy-balance approach demonstrates a powerful way to model physics of a system. In principle, one should be able to calculate the energies associated with various possible configurations of defects, atoms, impurities, etc., and solve for the distributions as a function of time and space. The solutions to very complex physical problems are solvable, in principle, given enough information about the material properties, external stresses, and computational resources. The limitation of physics-based reliability models is our ability to provide all the necessary information and for computers to handle all the data.

2.6   Chemical Potential

The chemical potential of an ensemble of particles is simply an energy term related to the temperature and concentration of these particles. Temperature is a measure of the average energy of a closed system. This energy is related to the molecular motion by Boltzmann's constant, k. The value for k is in energy per unit temperature, and the temperature scale is always absolute temperature in Kelvin. Boltzmann's constant,

$$k_B = 8.617 \times 10^{-5} \, eV/K \tag{2.48}$$

where electron volts are chosen as a common unit of energy when describing thermal activation as will become apparent later on.

The chemical potential, E_μ, of a gas is related to k by the natural log of the number of molecules in the ensemble:

$$E_\mu = k_B T \cdot ln\left(\frac{N}{N_0}\right) \tag{2.49}$$

where N is the number of molecules in the closed system and N_0 is a relative number for the arbitrary ground state of the system.

The chemical potential is only meaningful in relation to another system, in which case the ground state terms will cancel. In like manner to the gravity problem, the number of molecules in a closed system can be replaced by a density of these same molecules per unit volume:

$$n = \frac{N}{V} = P/RT \tag{2.50}$$

where P is the pressure, R is the ideal gas constant, and T is the absolute temperature in Kelvin.

The chemical potential is the energy associated with a concentration of a species of molecules. It is also a way we can benchmark the energy of diffusive species, defects, or any uncharged particle. Diffusion is governed by the energetics of a system of particles or molecules. If there is a concentration gradient, diffusion will proceed until equilibrium has been achieved and there is no more net flux of particles. The chemical potentials will also tend to become equal and the concentrations across any arbitrary boundary will tend to become equal. This is equivalent to saying that the energy of a system will approach a minimum as equilibrium is reached.

EXAMPLE: How can we determine the air pressure as a function of height, assuming equal temperature, in the atmosphere? As we said, the pressure is related to the density, so we can express the chemical potential as a function of the relative pressure,

$$E_\mu(z) = k_B T \cdot ln \frac{P(z)}{P_0} + E_g(z) \tag{2.51}$$

for any height z and gravitational energy E_g. We also know that the gravimetric energy is related to z as in the previous example:

$$E_g(z) = mgz \tag{2.52}$$

where E_g is the energy associated with the gravitational potential and m is the mass per molecule of air.

We now only need to solve for $P(z)$ under the condition of diffusive equilibrium and the chemical potential becomes equal for all z:

$$k_B T \cdot ln \frac{P(0)}{P_0} = k_B T \ ln \frac{P(z)}{P_0} + mgz \tag{2.53}$$

which can be solved for $P(z)$,

$$P(z) = P(0) \cdot e^{\frac{mgz}{k_B T}} \tag{2.54}$$

and the term, P_0, drops out, as will always happen since it is a relative ground state.

The chemical potential is the same for all identical molecules in a system. For noninteracting molecules, relative concentrations will interdiffuse based on the chemical potential of each other. In equilibrium, the chemical potentials become equal and the whole system can be described by a single energy state determined by the temperature.

If two dissimilar materials are in diffusive contact with each other, there will be a continual interdiffusion across the junction since equilibrium will never be reached until both materials have become saturated with the other. Fortunately, however, diffusion is thermally activated, and the diffusion rate is much slower for solids than gasses at room temperature. This is very important from a reliability point of view since extremely slow diffusive processes will speed up at higher temperatures and can result in eventual failures or degradation.

If a species is charged, such as an electron or an ion, the chemical potential is augmented with the electrical potential by adding the appropriate energy term related to the charge and surrounding field. In like manner, as expressed above, any other energetic considerations can simply be added to the energy Hamiltonian and solved.

2.6.1 Thermodynamics

Diffusion is irreversible. This means that once a species diffuses from one region to another in a closed system, the total energy is reduced by increasing the entropy of the universe. In the case of air diffusing into a previously evacuated chamber, it is obvious that one cannot just sit and let the air back out again. This is because the chemical potential of the total system reached equilibrium, and the net energy of the system was minimized. Additional energy will be required to extract the air again. Again, this is a consequence of the inherent increase in entropy associated with minimizing free energy resulting from the fundamental principle of thermodynamics.

When a system transfers its energy from one state to another (potential to kinetic), the net result is no loss of energy at best. This is the first law of thermodynamics. In the bouncing ball example given previously, the highest the ball can possibly bounce would be to its original position. However, the ball will never make it to that original position because there will always be some energy lost somewhere along the way, either from friction, hitting the ground and making a noise, or something else. This something else is the nonconservative imperative of nature which makes perpetual motion impossible. This is known as the second law of thermodynamics.

The third law of thermodynamics states that a closed system will always seek its minimum energy configuration. This is equivalent to saying that entropy will be maximized for a closed system in equilibrium. This is the property that assures that molecules will always be around to fill a vacuum. As long as there are no barriers preventing air from entering a room or a chamber, you can be very sure that you will be able to breathe wherever you go, as long as the door is open.

The third law of thermodynamics is often expressed in the following way:

$$E_f = \sum_i E_i + T\Delta S \tag{2.55}$$

where E_f is the final energy, E_i are all the initial energies of the subsystems, T is the absolute temperature, and ΔS is the increase in entropy. The units of entropy are in energy per temperature since the temperature and all the other thermodynamic properties are related to the entropy. The entropy, S, describes the degree of randomness of a system.

An irreversible process takes place because states are available for the particles to fill, and eventually, all possible states are equally probable. The increase in entropy is related to the energy that escapes a system in the form of work. So, any system that does work always increases the entropy of the universe and is, therefore, irreversible. Every process which does anything will increase the number of possible energy states available to the overall system, and entropy will increase as the randomness increases.

2.7 Thermal Activation Energy

In terms of statistical mechanics, the temperature of a system describes the average energy of every molecule in the system. If there is unrestricted diffusive contact within an entire system, it will eventually become uniform in temperature in isolation from all other systems. The average molecular thermal energy of the system is therefore exactly described by the absolute temperature multiplied by Boltzmann's constant, $k_B T$.

Classical massive particles are distributed in energy by Maxwell-Boltzmann statistics, which state that the probability of finding a particle with energy, E, is related to the temperature by the average thermal energy, $k_B T$, by the relation:

$$f(E) = e^{\frac{-E}{k_B T}} \tag{2.56}$$

which is an exponential distribution of energy. The average, or expected, energy is exactly $k_B T$. This relation is valid for a large ensemble of particles that have no spin. Gases and atoms, for example, follow this distribution. Electrons, phonons, and photons, on the other hand, are quantum mechanical particles and do not follow this law precisely. However, when $E \gg k_B T$, this is a very good approximation for nearly any energetic distribution of any type of particle.

The result of knowing the distribution is that if the relative energy associated with a given state is known, the relative occupancy of that state is exactly determined from these statistics. Conveniently, the absolute energies do not have to be known when the probability density function is strictly exponential as in Eq. (2.53). Only relative energy levels are needed, and the probabilities can all be normalized to any one energy level.

Boltzmann statistics are used in a very straightforward manner. If the energy of a system of particles is E_0 and the energy associated with a particular state is E_1, then the relative concentration of particles in energy state, E_1, is:

$$\frac{e^{\frac{-E_1}{k_B T}}}{e^{\frac{-E_1}{k_B T}} + e^{\frac{-E_0}{k_B T}}} \tag{2.57}$$

where the denominator is the total probability of both states being occupied. Stated in terms of statistics, this is the conditional probability that the particle is in state 1, given that it is either in state 1 or state 0.

This expression can be rearranged to describe the probability of particles with sufficient energy to surmount a relative energy barrier, which can be stated as being a relative activation energy, E_A, where:

$$E_A = E_1 - E_0$$

for a given system at a uniform energy. Thus, the probability of particles being in one state versus another is related to the difference by the following relation:

$$p(E_a) = \left(1 + e^{E_A/k_B T}\right)^{-1} \tag{2.58}$$

which allows us to determine, from first principles, the likelihood of an energetic event simply by knowing the relative energies and the temperature (Figure 2.10).

We can refer to the random walk formalism for diffusion along a single axis where the slots are equally spaced by A_x and the transitions occur at an average frequency, v. The probability of a transition is dependent on the energy associated with a diffusive event, E_A, and the temperature, T, by Eq. (2.55) which modifies the diffusivity as defined by Eq. (2.16) by the same amount. Hence, the diffusivity can be seen to equal:

$$D(T) = D_0 \cdot \left(1 + e^{E_A/k_B T}\right)^{-1} \tag{2.59}$$

where $D_0 = v \cdot \Delta x^2$ and the average transitional frequency is $1/\Delta t$. This is the exact diffusion relation which is valid for a single activation energy diffusive process for all E_A and $k_B T$.

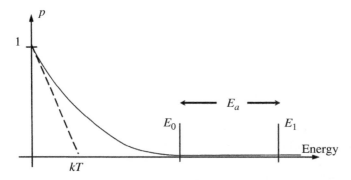

Figure 2.10 Probability function for energies of particles in a classical system.

When the activation energy is much larger than $k_B T$, which is true for most interesting phenomena, the absolute probability of occupancy is very small. This probability can be approximated by the Boltzmann relation since the exponential term will be much greater than 1.

$$p(E_a) \approx e^{\frac{-E_A}{k_B T}} (E_A \gg kT_B) \tag{2.60}$$

and the diffusivity of a particular species can be approximated by Eq. (2.18), but when the activation energy is much smaller than $k_B T$, the probability is nearly unity. As was stated earlier, diffusion is thermally activated and is generally found to be a single activation energy process where the activation energy is relatively large compared to $k_B T$. Thus, diffusion is a prime example of a process that can lead to failures which can be accelerated by raising the temperature of a system. This probability is precisely the conditional probability, A, that regulates a diffusion event as seen with respect to the random walk.

2.7.1 Arrhenius Relation

All thermally activated processes are ones where events occur probabilistically when enough events at a particular energy have occurred during some time interval. In the case of diffusion, if a particular piece of material has a fixed length, the time for a critical number of defects to diffuse that length is related to the time to fail, t_f as:

$$t_f = D^{-1} \cdot L_D^2 \tag{2.61}$$

where D is the diffusivity, which is related to the temperature by Eq. (2.18). Thus, the time to fail is related to the activation energy by what is known as the Arrhenius relationship:

$$t_f = t_0 \cdot e^{\frac{E_A}{k_B T}} \tag{2.62}$$

so that the relationship between temperature and expected lifetime is determined with very high confidence.

Any process which is regulated by energetic potentials and occupation probabilities is accelerated by increased temperature. Atoms and defects are always moving with their distributions in energy. The probability that a certain number of defects will occur over a given amount of time can be expressed as a rate constant that depends on an activation energy, E_A. If an error rate due to a single thermally activated process is determined to have a certain probability at a given temperature, then raising the temperature should raise the rate or decrease t_f as described by Eq. (2.62).

The error or failure rate is often plotted on a logarithmic scale versus $1/T$, where T is the absolute temperature. The slope of this rate constant is proportional to the activation energy and allows a way to predict behavior at lower temperatures by linear extrapolation. The Arrhenius relation is one of the only truly physics-based ways one can determine the mean time to fail (*MTTF*) of a single failure mode at operating conditions based on accelerated testing. If a single activation energy process is justified, then an Arrhenius plot can be an extremely powerful way to predict failures based on real experimental data.

An example of a thermally activated process is any chemical reaction. Most exothermic reactions occur because of a thermally activated process. Raising the temperature will accelerate mort chemical and surface reactions allowing the results to be extrapolated to a room temperature lifetime or to another operating condition. Thus, this tool is extremely powerful when determining how to model the physics of a system and design accelerated tests to failure.

Caution must be used, however, when utilizing the Arrhenius model on systems that have multiple failure modes or where the failure is not due to a thermally activated process. This method of extrapolating lifetimes is limited exclusively to situations where all the physics presented here is valid. Catastrophic failures due to shock or overstressing, for example, cannot be extrapolated by this relation. Furthermore, if more than a single failure mode is expected, this relation cannot predict lifetime.

2.7.2 Einstein Relation

Although the drift and diffusion are independent events, the former depending on electric field and the other depending on temperature and concentration gradients, they both are limited by more fundamental properties. Both are statistical processes which are thermally induced with a fixed activation energy. If we assume the mechanism behind increasing the probability for a transition in one direction is a lowering of the barrier from E_A to $E_A - \Delta E$, then we must assume an increase of ΔE in the opposite direction. This energy change is directly proportional to the field, E, and the distance between hops, Δx, along the direction of the field.

As we learned earlier, if there is diffusion in one dimension where there is an increased probability to move in one preferred direction, the velocity is found from Eq. (2.5) by taking the relative probability of moving to the right, p, minus the probability of moving to the left, $q = 1 - p$, times the transition rate, $\Delta x/\Delta t$. Thus, the velocity can be described as:

$$\overline{v_z} = \mu \cdot \overline{E} = \frac{\Delta x}{\Delta t}\left(1 - e^{q\Delta E/k_B T}\right) \tag{2.63}$$

assuming that: $\Delta E = \overline{E}\Delta x$.

This expression simplifies in the limit that $\Delta x \to 0$,

$$\mu \cdot \overline{E} = \frac{qE \cdot \Delta x^2}{k_B T \cdot \Delta t} \tag{2.64}$$

we get a simple relation between μ and D:

$$D = \mu \frac{k_B T}{q} \tag{2.65}$$

which is a very fundamental equally called the *Einstein Relation* and relates to all combined drift and diffusion processes in quasi-thermal equilibrium.

2.7.3 Magnitude of Energy

The average energy at room temperature is approximately 0.0259 eV. The size of an energy barrier must be in order to have any pronounced retarding effect must be at least several times $k_B T$. Hence, the smallest activation energy that will usually be observed to have a measurable effect is a few tenths of an eV. If the activation energy is smaller than this, or the order of $k_B T$, then the diffusivity saturates at its ultimate limit of D_0 and is not thermally dependent.

On the other hand, if we expect to see an effect in our lifetime, of the order of 10^9 s, we can figure out the largest energy that will have this kind of time constant. All failures that will concern people will be observed over the span of seconds to lifetimes. So, nine orders of magnitude is a very large estimate of the complete range of time frames of interest at room temperature:

$$\frac{E_{a_{max}}}{E_{a_{min}}} \approx ln\left(10^9\right) \approx 20 \tag{2.66}$$

Thus, if the smallest energies of interest are of the order 0.15 eV, the largest energy barriers that are meaningful will be around 3 eV. If a barrier is more than a few eV in energy, the temperature will have to be raised significantly to force the event to occur in a time span of less than a few hours or weeks.

2.8 Oxidation and Corrosion

Oxidation and corrosion are two names for what is basically the same process. Corrosion occurs because of surface oxidation which changes the nature of the original material over time. Iron rusts, silicon oxidizes to form SiO_2, aluminum forms alumina, etc. Nearly all metals, with very few exceptions such as gold, will oxidize in air at room temperature. What causes a material to oxidize is that there is an energetically favorable state available to the surface of a metal in the presence of oxygen molecules so that the intermetallic bonds break in favor of forming metal-oxygen bonds. The reaction rate depends on the surface chemistry, oxygen concentration, reactive species, and of course, temperature.

The generic chemical reaction is described for a metal, M, in the presence of oxygen, O_2, or water, H_2O forms a metallic oxide and energy:

$$M + \frac{x}{2}O_2 \rightarrow MO_x + \Delta H \tag{2.67}$$

where ΔH is the energy released as a result of the reaction, called the Enthalpy. This reaction only proceeds if $\Delta H > 0$. So, it is thermodynamically favorable. This reaction will also proceed in the presence of water:

$$M + xH_2O \rightarrow MO_x + H_2 + \Delta H \tag{2.68}$$

which is generally more energetically favorable. Consequently, materials oxidize, corrode, or rust (depending on the material) more quickly in a moist environment than in a dry one (Figure 2.11).

2.8.1 Reaction Rate

The surface reaction rate is directly proportional to the concentration of the oxidizing species (O_2 or H_2O) that reaches the metallic surface. This relates to the concentration that is able to diffuse through the oxide thickness that has already formed since the reaction occurs only at the oxide-metal interface. The three steps of oxide formation are illustrated by the following diagram, which shows the relative concentrations of oxidizing species at the two interfaces: the gas that reaches the surface and the concentration that diffuses the oxide to react at the oxide-metal interface.

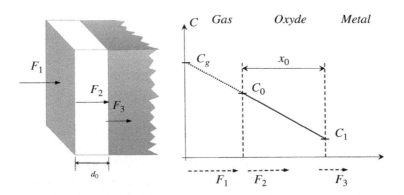

Figure 2.11 Representation of simple oxidation due to diffusion of oxygen molecules.

The molecular fluxes, F_1–F_3, represent the three events necessary for oxidation:

F_1—Transport to the surface from the atmosphere.
F_2—Diffusion through the existing oxide.
F_3—Surface reaction rate at the interface.

The first flux is linearly related to the concentration of the oxidant in the atmosphere just beyond the exposed interface, C_a, and the equilibrium concentration just inside the oxide, C_a, by the mass transfer coefficient, h_G,

$$F_1 = h_G(C_a - C_0) \tag{2.69}$$

The concentrations are directly proportional to the partial pressure of the oxidant in the atmosphere, which is related to the temperature. The transport term is thermally activated since it represents the susceptibility of the oxide to absorb oxygen or water.

The second flux is also thermally activated since it is the steady-state diffusion of oxidant through the existing oxide at any time,

$$F_2 = \frac{D(C_a - C_0)}{d_0} \tag{2.70}$$

where d_0 is the oxide thickness and C_i is the concentration at the interface, and D is the molecular diffusivity of the oxide. This is the only place where the oxide thickness, do, comes into the relation and it is the variable, which is solved for d_0, for all time.

The third flux is the surface reaction rate constant, k_s, which is proportional to the concentration at the interface,

$$F_3 = k_s C_i \tag{2.71}$$

which is related to the surface chemistry and the enthalpy of formation. This final constant is also thermally activated since it depends on the energetics of the oxidation process. The chemical reaction is determined by the total energy of the oxidant present at the metallic surface.

We can assume that the oxidation process takes a long time, and all three processes are in steady state. This condition simply says that all three fluxes become equal since the concentration profiles do not change on the time scale of the diffusion. This allows us to equate all three fluxes as our boundary condition and set them equal to the net flux, F,

$$F = F_1 = F_2 = F_3 \tag{2.72}$$

The flux alone is related to the rate of oxide growth rate which will allow us to solve for the oxide thickness, d_0, as a function of oxidant concentration, C_a, at a given temperature by:

$$F = N_s \frac{d(d_0)}{dx} \tag{2.73}$$

where N_s is the molecular surface concentration. This is the absolute number of bonds that can form an oxide at the surface per unit area. The algebra is straightforward and results in a quadratic relation for d_0:

$$d_0^2 + A \cdot d_0 = B(t + \tau) \tag{2.74}$$

where $A = 2D(k_s^{-1} - h_G^{-1})$ and: $B = \dfrac{2DC_0}{N_s}$ and the initial oxide thickness, x_i is related to an initial time, τ, where:

$$\tau = \frac{x_i + Ax_i}{B} \tag{2.75}$$

which gives us a relation that can be solved for oxide thickness at any time. The initial time, τ, can usually be ignored if one is concerned about the long-term oxidation or corrosion that occurs over years of exposure.

Eq. (2.75) describes all oxides grown on metals. The diffusion coefficient, D, is thermally activated, as are the surface rate constants, so the rate of oxide growth depends on temperature via the Arrhenius model. That is, raising the temperature alone, in the presence of the same oxidizing species will accelerate the oxidation process in a reliable manner. Also, the constants can be determined experimentally by performing successive oxidations and measuring the resulting thicknesses.

2.8.2 Limiting Time Scales

The oxidation rate constants, A and B have important physical meanings. These relate to the limiting factors associated with the oxidation growth process. The oxidation may be limited by the surface reaction rate or by the diffusion of oxidant through the existing oxide. At short times, or when the oxide is highly permeable (as in the case of rust on iron), the diffusion is effectively infinite for all time, and the rate of oxide growth is approximately constant:

$$d_o \approx \frac{B}{A}t, \;\; where \; t \ll \frac{A^2}{4B} \tag{2.76}$$

which is also true for any metal at very short times, when the oxide is very thin. When the oxide is much thicker, or if the diffusivity is very small, the oxidation slows down to a parabolic relation:

$$d_o \approx \sqrt{Bt}, \;\; where \; t \gg \frac{A^2}{4B} \tag{2.77}$$

where B is related to the diffusivity of the oxidant through the oxide. It should be of no surprise that this is the same form as the thermal diffusion length in Eq. (2.17).

2.8.3 Material Properties

In terms of reliability, oxide growth can be both beneficial and detrimental. For many metals, such as aluminum and chromium, the resulting oxide is very hard and resists further oxidation. These oxides form naturally at room temperature and result in a very thin protective coating for the metal itself. The oxide is also electrically insulating which is problematic if an electrical connection is desired across to the metal. Iron, on the other hand, forms a very porous oxide, rust, that has very poor mechanical stability. The oxidation process is accelerated by moisture. It was found, however, that if chromium is incorporated into the iron in an alloy, an impermeable chromium oxide forms to protect further oxidation of the iron. This invention is what leads to the development of stainless steel which does not rust the way iron does.

The major concern about an oxide layer which forms on a metal is the differences in material properties. Usually, the materials are completely different. The metal will be a deformable, malleable material that is electrically conductive, but the native oxide that forms is hard, brittle, and electrically insulating. The oxide is also usually thermally insulating which may introduce other reliability concerns. One inherent problem associated with a thin oxide layer on a structural piece of metal is a difference in thermal expansion. As a material heats up and cools down, there will be nonuniform stresses at the interfaces. The result can be excess local stress which could lead to premature fatigue and fractures simply due to the differences in hardness, strength, and thermal expansion coefficients.

2.9 Vibration

The final section on fundamental failure phenomena that are modeled by the first principles of physics deals with normal vibrational modes and the energies associated with them. This will not be a complete treatment of acoustics; rather it will demonstrate the essentials of physical vibrational modes, which may lead to fatigue in resonant structures.

Vibrations are acoustic waves which propagate in a material under various boundary conditions. The formalism for acoustic generation and propagation in a solid or a fluid is very straightforward. The propagation of nondispersive acoustic waves is described by the wave equation for propagation in a direction, x,

$$v_s^2 \frac{\partial^2}{\partial x^2} \overline{U} = \frac{\partial^2}{\partial t^2} \overline{U} \tag{2.78}$$

where v_s is the velocity of sound and U is a vector parameter that relates to the amplitude of the acoustic wave.

2.9.1 Oscillations

Energy is transmitted by acoustic waves; thus, we will use an energy-based formalism to describe the oscillatory behavior of acoustic oscillations. One can appreciate the fact that vibrations occur in all materials by virtue of having a mass and an elastic modulus (Young's modulus). The classic mass on a spring analogy has often been used to determine the characteristic frequency of a massive body that is fixed on one end and free on the other.

The force imposed on the mass by the spring is always balanced by the spring displacement. The force is always proportional to the acceleration, and the energy associated with the system is conserved. Thus, the Hamiltonian consists of the kinetic energy of the mass in motion and the potential energy of the spring:

$$H = \frac{1}{2m}p^2 + \frac{1}{2}Ku^2 \tag{2.79}$$

where u is the displacement from equilibrium, p is the instantaneous momentum of the mass, m, and K is the spring constant. The momentum is the mass times the velocity, which may be expressed as a derivative of the displacement:

$$p = m\frac{du}{dt} \tag{2.80}$$

which is a vector quantity in the direction of the velocity. Both position and momentum terms are squared to yield a scalar quantity for the energy (Figure 2.12).

In the simple example, where there is no friction, the total energy, H, is constant. So, once energy is introduced to the system, it is completely transferred from kinetic to potential and vice versa. One can plot the momentum, p, versus displacement, x, for a given energy on a single set of axes. The result of Eq. (2.81) is an ellipse describing the transfer of energy from potential energy (displacement) to

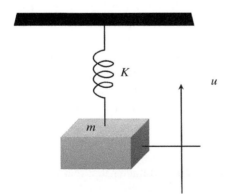

Figure 2.12 A mass, m, suspended from a rigid surface by a spring, K.

Figure 2.13 A constant energy ellipse representing the relationship between displacement, u, and momentum, p.

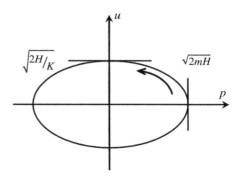

kinetic (momentum). The mass moves along the ellipse at a constant speed determined by the oscillatory frequency:

$$v_0 = \frac{1}{2\pi} \sqrt{\frac{K}{m}} \tag{2.81}$$

and is independent of energy, H. More energy in the system will increase the amplitude of oscillations, but not the frequency. This is the principle of a pendulum, where the restoring force, K, is determined by gravimetric acceleration instead of a material constant (Figure 2.13).

Given a physical system where there exists a single resonant frequency, it is quite straightforward to understand that energy introduced into that system at the resonant frequency will be absorbed by that system. If an ideal mass on a spring is excited by an oscillating source at precisely the resonant frequency, v, then that energy will be completely converted to motion and a large displacement will be achieved with a minimum of energy. What may not be obvious from this idealization is that there will be zero absorption of energy at any frequency other than v_0.

Because of this idealization, one can appreciate that the only reason that structures do vibrate at all is because they are excited with energetic components at the resonant frequencies. These vibrational components are easily achieved by transients and impulse shocks that contribute Fourier components across the frequency spectrum and can lead to vibrations.

2.9.2 Multiple Resonances

Most physical systems are not made up of a point mass attached to a massless spring. In practice, parts are made of massive members having a specific length. The next level of sophistication is a cantilever beam of length, L, and mass, m, that is held fixed at one end and is open on the other. This system has multiple resonant frequencies at harmonic intervals determined by the boundary conditions. There is a fundamental resonant frequency, ν_0, which is precisely the same as Eq. (2.72) where K is the spring constant of the whole member.

The additional factor of the mass being distributed along the member allows additional resonances at frequencies determined by the boundary conditions; zero displacement at $x = 0$ and a free surface at $x = L$. The result is that at frequencies of odd multiples of the fundamental resonance (3, 5, etc.), there will be higher-order modes. Therefore, any vibrational excitation will be accommodated by an appropriate superposition of these orthogonal modes. However, the fundamental resonant mode represents the largest displacement for a given driving energy.

2.9.3 Random Vibration

In any real product, the vibrational excitation is randomly applied. Without knowledge of all the local resonances and the operating environment, the possible vibrational excitations will be random. This is often described as a uniform frequency distribution. Conveniently, the Fourier transform of a uniform frequency distribution is an impulse in time. Hence, a truly random vibration pattern can be simulated by a purely impulsive mechanical excitation. This is essentially the "hit-it-with-a-hammer" approach to reliability testing (Figure 2.14).

Unfortunately, this test is all too realistic in simulating sharp impacts, for example, a drop test. A sudden impact is generally the most severe test for a piece of equipment since it excites all the harmonic vibrational modes simultaneously. Nonetheless, the drop test is a common way to specify to the user what is a safe impact that a device can withstand and not fail to operate.

Many sophisticated techniques have been developed to excite and detect mechanical resonances by applying controlled excitations and listening with an acoustic transducer. The individual resonances can be correlated and possible interactions of different members with each other may be determined. Furthermore, sinusoidally swept vibrations may be applied and absorption peaks can be detected acoustically. All the various techniques for measuring vibrational resonances have important uses in the design of mechanical systems.

Resonant elements can vibrate and fatigue at a much faster rate than would be expected under normal calculated loading. Furthermore, members that are next to each other may interfere and cause damage to each other. In applications of electronic components,

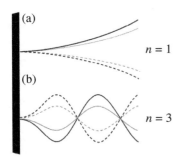

Figure 2.14 A cantilever oscillating at (a) the fundamental and (b) the 3rd harmonic.

resonant vibrations can result in short circuits. Vibrational resonances must be considered in any system design so that their effects can be minimized.

2.10 Summary

In summary, we have seen that there are a very small number of fundamental physical phenomena discussed in any nonlinear types of behavior such as fatigue and rupture, but rather concentrated on basic linear phenomena that can be used in numerical models. By incorporating basic ideas, including diffusion and entropy into a system, it becomes clear that many approximations can be made about the long-term reliability of a system. One can also use these concepts to develop physics-based accelerated life tests that target specific potential problems and allow for a more robust design.

This is the essence of the physics of failure approach to reliability. By understanding failure modes down to their most fundamental components, it is possible to model the propagation of these mechanisms and to design products better and faster. Now that computer models are relatively inexpensive and fast, it becomes feasible to model failure mechanisms in a product before it is even manufactured. The result of good modeling is a faster turnaround of a more robust and less expensive product for the consumer.

3

Physics-of-Failure-based Circuit Reliability

Microelectronics integration density is limited by the reliability of the manufactured product operated under expected operating conditions given a desired circuit density. Design rules, operating voltage, and maximum switching speeds are chosen to insure functional operation over the intended lifetime of the product. Thus, to determine the ultimate performance for a given set of design constraints, the reliability must be modeled for its specific operating condition.

Reliability modeling for the purpose of lifetime prediction is therefore the ultimate task of a failure physics evaluation. Unfortunately, all of today's industrial approaches to reliability evaluation fall short of predicting failure rates or wear-out lifetime of semiconductor products. This is attributed mainly to two reasons: the lack of a unified approach for predicting device failure rates and the fact that all commercial reliability evaluation methods rely on the acceleration of one, dominant, failure mechanism.

Over the last several decades, our knowledge about the root cause and physical behavior of the critical failure mechanisms in microelectronic devices has grown significantly. Confidence in the reliability models has led to more aggressive design rules that have been successfully applied to the latest nano-scale technology. One result of improved reliability modeling has been accelerated performance, beyond the expectation of Moore's Law. A consequence of more aggressive design rules has been a reduction in the weight of a single failure mechanism. Hence, in modern devices, there is no single failure mode that is more likely to occur than any other as guaranteed by the integration of modern failure physics modeling and advanced simulation tools in the design process. The consequence of more advanced reliability modeling tools is a new phenomenon of device failures resulting from a combination of several competing failure mechanisms. Therefore, a new approach is required for reliability modeling and prediction.

This chapter highlights the problematic areas of semiconductor reliability prediction including "single failure mechanism" criteria, "constant failure rate"

Reliability Prediction for Microelectronics, First Edition. Joseph B. Bernstein, Alain A. Bensoussan, and Emmanuel Bender.
© 2024 John Wiley & Sons Ltd. Published 2024 by John Wiley & Sons Ltd.

(CFR) approximation, and the controversial "accelerated testing." "Multiple failure mechanism" together with a novel testing method is suggested to overcome the inaccurate results of "zero failure" reported data. "Zero failure" criterion is the result of the "single failure rate" assumption and the related accelerated testing. Several case studies support the claim that our novel approach provides more accurate reliability prediction. We will show that the assumption of multiple competing failure mechanisms together with novel accelerated testing method from a CFR approximation for each mechanism for electronic system (consisting of millions of components) forms the basis of a more accurate approach to predicting system reliability.

3.1 Problematic Areas

One of the fundamentals of understanding a product's reliability requires an understanding of the calculation of the failure rate. The calculation of failure rates is an important metric in assessing the reliability performance of a product or process. This data can be used as a benchmark for future performance or an assessment of past performance, which might signal a need for product or process improvement. Reliability data are expressed in numerous units of measurement. However, most documents use the units of FITs (failures-in-time), where one FIT is equal to one failure occurring in 10^9 device-hours. More literally, FIT is the number of failures per 10^9 device hours. It is typically used to express the failure rate. Similarly, it can be defined as 1 ppm per 1000 hours of operation or one failure per 1000 devices run for one million hours of operation.

The traditional method of determining a product's failure rate is by using accelerated high-temperature operating life (HTOL) tests performed on a sample of devices randomly selected from its parent population. The failure rate obtained on the life test sample is then extrapolated to end use conditions by means of predetermined statistical models to give an estimate of the failure rate in the field application. Although there are many other stress methods employed by semiconductor manufacturers to fully characterize a product's reliability, the data generated from operating life test sampling is the principal method used by the industry for estimating the failure rate of a semiconductor device in field service.

The semiconductor industry provides an expected FIT for every product that is sold based on operation within the specified conditions of voltage, frequency, heat dissipation, etc. Hence, a system reliability model is a prediction of the expected mean time between failures (MTBF) or mean time to failure (MTTF) for an entire system as the sum of the FIT rates for every component. A FIT is defined in terms of an acceleration factor, AF, as

$$FIT = \frac{\#\,failures}{\#tested \cdot hours \cdot AF} \cdot 10^9 \qquad (3.1)$$

where #failures and #tested are the number of actual failures that occurred as a fraction of the total number of units subjected to an accelerated test. The acceleration factor, AF, is given by the manufacturer. More details about the acceleration factor are given in Section 3.1.2.

The HTOL or steady-state life test is performed to determine the reliability of devices under operation at high temperature conditions over extended time. It consists of subjecting the parts to a specified bias or electrical stress for a specified amount of time and at a specified high temperature (essentially just a long-term burn-in). Unlike production burn-in, which accelerates early life failures, HTOL testing is applied to assess the potential operating lifetimes of the sample population. It is therefore more concerned with acceleration of wear-out failures. As such, life tests should have sufficient durations to assure that the results are not due to early life failures or infant mortality. The HTOL qualification test is usually performed as the final qualification step of a semiconductor manufacturing process. The test consists of stressing some number of parts, usually about 100 parts, for an extended time, usually 1000 hours, at an accelerated voltage and temperature. Several aspects, however, shed doubt on the accuracy of this procedure [33].

In many cases, the individual components in an engineering system may have suffered zero or perhaps only a single failure during their operating history. Such a history may be that of continuous operation over operational time or it may be over several discrete periods. Probabilistic safety assessment provides a framework in which to analyze the safety of large industrial complexes. Therefore, the lack of data has encouraged either the elicitation of opinion from experts in the field (i.e. "expert judgment technique") or making a confidence interval. Several articles show the history of failure rate estimation for low-probability events and the means to handle the problem [34–36].

In general, apart from the Bayesian approach, when a reliability test ends in zero units having failed, traditionally reliability calculations suggest that the estimated failure rate is also zero, assuming an exponential distribution. However, obviously, this is not a realistic estimate of a failure rate, as it does not account for the number of units. In such cases, the first approximation is to select a failure rate that makes the likelihood of observing zero failures equal to 50% [37–39]. In other words, a failure rate that carries a high probability of observing zero failures for a given reliability test is selected. An upper $100(1 - \alpha)$ confidence limit for λ is given by:

$$\lambda_{100(1-\alpha)} = \frac{\chi^2_{2;\,100(1-\alpha)}}{2nT} = \frac{-\ln\alpha}{nT} \qquad (3.2)$$

where n is the total number of devices, T is the fixed time end of test, $\chi^2_{2;\,100(1-\alpha)}$ is the upper $100(1-\alpha)$ percentile of the chi-square distribution with 2 degrees of freedom. In practice, α is often set to 0.5, which is referred to as 50% zero failures estimate of λ. However, α can theoretically be set to any probability value desired. λ_{50} is the failure rate estimate that makes the likelihood of observing zero failures in a reliability test equal to 50%.

This is the basis for calculating the semiconductor failure rates. In the case of semiconductor devices, it is said that since failure rates calculated from actual HTOL test data are not the result of predictions (such as those contained in MIL-HDBK-217), they are calculated from the number of device hours of test, the number of failures (if any) and the chi-square statistic at the 60% confidence level. The formula for failure rate calculation at a given set of conditions is as follows:

$$FIT = \frac{\chi^2_{(2C+2)}}{2nT} \cdot 10^9 \text{ hours} \tag{3.3}$$

With $\chi^2_{(2C+2)}$ = CHI-square distribution factor with an erroneous $2C+2$ degrees of freedom (taken from chi-square tables), C = total number of failures, n = total number of devices tested, T = test duration for each device at the given conditions. In fact, there is no justification for adding these 2 degrees of freedom according to any understanding of statistical significance. In fact, the degrees of freedom in a chi-square test are $2C-2$. Nonetheless, this problematic approach for justifying reporting a "maximum" failure rate based on zero data has become an industry standard, from which the FIT calculation, FR, is determined.

$$MTTF = \frac{10^9}{FIT} \tag{3.4}$$

The calculation of the failure rates and $MTTF$ values at a given condition is accomplished through the determination of the accelerating factor for that condition [40, 41].

This whole approach is not correct. The problem is that the "zero failure rates" criterion is based on the inaccurate (and even incorrect) assumption of "single failure mechanism," on one hand, and the "confidence interval" which is built upon the mere zero (or at most) one data point, on the other hand. Unfortunately, with zero failures no statistical data is acquired. The other feature is the calculation of the acceleration factor, AF. If the qualification test results in zero failures, which allows the assumption (with only 60% confidence!) that no more than 1 failure occurred during the accelerated test. This would result, based on the example parameters, in a reported FIT = 5000/AF, which can be almost any value from less than 1 FIT to more than 500 FIT, depending on the conditions and model used for the voltage and temperature acceleration.

The accepted approach for measuring FIT could, in theory, be reasonably correct if there is only a single, proven dominant failure mechanism that is excited equally by either voltage or temperature. For example, electromigration (EM) is known to follow Black's equation (described later) and is accelerated by increased stress current in a wire or by increased temperature of the device. But, if there is only a single mechanism, then that mechanism will be characterized by a very steep (high β) Weibull slope and the constant rate model will be completely invalid. If, however, multiple failure mechanisms are responsible for device failures, each failure mechanism should be modeled as an individual "element" in the system and the component survival is modeled as the survival probability of all the "elements" as a function of time.

If multiple failure mechanisms, instead of a single mechanism, are assumed to be time-independent and independent of each other, FIT (CFR approximation) should be a reasonable approximation for realistic field failures. However, it would not be characterized by a single failure rate model or activation energy. Under the assumption of multiple failure mechanisms, each will be accelerated differently depending on the physics that is responsible for each mechanism. If, however, an HTOL test is performed at a single arbitrary voltage and temperature for acceleration based only on a single failure mechanism, then only that mechanism will be accelerated. In that circumstance, which is generally true for most devices, the reported FIT (especially one based on zero failures) will be meaningless with respect to other failure mechanisms. Table 3.1 gives definitions of some of the terms used to describe the failure rate of semiconductor devices [42].

3.1.1 Single-Failure Mechanism Versus Competing-Failure Mechanism

Accelerated stress testing has been recognized to be a necessary activity to ensure the dependability of high-reliability electronics. The application of enhanced stresses is usually for the purpose of (i) ruggedizing the design and manufacturing process of the package through systematic step-stress and increasing the stress margins by corrective action (reliability enhancement testing); (ii) conducting highly compressed/accelerated life tests (ALTs) in the laboratory to verify in-service reliability (ALTs); and (iii) eliminating weak or defective populations from the main population (screening or infant mortality reduction testing).

In general, accelerated life testing techniques provide a shortcut method to investigate the reliability of electronic devices with respect to certain dominant failure mechanisms occurring under normal operating conditions. Accelerated test are usually planned on the assumption that there is a single dominant failure mechanism for a given device. However, the failure mechanisms that are dormant under normal use conditions may start contributing to device failure under

Table 3.1 Definitions of terms used to describe the failure rate of semiconductor devices.

Terms	Definitions/Descriptions
Failure rate (λ)	Measure of failure per unit of time. The useful life failure rate is based on the exponential life distribution. The failure rate typically decreases slightly over early life, then stabilizes until wear-out which shows an increasing failure rate. This should occur beyond useful life
Failure in time (FIT)	Measure of failure rate in 10^9 device hours; e.g. 1FIT = 1 failure in 10^9 device hours
Total device hours (TDH)	The summation of the number of units in operation multiplied by the time of operation
Mean-time to failure (MTTF)	Mean of the life distribution for the population of devices under operation or expected lifetime of an individual, MTTF = $1/\lambda$, which is the time when 63.2% of the population has failed. Example: For $\lambda = 10$ FITs, MTTF = $1/\lambda = 100$ million hours
Confidence level or limit (CL)	Probability level at which population failure rate estimates are derived from sample life test. The upper confidence level interval is used
Acceleration factor (AF)	A constant derived from experimental data which relates the time to failure at two different stresses. The AF allows extrapolation of failure rates from accelerated test conditions to use conditions

accelerated conditions and the life test data obtained from the accelerated test would be unrepresentative of the reality of device usage. But the reverse is also true, the failure mechanisms that are dormant under accelerated conditions condition may contribute significantly to device failure under normal use. Thus, the life test data obtained from the accelerated test would be unrepresentative of the failure rate under actual usage. Moreover, accelerated stress accelerates various failure mechanisms simultaneously and the dominant failure mechanism is the one which gives the shortest predicate life [41].

A model which accommodates multiple failure mechanisms considers a system with k failure mechanisms, each of which is independent of others and follows the exponential distribution. The lifetime of such a system is the smallest of k failure mechanism lifetimes. If the variable X denotes the system lifetime, then:

$$X = min(X_1, X_2, ..., X_k) \tag{3.5}$$

This model is often called competing mechanism model. The probability of a system surviving failure type I at time t is:

$$R_i(t) = P(X_i > t) = 1 - G_i(t) \tag{3.6}$$

where $G_i(t)$ is the distribution function of lifetime for failure type i. From the assumption that the failure mechanism develops independently of one another, the probability of surviving all k failure mechanisms at time t is:

$$P(X_i > t | i = 1, 2, ..., k) = \prod_{i=1}^{k} (1 - G_i(t)) \tag{3.7}$$

then the distribution function of system failure is:

$$F(t) = 1 - \prod_{i=1}^{k} (1 - G_i(t)) \tag{3.8}$$

We may consider each failure mechanism as an "element" in the system. The system survives only if all the "elements" survive, just like a series system reliability model. Then the probability of survival of the system is:

$$R(t) = R_1(t) \cdot R_2(t) \cdot ... \cdot R_k(t) \tag{3.9}$$

If each failure mechanism has a time-variant reliability distribution, then the system reliability distribution is also time-dependent and rather complex. However, simulations show that the exponential distribution for each failure distribution would result in reasonable approximation (see Section 3.2.1). Applying CFR to the assumptions, the system reliability distribution also follows the exponential distribution. For each failure mechanism:

$$R_i(t) = exp(\lambda_i t) \tag{3.10}$$

Then:

$$R(t) = \prod_{i=1}^{k} exp(\lambda_i t) = exp(\lambda t) \tag{3.11}$$

where $\lambda = \lambda_1 + \lambda_2 + ... + \lambda_k$ is the sum of all failure rates. The above-competing model provides the basis to find the more accurate form of acceleration factors of complex systems with multiple failure mechanism.

In modern systems including electronics, we can assume that throughout its operating range, no single failure mechanism is more likely to occur than any other as guaranteed by the integration of modern failure and modern simulation tools in the design process. The consequence of more advanced reliability modeling tools is a new phenomenon of device failures resulting from a combination of several competing failure mechanisms. It seems that the multiple failure mechanism realm is even found in standards, i.e. Joint-Electron-Device-Engineering-Council (JEDEC) Standards JEP122 [43] and JESD85 [44]. These handbooks give instructions to calculate multiple activation energy procedures for CFR distributions. These references suppose that devices fail due to several different failure mechanisms which can be assigned to different activation energies [14, 42]. In fact, JEP122H states explicitly, *"When multiple failure mechanisms and thus multiple acceleration factors are involved, then a proper summation technique, e.g. sum-of-the-failure rates method, is required."* Therefore, it is appropriate to use a proper linear superposition of failure rates.

3.1.2 Acceleration Factor

Accelerated life testing of a product is often used to reduce test time by subjecting the product to higher stress. The resulting data is analyzed and information about the performance of the product at normal usage conditions is obtained. Accelerated testing usually involves subjecting test items to conditions more severe than encountered in normal use. This should result in shorter test times, reduced costs, and decreased mean lifetimes for test times. In engineering applications, accelerated test conditions are produced by testing items at higher-than-normal temperatures, voltage, pressure, load, etc. The data collected at higher stresses are used to extrapolate to some lower stress where testing is not feasible. In most engineering applications, a usage or design stress is known. The time to fail under accelerated conditions is extrapolated to the usage condition based on a model for that mechanism which is supposed to be accelerated by the applied conditions.

In traditional approaches, most models will calculate the acceleration factor compared to actual use conditions but do not include any interaction terms with respect to other failure mechanisms. Hence, the relative change in lifetime acceleration factors due only to one applied stress does not consider the effect of other mechanisms whose acceleration also depends on the same stress. The general Eyring model is almost the only model to include terms that have stress and temperature interactions (in other words, the effect of changing temperature varies, depending on the levels of other stresses).

In multiple models with no interaction, one can compute acceleration factors for each stress and multiply them together. This would not be true if the physical mechanism required interaction terms. However, it could be used as the first approximation. If, instead, the temperature and voltage acceleration effects degradation independently, the overall acceleration factor can be obtained by multiplying the temperature acceleration factor by the voltage acceleration factor.

The traditional form for calculating acceleration factor is:

$$AF_{sys} = \prod_{i=1}^{k} AF_i \tag{3.12}$$

In general, the acceleration factors are provided by the manufacturers since only they know the failure mechanisms that are being accelerated in the HTOL and it is generally based on a company-proprietary variant of the MIL-HDBK-217 approach for accelerated life testing. The true task of reliability modeling, therefore, is to choose an appropriate value for *AF* based on the physics of the dominant device failure mechanisms that would occur in the field. More generally, handbooks speak of an average or effective activation energy if they intend to consider multiple mechanism. Of course, activation energies cannot be combined due to the highly nonlinear relation of temperature and failure rate based on the Arrhenius model. Hence, there is no justification for a single model representing multiple mechanisms having multiple activation energies.

The qualification of device reliability, as reported by a *FIT* rate, must be based on an acceleration factor, which represents the failure model for the tested device at the tested stress level. If we assume that there is no failure analysis (FA) of the devices after the HTOL test, or that the manufacturer will not report FA results to the customer, then a model should be made for the acceleration factor, *AF*, based on a *combination* of competing mechanisms, i.e. a sum-of-rates model. This will be explained by way of example. Suppose there are two identifiable, constant-rate competing failure modes (assume an exponential distribution). One failure mode may be accelerated only by temperature, for example. We denote its failure rate as $\lambda_1(T)$. The other failure mode may be only accelerated by voltage, and the corresponding failure rate is denoted as $\lambda_2(V)$. By performing the acceleration tests for temperature and voltage separately, we can get the failure rates of both failure modes at their corresponding stress conditions. Then we can calculate the acceleration factor of the mechanisms. If for the first failure mode we have $\lambda_1(T_1)$, $\lambda_1(T_2)$, and for the second failure mode, we have $\lambda_2(V_1)$, $\lambda_2(V_2)$, then the temperature acceleration factor is:

$$AF_T = \frac{\lambda_1(T_2)}{\lambda_1(T_1)}, T_1 < T_2 \tag{3.13}$$

and the voltage acceleration factor is:

$$AF_V = \frac{\lambda_2(V_2)}{\lambda_2(V_1)}, V_1 < V_2 \tag{3.14}$$

The system acceleration factor between the stress conditions of (T_1, V_1) and (T_2, V_2) is:

$$AF = \frac{\lambda_1(T_2, V_2) + \lambda_2(T_2, V_2)}{\lambda_1(T_1, V_1) + \lambda_2(T_1, V_1)} = \frac{\lambda_1(T_2) + \lambda_2(V_2)}{\lambda_1(T_1) + \lambda_2(V_1)} \tag{3.15}$$

The above equation can be transformed to the following two expressions:

$$AF = \frac{\lambda_1(T_2) + \lambda_2(V_2)}{\lambda_1(T_1)/AF_T + \lambda_2(V_1)/AF_V} \tag{3.16}$$

or

$$AF = \frac{\lambda_1(T_1)AF_T + \lambda_2(V_1)AF_V}{\lambda_1(T_1) + \lambda_2(V_1)} \qquad (3.17)$$

These two equations can be simplified based on different assumptions. When $\lambda_1(T_1) = \lambda_2(V_1)$ (i.e. equal probability at-use condition):

$$AF = \frac{AF_T + AF_V}{2} \qquad (3.18)$$

Therefore, unless the temperature and voltage are carefully chosen so that AF_T and AF_V are very close, within a factor of about 2, then one acceleration factor will overwhelm the failures at the accelerated conditions. Similarly, when $\lambda_1(T_2) = \lambda_2(V_2)$ (i.e. equally likely during accelerated test condition), AF will take this form:

$$AF = \frac{2}{\dfrac{1}{AF_T} + \dfrac{1}{AF_V}} \qquad (3.19)$$

and the acceleration factor applied to *at-use* conditions will be dominated by the individual factor with the greatest acceleration. In either situation, the accelerated test does not accurately reflect the correct proportion of acceleration factors based on the understood physics-of-failure (PoF) mechanisms.

This discussion can be generalized to incorporate situations with more than two failure modes. In the case that a device has n independent failure mechanisms, and λ_{LTFMi} represents the ith failure mode at accelerated condition, λ_{useFMi} represents the ith failure mode at normal condition, then AF can be expressed by:

$$AF = \frac{\lambda_{useFM(1)} \cdot AF_1 + \lambda_{useFM(2)} \cdot AF_2 + \ldots + \lambda_{useFM(n)} \cdot AF_n}{\lambda_{useFM(1)} + \lambda_{useFM(2)} + \ldots + \lambda_{useFM(n)}} = \frac{\sum\limits_{i=1}^{n} AF_{(i)}}{n} \qquad (3.20)$$

for a device which has failure modes with equal frequency of occurrence during the use conditions and:

$$AF = \frac{\lambda_{LTFM(1)} + \lambda_{LTFM(2)} + \ldots + \lambda_{LTFM(n)}}{\lambda_{LTFM(1)} \cdot AF_1^{-1} + \lambda_{LTFM(2)} \cdot AF_2^{-1} + \ldots + \lambda_{LTFM(n)} \cdot AF_n^{-1}} = \frac{n}{\sum\limits_{i=1}^{n} \dfrac{1}{AF_{(i)}}} \qquad (3.21)$$

Only if acceleration factors for each mode are almost equal, i.e. $AF_1 \approx AF_2$, the total acceleration factor will be $AF = AF_1 = AF_2$, and certainly not the product of the two (as is currently the model used by industry). However, if the acceleration of one failure mode is much greater than the second, the standard FIT calculation (Eq. (3.12)) could be incorrect by many orders of magnitude.

Due to the exponential nature of acceleration factor as a function of voltage, current, frequency, or temperature, if only a single parameter is changed, then it is not likely for more than one mechanism to be accelerated significantly compared to the others for any given stress. Since at least four mechanisms are generally included, various voltage and temperature dependencies must be considered to make a reasonable reliability model for electron devices, packages, and systems. Since the individual failure rates can be combined linearly, a matrix can combine the relative proportions since they are linearly related to the final acceleration factor. The rates are linearly summed although the rate functions themselves are highly nonlinear exponential and power-law factors of the related physics models [33].

3.1.3 An Alternative Acceleration Factor Calculation – Matrix Method

Semiconductor voltage, current, and temperature acceleration factors have been studied in detail. The key factor in all the models is that the single activation energy Arrhenius relationship is unique for each mechanism, and it is not appropriate to combine or weigh the activation energies when combining multiple failure mechanisms. Thereafter, an alternate model is proposed to estimate the system reliability. Unlike the single dominant mechanism realm, multiple-failure mechanisms require a method to separate and detect different failure mechanisms. The result could be summarized as saying that time-dependent dielectric breakdown (TDDB) or bias temperature instability (BTI) dominates under high temperature with high voltage, hot carrier injection (HCI) dominates under high voltage and low temperature, and EM dominates under high temperature with low voltage. So, the basic idea is that different mechanisms are accelerated under different kinds of conditions. This section compares the results obtained under single failure mechanism assumption and those obtained in the multiple-failure mechanism realm; in the first step, the failure rate of three failure mechanisms (EM, TDDB, and HCI) are calculated under the single-mechanism assumption and then they are recalculated under multiple-failure mechanism condition; the results allow us to compare results to those of use condition.

Generally, for multiple mechanisms, the system acceleration factor can be expressed based on those of the mechanisms. For n independent mechanisms, it can be written:

$$\frac{\lambda_{s-test}}{AF_s} = \lambda_{s-use} = \sum_{i=1}^{n} \lambda_{i-use} \tag{3.22}$$

Then:

$$\frac{\lambda_{s-test}}{AF_s} = \sum_{i=1}^{n} \frac{\lambda_{i-test}}{AF_i} \tag{3.23}$$

$$\frac{\lambda_{s-test}}{AF_s} = \sum_{i=1}^{n} \frac{\alpha_i \lambda_{s-test}}{AF_i} \tag{3.24}$$

$$\frac{1}{AF_s} = \sum_{i=1}^{n} \frac{\alpha_i}{AF_i} \tag{3.25}$$

where: λ_{s-test} is the system failure rate in test condition, λ_{i-test} is the ith mechanism failure rate in test condition, AF_s is the system acceleration factor, AF_i is the ith mechanism acceleration factor, and α_i is the weight of the ith mechanism failure rate in the system failure rate.

A failure rate matrix (Figure 3.1) would be an appropriate form to show the possibility of getting optimum test results for multiple failure mechanisms assumption; λ_{p1}, λ_{p2}, ... λ_{pq} are failure rates with fixed high voltage but a variety of temperatures, while λ_{1q}, λ_{2q}, ..., $\lambda_{(p-1)q}$ are those related to fixed high temperatures but different voltages.

Figure 3.1 Failure rates from the ALT.

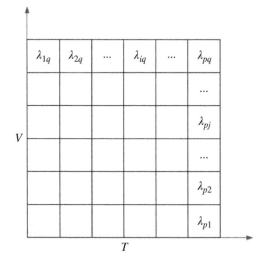

3.1.4 Single-Failure Mechanism Assumption: Conventional Approach

In a case study, three mechanisms, TDDB, HCI, and EM, are selected to model the ALT for 90 nm node deep submicron (DSM) technology. For the TDDB, gate oxide lifetime dramatically shortens with the increase in direct tunneling current, and the dependence on lifetime t_f on gate voltage V_g is given by:

$$t_f = A_1 V_g^c \, exp\left(\frac{E_{a1}}{k_B \cdot T}\right) \tag{3.26}$$

where A_1 is the technology constant and k_B is the Boltzmann constant. Then:

$$\lambda_{\text{TDDB}} = A_1 V_g^c \, exp\left(-\frac{E_{a1}}{k_B \cdot T}\right) \tag{3.27}$$

The typical range of E_{a1} is from 0.6 to 0.9 eV, and the range of c is from 38 to 45. In this case (90 nm node), the use condition is assumed to be 1.2 V (voltage) and 20°C (temperature). Furthermore, it is assumed that $E_{al} = 0.65$ eV and $c = 40$, based on the failure rate being 0.5 FIT at the use condition (for TDDB), which leads to $A_1 = 51.1 \cdot 10^7$.

Similarly, the failure rate of HCI can be modeled as:

$$\lambda_{\text{HCI}} = A_2 \, exp\left(-\frac{\gamma}{V_d}\right) exp\left(-\frac{E_{a2}}{k_B \cdot T}\right) \tag{3.28}$$

E_{a2} is reported to be in the range of -0.1 to -0.2 eV, and the typical range of value of γ is reported as 30 to 80. If $E_{a2} = -0.15$ eV, and $\gamma_2 = 46$, which leads to $A_2 = 3.51 \cdot 10^{15}$.

The model for EM could be given by:

$$\lambda_{\text{EM}} = A_3' \cdot J^n \cdot exp\left(-\frac{E_{a2}}{k_B \cdot T}\right) \tag{3.29}$$

Generally, J can be estimated by:

$$J = \frac{C_{int}}{W \times H} \cdot f \cdot \gamma \tag{3.30}$$

where C_{int} is the interconnect capacitance of a specific node, V is the voltage drop across the interconnect segment, W is the interconnect width, H is the interconnect thickness, f is the current switching frequency and γ is the probability that the line switches in one clock cycle.

Then, it could be possible to derive:

$$\lambda_{\text{EM}} = A_3' \cdot \left(\frac{C_{int}}{W \cdot H} \cdot f \cdot \gamma\right)^n \cdot exp\left(-\frac{E_{a2}}{k_B \cdot T}\right) = A_3 \cdot V^n \cdot exp\left(-\frac{E_{a2}}{k_B \cdot T}\right) \tag{3.31}$$

The typical value of $n = 2$ and the range of E_{a2} is 0.9 to 1.2 eV.

Assume that $n = 2$ and $E_{a2} = 0.9$, and the failure rate due to EM is 20 FIT under use conditions. If, we consider these to be independent failure mechanisms, the failure rate at 1.7 V and 140°C would be:

$$\lambda_{1.7\,V,140°C} = \lambda^{TDDB}_{1.7\,V,140°C} + \lambda^{HCI}_{1.7\,V,140°C} + \lambda^{EM}_{1.7\,V,140°C} = 8.93 \cdot 10^7 FIT \qquad (3.32)$$

Similarly, the failure rates in the other tested conditions are shown in Figure 3.2, which can be obtained by the manufacturers in the ALT.

The traditional method used by the manufacturers considers a dominant failure mechanism for a system, and therefore the dominant failure mechanism model would have been used to extrapolate the failure rate from test conditions:

$$\lambda = A_4 \cdot V_g{}^c \cdot exp\left(-\frac{E_A}{k_B T}\right) \qquad (3.33)$$

where A_4, c, and E_A are estimated from matrix Figure 3.2.

To find c, one could write:

$$\lambda = A_4 V_g{}^c exp\left(-\frac{E_A}{k_B T}\right) \qquad (3.34)$$

then by getting logarithm from both sides:

$$ln\,\lambda = clnV_g - \frac{E_A}{k_B T} + ln\,A_4 \qquad (3.35)$$

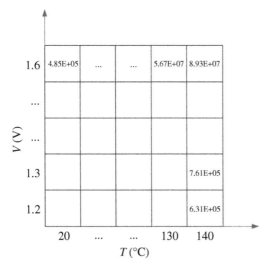

Figure 3.2 The values of failure rates obtained in the case study.

Table 3.2 The traditional test conditions under high voltages and high temperature (140°C).

	1.52 V, 140°C	1.54 V, 140°C	1.56 V, 140°C	1.58 V, 140°C	1.6 V, 140°C
System failure rate	1.24×10^7	2.02×10^7	3.31×10^7	5.44×10^7	8.93×10^7

When the temperature is fixed, there is a linear relation between $ln\,\lambda$ and V_g, and c is the slope. In practice, due to the long test time, the manufacturers often start the test from relatively high voltage instead of the use voltage. For instance, the voltage test condition in 40°C starts from 1.52 V instead of 1.2 V (as shown in Table 3.2). The voltage factor c is dominated by TDDB, and is estimated as 38.6, close to 40 which is related to that of TDDB (Figure 3.3).

In order to estimate E_a, one can write:

$$ln\,\lambda = -\frac{E_A}{k_B} \cdot \frac{1}{T} + ln\,A_4 + c\,ln\,V_g \tag{3.36}$$

When the voltage is fixed, there is a linear relation between $ln\,\lambda$ and $\frac{1}{T}$, and $-\frac{E_A}{k_B}$ is the slope. Similarly, in the case of fixed voltage, the temperature test starts from relatively high temperature instead of use temperature. In 1.6 V, temperature test

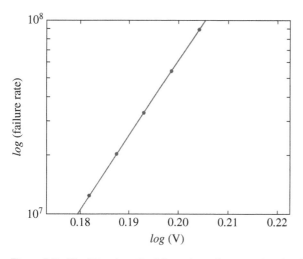

Figure 3.3 Traditional method for voltage factor estimation (c).

Table 3.3 The traditional test conditions under high voltage (1.6 V) and low temperatures.

	1.6 V, 120°C	1.6 V, 125°C	1.6 V, 130°C	1.6 V, 135°C	1.6 V, 140°C
System failure rate	3.52E+07	4.48E+07	5.67E+07	7.14E+07	8.9E+07

conditions start from 120°C instead of 20°C. The test data are shown in the following Table 3.3 (only sum can be obtained by the manufacturers in the test).

Then, $-\dfrac{E_a}{k_B}$ can be estimated as -7558.8 and E_A is equal to 0.651 eV.

Since it is supposed that TDDB is the dominant failure mechanism, the estimation of E_A is dominated by TDDB and E_A is approximately equal to that of TDDB (Figure 3.4).

Then the failure rate under use condition (1.2 V and 20°C) can be extrapolated from the data provided under the test condition (1.6 V and 140°C). The system acceleration factor can be calculated as:

$$AF_s = AF_t \cdot AF_V = \frac{A_4 V_{g_{test}}^c \; exp\left(-\dfrac{E_A}{k_B \cdot T_{tset}}\right)}{A_4 V_{g_{use}}^c \; exp\left(-\dfrac{E_A}{k_B \cdot T_{use}}\right)} = 1.02 \cdot 10^8 \tag{3.37}$$

Therefore:

$$\lambda_{1.2 \, V, 20°C} = \frac{\lambda_{1.7 \, V, 140°C}}{AF_s} = \frac{8.93 \cdot 10^7}{1.20 \cdot 10^8} = 0.75 \text{ FITs.} \tag{3.38}$$

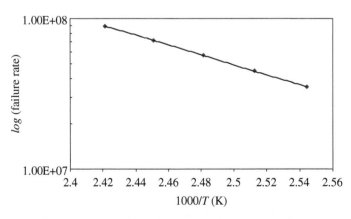

Figure 3.4 Estimation of E_A.

The failure rate estimated from the test data (and provided by manufacturers) is much smaller than that calculated from the real failure rate under use condition (50.5 FITs). In this case, since TDDB is supposed to be the mere dominant failure mechanism, the related acceleration factor and failure rate are calculated under high voltage conditions and those of others are completely ignored. Therefore, the failure rate extrapolated from high temperature and voltage is essentially that of TDDB under the use condition, which is only a piece of the puzzle. The whole picture would appear when data comes from high voltage/high temperature, high voltage/low temperature, and high temperature/low voltage test conditions.

3.1.5 Failure Rate Calculations Assuming Multiple-Failure Mechanism

To show the difference, the previous case study is re-calculated with the multiple failure mechanism approach together with the suggested test matrix; the model and parameters of each mechanism are the same as those in the traditional approach. All the failure rates of TDDB, HCI, and EM under use conditions are also assumed to be 20 FITs.

As mentioned above, TDDB dominates under fixed high temperatures with increasing high voltage, and the data in this field can be regarded as the TDDB failure rate approximately. Its estimation process, including the data, is the same as that in previous sections. Then, TDDB failure rate would be $\lambda_{1.2\,V,20°C}^{TDDB} = 0.75\ FITs.$

For HCI, using the previous equation:

$$\lambda_{HCI} = A_2\,exp\left(-\frac{\gamma_2}{V_d}\right)exp\left(-\frac{E_{a2}}{k_BT}\right) \tag{3.39}$$

then:

$$ln\,\lambda_{HCI} = -\frac{\gamma_2}{V_d} - \frac{E_{a2}}{k_BT} + ln\,A_2 \tag{3.40}$$

When the temperature is fixed, there is linear relation between $ln\,\lambda_{HCI}$ and $\frac{1}{V_d}$, and $-\gamma_2$ is the slope. HCI dominates under fixed high voltage and low temperature, therefore the related data could be regarded as the HCI failure data as shown in Figure 3.5, i.e. the voltage factor is $\gamma_2 = 47.313$ (see Table 3.4).

Since the acceleration test data has the same temperature as the one under use condition, only voltage acceleration factor is needed to be calculated for HCI acceleration factor:

$$AF_{HCI} = \frac{A_2 \cdot exp\left(\dfrac{\gamma}{V_{d_{test}}}\right) \cdot exp\left(-\dfrac{E_{a2}}{k_B \cdot T_{test}}\right)}{A_2 \cdot exp\left(\dfrac{\gamma}{V_{d_{use}}}\right) \cdot exp\left(-\dfrac{E_{a2}}{k_B \cdot T_{use}}\right)} = 1.91 \cdot 10^4 \tag{3.41}$$

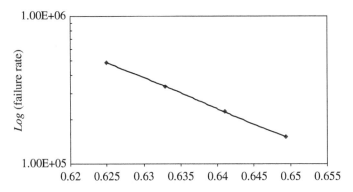

Figure 3.5 Estimation of voltage acceleration factor γ_2 (HCI).

Table 3.4 Failure rate with fixed low temperature and high voltages (HCI).

	1.52 V, 20°C	1.54 V, 20°C	1.56 V, 20°C	1.58 V, 20°C	1.6 V, 20°C
Failure rate	7.03×10^5	1.53×10^5	2.27×10^5	3.33×10^6	4.85×10^6

and

$$\lambda_{1.2\,V,20°C} = \frac{\lambda^{HCI}_{1.6\,V,20°C}}{AF_{HCI}} = 25.4\ \text{FITs} \tag{3.42}$$

To estimate the activation energy of EM, one can use:

$$\lambda_{EM} = A_3 \cdot V^n \cdot exp\left(-\frac{E_{a3}}{k_B \cdot T}\right) \tag{3.43}$$

Then by getting logarithm from both sides:

$$ln\,\lambda_{EM} = -\frac{E_{a3}}{k_B T} + nlnV + ln\,A_3 \tag{3.44}$$

When the voltage is fixed, there is a linear relation between $ln\,\lambda_{EM}$ and $1/T$ and $-\dfrac{E_{a3}}{k_B}$ is the slope. Since EM dominates under fixed high temperatures and low voltages, the related could be regarded as the EM failure rate approximately. Then the voltage factor would be $E_{a3}/k_B = -10{,}439$, which leads to $E_{a3} = 0.9\,eV$ (Table 3.5 and Figure 3.6).

Table 3.5 Failure rate with fixed low temperature and high voltages.

	1.2 V, 120°C	1.2 V, 125°C	1.2 V, 130°C	1.2 V, 125°C	1.2 V, 140°C
Failure rate	1.74E+05	2.43E+05	8.41E+05	3.371E+05	6.31E+05

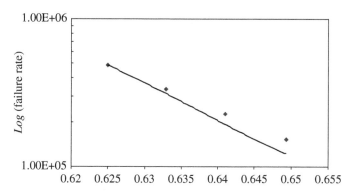

Figure 3.6 Estimation of E_{a3} (EM).

However, since the acceleration test data has the same voltage as the data under use conditions, only E_{a3} is needed to be calculated for EM acceleration factor:

$$AF_{EM} = \frac{A_3 V_{test}^n \, exp\left(\dfrac{-E_{a3}}{k_B T_{test}}\right)}{A_3 V_{use}^n \, exp\left(\dfrac{-E_{a3}}{k_B T_{use}}\right)} = 3.15 \cdot 10^4 \qquad (3.45)$$

which leads to:

$$\lambda_{1.2\,V,20°C} = \frac{\lambda_{1.2V,14°C}^{EM}}{AF_{EM}} = 20.0 \, FITs \qquad (3.46)$$

Therefore, the system failure rate could be calculated from the following equation:

$$\lambda_{1.2\,V,20°C}^{S} = \lambda_{1.2\,V,20°C}^{TDDB} + \lambda_{1.2\,V,20°C}^{HCI} + \lambda_{1.2\,V,20°C}^{EM} = 46.15 \, FITs \qquad (3.47)$$

The result of this method is by far closer to those of use conditions (50.5 FITs), which shows that it is more accurate than the one given by traditional single failure mechanism approach together with the related test data.

3.1.6 Constant-Failure-Rate Approximation/Justification

The Achilles' heel of MIL Handbook 217F is the assumption of CFR for the new generation of electronic devices. It was said that the assumption of CFR could only be valid in the days of vacuum tubes not for solid-state components; the failure rate of solid-state components not only decreases but also approaches zero. Moreover, the use of CFR gives a pessimistic result for the new devices.

Recent studies based on the PoF not only revealed different failure mechanisms for microelectronic devices and circuits but also modeled their failure distributions. It was believed that almost all the failure mechanisms were in the wear-out part of bathtub curve. However, as technology progressed, the new era of "constant rate processes" approximation is dominant for the reliability of microelectronic devices and circuits. The "constant rate processes" assumption is the result of failure mechanisms shifting from wear-out part to CFR part due to deep submicron (DSM) technologies as well as the shrinkage of transistor geometries. TDDB Weibull slope approaches unity as one of the significant signs of this shift.

However, the debate over the last few decades concluded the disadvantages and uselessness of MIL Handbook 217F make the "constant rate processes" assumption almost unusable. On one hand, experts do not want to build the new handbook on the weak bricks of MIL Handbook 217F; on the other hand, the memory-lessness property of CFR is an obstacle in the way of accepting the new approach. The industry is willing to discard the good with the bad.

Since in the absence of a dominant failure mechanism in microelectronic chip/system reliability, all failure distributions could be mixed with the same wear-out drivers such as voltage, temperature, and frequency. It is useful to review the results of distribution mixtures and their properties. The counterintuitive result and the interpretations of distribution mixture properties would help retain the "constant rate process" approach and its memoryless-ness characterization. The decreasing failure rate (DFR) of a bunch of increasing failure distributions mixture together with "negative aging" property, gives a new kind of perspective to the reliability issues.

Drenick's theorem [55] is another way of justifying the CFR approximation for a system consisting of a very large number of components, like a microelectronic chip.

In microelectronic system reliability, one may define a system as an integrated collection of subsystems, modules, or components, which function together to attain specific results. Systems fail because some or all its components do. It is obvious that there is a connection between the reliability of components of a system and that of the whole system. The system reliability is given as a function of the reliability of its components. It is not easy or simple to formulate this functional dependence. Therefore, it is common to simplify the case as a system with independence components time to failure, and simple coherent structure functions, like those of series, parallel, and cross-linked. In practice, however, exact evaluation of system reliability is extremely difficult and sometimes impossible. Once one obtains the expression for the structure function, the system reliability computations become straightforward.

This section claims that the "constant failure rate" approach could be a good approximation to predict the microelectronic system reliability. However, like any other statistical analysis, it needs to be justified by gathering more field data; comparing the results of simulation with the actual results would be the best justification for the new approach.

3.1.7 Exponential Distribution and Its Characterization

The exponential distribution is one of the key distributions in the theory and practice of statistics. It possesses several significant properties – most notably, its characterization through the lack of memory, yet it exhibits great mathematical tractability. Consequently, there is a vast body of literature on the theory and applications of exponential distributions. Exponential distributions are encountered as life-time distributions with constant hazard rates. If the cumulative distribution function of X, a random variable representing lifetime, is $F_X(x)$ and the constant hazard rate is λ, then:

$$-\frac{d \, log(1 - F_X(x))}{dx} = \lambda \, (x > 0) \tag{3.48}$$

or, in terms of the survival function, $\overline{F}_X(x) = 1 - F_X(x)$,

$$-\frac{d \, log \, \overline{F}_X(x)}{dx} = \lambda \, (x \geq 0) \tag{3.49}$$

Solving the differential equation by applying $\overline{F}_X(0) = 1$:

$$\overline{F}_X(x) = e^{-\lambda x} \, (x \geq 0) \tag{3.50}$$

and the corresponding probability density function of X as:

$$f_X(x) = \lambda e^{-\lambda x} (x \geq 0) \tag{3.51}$$

If $\lambda = 1$, then the standard exponential distribution with the survival function:

$$\overline{F}_X(x) = e^{-x} \, (x \geq 0) \tag{3.52}$$

and the probability density function:

$$f_X(x) = e^{-x} \tag{3.53}$$

For $y > 0$ and $x > 0$:

$$P[X > x + y | X > x] = P[X > y] \tag{3.54}$$

That is, the future lifetime distribution, given survival to age x, does not depend on x. This property, termed lack of memory, or, equivalently, lack of aging, characterizes exponential distributions, as can easily be seen. The last equation can also be written equivalently as:

$$S_X(x + y)/S_X(x) = S_X(y) \tag{3.55}$$

That is:

$$\log S_X(x + y) = \log S_X(x) + \log S_X(y) \tag{3.56}$$

Differentiating the above equation with respect to y:

$$-\frac{d \log S_X(x + y)}{dy} = -\frac{d \log S_X(y)}{dy} = \lambda_X(y) \tag{3.57}$$

Since

$$-\frac{d \log S_X(x + y)}{dy} = -\frac{d \log S_X(x + y)}{d(x + y)} = \lambda_X(x + y) \tag{3.58}$$

The hazard rate y is the same as $(x + y)$ for any x and y, that is, it is constant.

This lack of memory properly is a critical part of many of the statistical analyses based on the exponential distribution. Exponential distributions are commonly employed in the formation of models of lifetime distributions, and stochastic processes in general. Even when the simple mathematical form of the distribution is inadequate to describe real-life complexity, it often serves as a benchmark with reference to which effects of departures to allow for specific types of disturbance can be assessed [45].

The exponential distribution is remarkably friendly. Closed-form expressions exist for their density, distribution, moment generation function, mean residual life function, failure rate function, moments of order statistics, record values, etc. So, everyone introducing a new concept or functional to classify or organize distributions inevitably uses the exponential distribution as one of their examples. The exponential distribution is often the simplest example or the most analytically tractable.

The most important characterizations of exponential distribution could be classified as:

- Lack of memory (no degradation);
- Distributional relations among order statistics;
- Independence of functions of order statistics;

- Moments of order statistics;
- Geometric compounding;
- Record value concepts.

Mixtures of distributions of lifetimes occur in many settings. In engineering applications, it is often the case that populations are heterogeneous, often with a small number of subpopulations. The concept of a failure rate in these settings becomes a complicated topic especially when one attempts to interpret the shape as a function of time. Even if the failure rates of the subpopulations of the mixture have simple geometric or parametric forms, the shape of the mixture is often not transparent. Recent results focus on general results to study whether it is possible to provide approximation for a system.

3.2 Reliability of Complex Systems

Mixtures are a common topic in most areas of statistics. They also play a central role in reliability and survival analysis. However, the failure rate of mixed distributions is a source of much confusion.

The density function of a mixture from two subpopulations with density functions f_1 and f_2 is simply given by

$$F(t) = pf_1(t) + (1-p)f_2(t) \quad t \geq 0, 0 \leq p \leq 1 \tag{3.59}$$

Thus, the survival function of mixture is also a mixture of the two survival functions, that is

$$\overline{F}(t) = p\overline{F}_1(t) + (1-p)\overline{F}_2(t) \tag{3.60}$$

The mixture failure rate $r(t)$ obtained from failure rates $r_1(t)$ and $r_2(t)$ associated with f_1 and f_2, respectively, can be expressed as:

$$r(t) = \frac{pf_1(t) + (1-p)f_2(t)}{p\overline{F}_1(t) + (1-p)\overline{F}_2(t)} \tag{3.61}$$

where $f_i(t)$, $\overline{F}_i(t)$ are the probability density and survival function of the distribution having failure rate $r_i(t)$, $i = 1,2$:

$$r(t) = h(t)r_1(t) + (1-h(t))r_2(t) \tag{3.62}$$

where

$$h(t) = \frac{1}{(1+g(t))}, \quad g(t) = \frac{(1-p)\overline{F}_2(t)}{p\overline{F}_1(t)} \tag{3.63}$$

clearly $0 \leq h(t) \leq 1$.

The above equation can easily be generalized to accommodate mixtures of k subpopulations giving

$$r(t) = \frac{\sum_{i=1}^{k} p_i f_i(t)}{\sum_{i=1}^{k} p_i \overline{F}_i(t)} \tag{3.64}$$

where $i = 1, 2, 3, ..., k, 0 < p_i < 1, \sum_{i=1}^{k} p_i = 1, k \geq 2$.

It has been proven that a mixture of two DFR distributions is again DFR.

As mentioned before, the first kind of counterintuitive case is that studied by Proschan [46]. In 1963, the team worked on the pooled data for airplane air conditioning systems whose lifetimes are known to be exponential and exhibit a DFR. Because DFRs are usually associated with systems that improve with age, this was initially thought to be counterintuitive [45]. Pooled data, on the time-scale of successive failures of the air conditioning systems of a fleet of jet airplanes, seemed to indicate that the life distribution had the DFR. More refined analysis showed that the failure distribution for each airplane separately was exponential but with a different failure rate. Using the theorem that a mixture of distributions each having a nonincreasing failure rate (a mixture of exponential distribution, for instance) itself has a nonincreasing failure rate, the apparent DFR of the pooled air conditioning life distribution was satisfactorily explained. This has implications in other areas, where an observed DFR may well be the result of mixing exponential distributions having different parameters.

Proschan brings two hypotheses called H_0 and H_1 as follows: if all the planes under investigation had the same failure rate, then the failure intervals pooled together for the different planes would be governed by a single exponential distribution (H_0 hypothesis). On the other hand, if to each plane there corresponded different failure rates, it would then follow that distribution of the pooled failure intervals would have a DFR (H_1 hypothesis). To validate this inference, he invokes the following theorem from Barlow et al. [47].

The theorem says: if $F_i(t)$ has a DFR, $i = 1, 2, 3..., n$, then

$$G(t) = \sum_{i=1}^{n} p_i F_i(t) \tag{3.65}$$

has a DFR, where each

$$p_i > 0, \sum_{i=1}^{n} p_i = 1 \tag{3.66}$$

After proving the theorem, he rejects H_0 in favor of H_1 and concludes that the pooled distribution has a DFR, as would be expected if the individual airplanes each displayed a different CFR. Another anomaly, at least to some, was that mixtures of lifetimes with increasing failure rates (IFRs) could be decreasing at certain intervals. A variant of the above is due to Gudand and Sethuraman [48], which gives examples of mixtures of very rapidly IFRs that are eventually decreasing.

We recall that the mixture of exponential distributions (which have CFR) will always have the DFR property (shown by Proschan in 1963). We can, thus, refer to a converse result due to Gleser [49] who demonstrates that any Gamma distribution with shape parameter less than 1, and therefore has a DFR distribution, can be expressed as a mixture of exponential distributions. They tried to show that it is reasonable to expect that mixture of IFRs distributions that have only moderately IFRs can possess the DFR property and gave some examples as well. A detailed study of classes of IFR distributions, whose mixtures reverse the IFR property, was given in 1993. Besides, they studied a class of IFR distributions that, when mixed with an exponential, becomes DFR. A few examples of distributions from this class are the Weibull, truncated extreme, Gamma, truncated normal, and truncated logistic distributions [48].

By considering that the mixtures of DFR distributions are always DFR, and some mixtures of IFR distributions can also be ultimately DFR, Gudand and Sethuraman studied various types of discrete and continuous mixtures of IFR distributions and developed conditions for such mixtures to be ultimately DFR [46].

They showed the unexpected results that mixture of some IFR distributions, even those with very rapid IFRs (such as Weibull, truncated extreme), become ultimately DFR distribution. In practice, data from different IFR distributions are sometimes pooled, to enlarge sample size, for instance. These results serve as a warning note that such pooling may reverse the IFR property of the individual samples to an ultimately DRF property for the mixture. This phenomenon is somewhat reminiscent of Simpson's paradox, wherein a positive partial association.

Between two variables may exist at each level a third variable, yet a negative overall unconditional association holds between the two original variables.

They provide the condition on mixture of two IFRs to show that the result has DFR, while the mixture of IFR distribution functions is given by:

$$F_P(t) \equiv p_1 F_1(t) + p_2 F_2(t) \tag{3.67}$$

where $0 \le p_1, p_2 \le 1$ and $p_1 + p_2 = 1$.

The provided conditions lead to interesting results that certain mixtures of IFR distributions, even those with very rapid IFRs become DFR distributions [48].

When assessing reliability of systems, an IFR seems to be very reasonable as age, use, or both may cause the system to wear out over time. The reasons for DFR, that is, a system improving as time goes by are less intuitive. Mixtures of lifetime distributions turn out to be the most widespread explanation for this "positive aging." The more paradoxical case corresponds to the exponential distribution that shows a CFR while exponential mixtures belong to the DFR class. Proschan found that a mixture of exponential distributions was the appropriate choice to model the failures in the air-conditioning systems of planes. Such mixing was the reason for the DFR that the aggregated data exhibited. The DFR as well as the DFR average (DFRA) classes are closed under mixtures. A similar result does not hold for the IFR class; however, many articles focus on mixtures of IFR distributions that reverse this property over time and exhibit a DFR as time elapses.

By using a Cox proportional hazard (PH) rate model [50], a non-negative random variable and the conditional failure rate are defined; the conditional failure rate is in the form of, could be, a Gamma distribution applied to an increasing Weibull distribution. The result shows that the greater the mean of mixing distribution, the sooner the IFR average (IFRA) property is reversed.

In the survival analysis literature, it is known that if an important random covariate in a Cox model is omitted, the shape of the hazard rate is drastically changed. Other types of articles mention that in many biological populations, including humans, lifetime of organisms at extremely old age exhibits a decreasing hazard rate. A natural question to ask is whether this means that some of the individuals in the population are improving or not, which leads to misconceptions and misunderstandings.

In most systems, the population of lifetimes is not homogeneous. That is, all the items in the population do not have the same distribution; there is usually a percentage of the lifetimes which are of a type different from the majority. For example, in most industrial populations, there is often a subpopulation of defective items. For electronic components, most of the population might be exponential, with long lives, while a small percentage often has an exponential distribution with short lives. Even though each of the subpopulations has CFR, the mixed population does not. Proschan [46] observed that such a population has DFR. An intuitive explanation is that there is a stronger (i.e. lower failure rate) and a weaker (i.e. higher failure rate) component and as time goes by, the effect of the weaker component dissipates and the stronger takes over. Another way of saying this is that the failure rate of the mixture approaches the stronger failure rate. Although this had been observed in various special cases, one of the first general results appears in Block, Mi, and Savits [51, 52]. Mi discusses this for mixtures of discrete life distributions. Gudand and Sethuraman observed that mixtures of components which had rapid IFRs could still turn out to have eventually DFRs when mixed. It is important to know the behavior which occurs when populations are pooled. This pooling can occur naturally or can be done by statisticians to increase sample size. Besides, for modeling purposes, it is often useful to have available a distribution which has a particular failure shape. For example, it is useful to know how to pool distributions to obtain the important bathtub-shaped failure rate. Moreover, the tail behavior of the failure rate of mixtures with different lifetime distributions could drive one to conclude that if the failure rate of the strongest component of the mixture decreases, then the failure rate of the mixture decreases to the same limit. For a class of distributions containing the gamma distributions, this result can be improved in the sense that the behavior of the failure rate of the mixture asymptotically mirrors that of the strongest component in whether it decreases or increases to a limit [53].

Some general rules of thumb concerning the asymptotic behavior of the failure rate when several distributions are mixed were already found. First, the failure rate of the mixture will approach the stronger (lowest) failure rate so that there is a downward trend. However, if the strongest failure rate is eventually increasing, the mixture will become increasing. If one of the mixture probabilities is close to one, the mixture failure rate will initially behave like that component. If the component with probability close to one becomes the strongest component, then the mixture will eventually behave like that component. If the failure rates cross, then the point of intersection is also a factor. Differences in the y-intercepts and the ratio of the slopes also play a role [54].

The study of the lives of human beings, organisms, structures, materials, etc., is of great importance in actuarial, biological, engineering, and medical science. Research on aging properties is currently being pursued. While positive aging concepts are well understood, negative aging and concepts (life improved by age) are less intuitive. As mentioned before, there have been cases reported by several authors where the failure rate functions decrease with time. Sample examples are the business mortality, failures in the air conditioning equipment of a fleet of a Boeing 727 aircraft or in semiconductors from various lots combined, and the life of integrated circuit modules. In general, a population is expected to exhibit DFR when its behavior over time is characterized by "work hardening" in engineering terms, or "immunity" in biological terms. Modern phenomenon of DFR includes reliability growth in software reliability.

3.2.1 Drenick's Theorem

Drenick published a paper [55] in which he proved that, under certain constraints, systems which are composed of a "large" quantity of nonexponentially distributed subcomponents tend toward being exponentially distributed. This profound proof allows reliability practitioners to disregard the failure distributions of the pieces of the system since it is known that the overall system will fail exponentially. Given that most systems are composed of a large number of subcomponents, it would seem that Drenick's Theorem is a reliability analysis godsend. The usefulness of this theorem, however, lays in the applicability of the proof's constraints.

Kececioglu [56] delineated the constraints of Drenick's theorem quite well as follows:

1) The subcomponents are in series.
2) The subcomponents fail independently.
3) A failed subcomponent is replaced immediately.
4) Identical replacement subcomponents are used.

If the four conditions above are met, then as the number of subcomponents and the time of operation tend toward infinity, system failures tend toward being exponentially distributed regardless of the nature of the subcomponents' failure distributions [57].

In his paper, Drenick emphasized the role of many-component system and said, "In theoretical studies of equipment reliability, one is often concerned with systems consisting of many components, each subject to an individual pattern of malfunction and replacement, and all parts together making up the failure pattern of the equipment as a whole." His work is concerned with that overall pattern and more particularly with the fact that it grows, statically speaking, the more complex the equipment. Under some reasonably general conditions, the distribution of time between failure tends to be exponential as the complexity and the time of the operation increase; and somewhat less generally, so does the time up to the first failure of the equipment. The problem is in the nature of the probabilistic limit theorem and, accordingly, the addition of independent variables and the central limit theorem is a useful prototype.

It may be useful to add several comments, some concerned with the practice of reliability work and others with the theory.

- It has been assumed that the failure incidents in one component of a piece of equipment are statistically independent of the rest. The independence among failures is more likely to be satisfied in well-designed than in poorly designed equipment.
- It always makes good sense to lump the failures of many, presumably dissimilar, devices into one collective pattern.
- In the long run, the failure pattern of a complex piece of equipment is determined essentially by the mean lives of its components.
- There is a distinction between (initial) survival probability and the residual one; although both are exponential, they have different means. Therefore, despite the similarity of two distributions, one should not use them interchangeably. Otherwise, it gives a pessimistic kind of result, for instance.
- Mathematically, even the weak form of limit law condition is sufficient for the asymptotic convergence of one distribution to the exponential one.

Drenick's assumption would be considered a mathematical justification for microelectronic system CFR approach [55, 57]. We agree with the MIL Handbook approach that a constant rate assumption is best for a full system. We only suggest that the failure rate should be seen as a combination of interacting or competing physical phenomena whose effects can be properly summed as a weighted set of failure rate processes. Each failure rate is determined by its own Physics, including its own Arrhenius and non-Arrhenius factors based on the physical mechanisms that describe its probability of failure over time.

3.3 Physics-of-Failure-based Circuit Reliability Prediction Methodology

Reliability prediction has an important role in business decisions like system design, parts selection, qualification warranties, and maintenance. Nowadays electronic system designers have their own industry-specific reliability prediction tools, such as the well-known MlLHDBK-217, Society of Automobile Engineers (SAE) reliability prediction method, Telcordia SR-332, and prediction of reliability, integrity, and survivability of microsystems (PRISM) [10, 58]. Many of those methods are empirically based and were built upon field data and extrapolations such as the "parts count" and the "parts stress" methods with various kinds of pre-factors [1]. One very important disadvantage of those empirical-based methods is the lack of integration of the PoF models because of the complexity and difficulty for the system designers to get detailed technology and microcircuit data. Prediction accuracy is diminished without those PoF models, and the situation is becoming worse with technological advancement. Today's microelectronic devices, featuring ultra-thin gate oxide and nanometer-scaled channels, suffer various detrimental failure mechanisms, including TDDB [59], negative bias temperature instability (NBTI) [60], hot carrier degradation (HCD) [61] and EM [62] as the nonideal voltage scaling brings higher field and current density. As mentioned already, each failure mechanism has a unique dependence on voltage and temperature stresses, and all can cause device failure. The traditional prediction methods are simply not as applicable as before, considering the multiple failure mechanisms' effect and the difficulty of obtaining enough up-to-date field data.

Device manufacturers face the same challenge to maintain and further improve reliability performance of advanced microelectronic devices, despite all kinds of difficulties from technology development, system design, and mass production. Conventional product reliability assurance methods, such as burn-in and HTOL are gradually losing competitiveness in cost and time because the gap between normal operating and accelerated test conditions is continuing to narrow. Increased device complexity also makes sufficient fault coverage tests more expensive. An accurate reliability simulation and prediction tool are greatly needed to guide the manufacturers to design and deploy efficient qualification procedures according to customers' needs and help the designers get in-time reliability feedback to improve the design and guarantee the reliability at the very first stage [3].

The need for accurate reliability prediction from both device and system manufacturers requires integration of PoF models and statistical models into a comprehensive product reliability prediction tool that takes the device and application details into account. A new PoF-based statistical reliability prediction methodology is proposed in this chapter. The new methodology considers the needs of both the device manufacturer and the system supplier by taking application and design into account. Based on circuit-level operation-oriented PoF analysis, this methodology provides an application-specific reliability prediction which can be used to guide qualification and system design.

Prediction methods based on handbook approaches usually provide conservative failure rate estimation [63]. Many adjustments must still be made to improve the prediction accuracy. The two important considerations are:

- Integration of PoF analysis and modeling. This is a prominent issue because advanced microelectronic devices are vulnerable to multiple failure mechanisms. These failure mechanisms have unique voltage and temperature dependence. No unified handbook based lifetime model can take these into account.
- Integration of failure mechanism lifetime distribution. CFR assumption might give fast and cost-effective failure rate estimation at the system level because the inherent inaccuracy is so obvious without justification from detailed PoF analysis. It is also statistically impossible to combine non-CFR models together into a single distribution.

3.3.1 Methodology

The PoF statistical approach is a device reliability modeling method that considers all intrinsic failure mechanisms under static and dynamic stresses. Since today's microelectronic devices integrate millions or even billions of transistors, running a full spectrum simulation is too resource-intensive. To simplify the simulation and reduce computation load, the PoF statistical methodology takes four unique approaches by considering the repetitive characteristics of complimentary metal-oxide-semiconductor (CMOS) circuits:

- Cell-Based Reliability Characterization: Standard cells (inverter, NOR, NAND, etc.) are the fundamental building blocks in modern very large-scale integration (VLSI) circuit design. Cells in the same category have similar structures and operation profiles. By using cell-based reliability characterization, designers can save time in both circuit design and simulation. Cell schematics and layouts are readily available from design kits, eliminating the need for system designers to understand the circuits from the ground up. This approach categorizes cells and provides an understanding of their reliability character, saving time and resources in simulation. Furthermore, since cell-level simulation requires fewer parameters than full device simulation, it provides a more efficient and accurate option for identifying the stress profile of transistors.

- Equivalent stress factors (ESF):
  - ○ Convert Dynamic Stresses to Static Stresses: The ESF is a powerful tool that can be used to convert dynamic stresses to static stresses that have the same degradation effect. It's a game-changer for lifetime modeling, which is typically built upon highly accelerated static voltage and temperature stresses. However, because a transistor in real operation experiences a dynamic stress profile, the static PoF models cannot be applied directly. ESFs are obtained through cell reliability characterization, and then applied in device reliability prediction.
  - ○ Specific Factors for Each Component: ESFs are specified to each cell, transistor, operation, and failure mechanism, making them highly precise. To estimate cell reliability in a real application, the cell operation profile is determined first. Then, ESFs are used to calculate the "effective" stress time for each failure mechanism of each transistor.
  - ○ Accumulate Degradation: Degradation under different stress conditions is accumulated and converted to an equivalent-total-stress-time (ETST) under a specified static stress condition by utilizing appropriate acceleration models. The ESF is the ratio of the ETST to the real stress time. This allows for accurate estimation of cell reliability as a series system, in which each transistor inside the cell corresponds to a component.
- Best-Fit Lifetime Distribution: To improve the prediction accuracy, the best-fit lifetime distribution for each failure mechanism should be taken instead of using the CFR model without justification. In the PoF statistical approach, Weibull distribution is used to model failure distributions, as will be described in more detail in the following chapters.
- Time-Saving Chip-Level Interconnects EM Analysis: Chip-level EM analysis is focused on the power network since it carries a large current density and is the weakest link. Although EM becomes more serious in submicron designs, it is limited to the power distribution network in most cases [64]. In the PoF statistical approach, chip-level EM analysis is limited to the power distribution network and provides a good approximation without running full-detailed interconnect network EM analysis. To optimize the EM resistance, lower-level interconnects are designed to the EM-failure-free by considering the Blech effect [65]. The power network becomes the weakest link because of the large current

density it carries and the local Joule heating effect. This has been verified by acceleration test results [66].

In the PoF statistical approach, chip-level EM analysis is focused on the power network since all designs should pass the design rules check, and final products must survive the high-temperature, high-voltage defect screening. This provides a good approximation without running a detailed interconnect network EM analysis.

A flowchart of the PoF-based statistical method is shown in Figure 3.7 (general framework) and Figure 3.8 (detailed paradigm). A comprehensive description of each step of the procedure is discussed below.

The PoF Flowchart Framework (Figure 3.7) is a 5-step sequence that outlines the processes for assessing asset reliability. The first step involves defining the assembly, materials and processes, and packaging of the asset. The second step involves evaluating the external environment the asset will be exposed to. In the third step, knowledge of PoF and failure mechanisms is applied to the asset under the given external conditions. The fourth step involves identifying and describing the reliability models. The final step is the device reliability estimation, which refines the

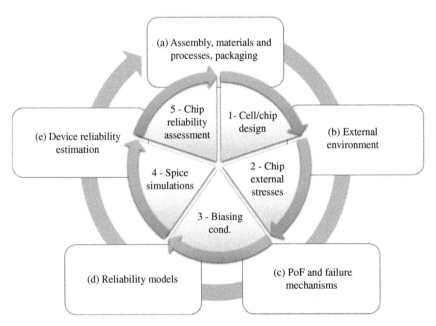

Figure 3.7 Flowchart framework of the physics-of-failure-based statistical reliability estimation.

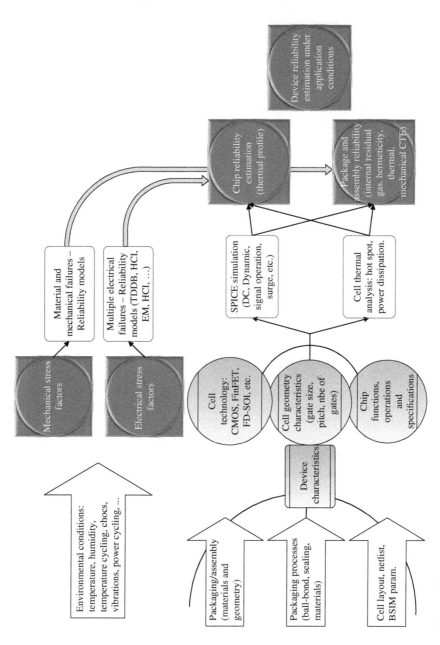

Figure 3.8 Physics-of-failure to comprehend reliability paradigm.

overall flow. The figure presented outlines a 5-step process, depicted by clock arrows, that aims to provide a detailed description of the cell/chip design content (tread 1). This process also highlights how external stresses impact chip stresses (tread 2), including which chip parameters are affected. Additionally, the process accounts for biasing conditions (tread 3), which can modify the applied stress using Spice simulations (tread 4), and aids in assessing chip reliability (tread 5). Each step is described in more detail in the next paragraphs.

Figure 3.8 presents the reliability paradigm, which offers a comprehensive view of the subelements that must be considered when estimating the device reliability of an asset subjected to operational conditions. These subelements include the packaging materials, geometry, and processes, as well as cell layout. Additionally, the paradigm involves SPICE simulation and thermal analysis to evaluate chip and device reliability under various environmental conditions that may introduce mechanical and electrical stress factors.

3.3.2 Assembly, Materials and Processes, and Packaging

When designing a new electronics system, it is crucial to consider reliability during the design phase. This is because it is the most cost-effective time to make changes to ensure a robust and reliable design. A reliable design can tolerate environmental stress and potential misuse of the product while allowing for variations in the manufacturing process. At each level of manufacturing, variations in the quality and consistency of materials and processes can introduce latent defects that can cause electronics systems to fail. Some of these latent defects include wire bond corrosion, die delamination, lead frame, and encapsulate defects at the IC level. At the Printed Wiring Board Assembly level, solder cracks, VIAs cracks, solder shorts, and cracks in Al metallization can cause failures. At the system level, loose electrical connectors, loose fastener hardware causing shorts, and fretting corrosion are common issues.

One can ask what are the differences between a defect due to latent failures and failures due to normal aging of good devices. Are they a manifestation of the manufacturing device inhomogeneity or normal behavior of the main lot? Consequently, why not consider failures due to latent defects being the result of the tails of a failure distribution with the same statistics as the mainstream lot?

Despite the importance of understanding the life cycle environmental profile (LCEP) is nearly impossible to know fully in a realistic application. Even when the LCEP is determined, new applications can have significantly different environmental conditions. Reliability predictions from failure prediction guides such as MIL-HDBK or FIDES have been based on the invalid assumption that the

Arrhenius equation applies to many wear-out modes in semiconductors and other electronics components. This has resulted in unnecessary costs in additional cooling and the belief that thermal derating during design provides longer life. There is no evidence to support this belief, and testing with steady-state elevated temperature may not provide a quantifiable acceleration of intrinsic wear-out mechanisms in electronics assemblies. In conclusion, it is crucial to prioritize reliability during the design phase of electronics systems, understanding that latent defects can occur at each level of manufacturing. Reliability predictions should be based on data and evidence and not on invalid assumptions. Therefore, the best approach to describe the PoF statistical reliability prediction methodology is to use a data-driven approach that considers the effects of environmental stress and latent defects at each level of manufacturing.

The rapid evolution of electronic materials and manufacturing methods creates a challenge to analyze and model intrinsic wear-out failure mechanisms from new materials and processes.

3.3.3 External Environment

The field of electronics is subject to rapid changes, both in the materials used and in the manufacturing methods. But in addition to these evolutions, some are driven by environmental concerns, such as the restriction of certain hazardous substances in electronics as defined by Directive 2002/95/EC related to new electrical and electronic equipment put on the market "after 1 July 2006 shall not contain lead or other hazardous materials." This directive has led to changes in the materials used in electronics, such as the switch to lead-free solders and the restriction of flame retardants. However, the intrinsic wear-out modes of active devices have much longer timescales than the technological obsolescence of systems. This means that wear-out mechanisms in electronics systems are rarely observed.

Acceleration life testing of a product is often used to reduce test time by subjecting the product to higher stress. The resulting data are analyzed and information about the performance of the product at normal usage condition is obtained. Acceleration testing usually involves subjecting test items to conditions more severe than encountered in normal use. This results in shorter test times, reduced costs, and decreased mean lifetimes for test times. In engineering applications, acceleration test conditions are produced by testing items at higher-than-normal temperatures, voltage, pressure, load, etc. The data collected at higher stresses are used to extrapolate to some lower stress where testing is not feasible. In most engineering applications, a usage or design stress is known.

3.3.4 PoF and Failure Mechanisms

For electronics manufacturers, most of the costs of failures occur in the first few years of the product's life, as shown by the declining failure rate on the left side of the bathtub curve. The challenge in building a system lifetime model is the complexity of fitting all temperatures and voltages when there are multiple failure mechanisms at work. Moreover, most common failure mechanisms in electronics are accelerated by more than one stress, making it challenging to develop a reliability prediction methodology that works for all scenarios.

Reliability prediction methodologies, such as MIL-HDBK 217 or FIDES, are often controversial in their application. These methodologies do not consider mixed factors such as the mounting and support of the device or the impact of environmental stresses such as temperature cycling, humidity cycling, and mechanical shock throughout the components' life cycles.

The systematic approach known as the PoF methodology is commonly used in engineering to understand the physical processes that lead to product failure. This involves identifying potential failure mechanisms, such as physical, mechanical, chemical, electrical, structural, or thermal processes, and likely failure sites on the product. Accelerated stresses are then applied to the product to simulate expected lifetime conditions, and the resulting data is analyzed to determine the root cause of failure. From this, a mathematical model is developed to explain the dominant failure mechanism and a statistical distribution such as Weibull or lognormal is used to predict the time to failure (TTF) under use conditions.

PoF methodology helps engineers design more reliable and durable products by identifying the underlying physical mechanisms that lead to failure and developing models that predict the expected lifetime of the product under use conditions. The goal of the PoF is to know the physics behind the root cause of a failure mechanism. The PoF approach can be used to design reliability assessments, testing, screening, and stress margins that prevent product failures. The PoF approach involves identifying potential failure mechanisms, failure sites, and failure modes, selecting appropriate failure models and input parameters, determining the variability for each design parameter, and computing the effective reliability function. Ultimately, the objective of any PoF analysis is to determine or predict when a specific end-of-life failure mechanism will occur for an individual component in a specific application.

An accurate quantitative acceleration test requires defining the anticipated failure mechanisms in terms of the materials used in the product to be tested, and determining the environmental stresses to which the product will be exposed when operating and when not operating or stored. When choosing a test or combination of tests on which a failure mechanism is based, assuming mechanisms that are anticipated to limit the life of the product, proper acceleration models must be considered.

These generally include:

- Arrhenius Temperature Acceleration for temperature and chemical aging effects
- Inverse Power Law for any given stress
- Miner's Rule for linear accumulated fatigue damage
- Coffin-Manson nonlinear mechanical fatigue damage
- Peck's Model for temperature and humidity combined effects
- Eyring/Black/Kenney models for temperature and voltage acceleration.

3.3.5 Key Considerations for Reliability Models in Emerging Technologies

The wear-out reliability of DSM components has been extensively studied and documented and many unanswered questions remain. Recent studies have shown that while wear-out failures occur prematurely and significantly reduce the useful life of components [67], there is a narrowing of the early life of components but a shortening of the useful life. The duration of the useful life period is defined as the component lifetime, which ends when the wear-out failure rate becomes higher than the random failure rate, as per the bathtub curve model. However, manufacturers provide little quantitative information on the reliability or the lifetime of their components, with reliability reports based only on HTOL test results. This traditional single-model HTOL approach leads to an expected FIT value that is unrealistically low, resulting in customers experiencing much higher reported failure rates in their applications.

Additionally, the HTOL test does not accurately accelerate all degradation mechanisms, and it does not consider certain parameters such as transistor size and material used that significantly impact failure mechanisms. The assumption that the activation energy is equal to 0.7 eV has not evolved since older generations of components and does not account for the fact that wear-out now consists of multiple degradation mechanisms instead of a single failure. As a result, it is difficult to determine the exact representativeness of the failure rate calculated based on the HTOL test or the suitability of 1000 hours of HTOL stress for a particular usage profile.

Moreover, environmental factors such as temperature, supply voltage, operating frequency, and clock duty cycle can accelerate failure mechanisms and reduce the device's lifetime, which poses a concern for the aeronautic market. Manufacturers do communicate information on lifetime under NDA, but several uncertainties and knowledge gaps still need to be addressed to predict accurately and manage the useful lifetime of these components. JEDEC's publication JEP122 acknowledges two issues with the current HTOL approach. First, multiple failure mechanisms compete for dominance in modern electronic devices. Secondly, each mechanism has significantly different voltage and temperature acceleration factors depending on device operation. To address these issues, JEDEC recommends a sum-of-failure-rates approach that accounts for multiple mechanisms. It is agreed that a single-point HTOL test with no failures cannot adequately consider the complexity of competing mechanisms.

3.3.6 Input Data

To carry out reliability simulation and prediction, all the following information needs to be gathered: device structure, cell function profile, cell thermal conditions, cell schematic and layout, SPICE simulation models, failure mechanism lifetime model, and statistical distribution.

- Application profile. The device application profile can be broken down into operation phases with distinguishable environmental factors. The PoF statistical approach deals with intrinsic failure mechanisms only, and the input data of each operation phase should include the ambient temperature (T_A) and the operating status power on hours. Other factors such as humidity and vibration can be considered.
- Device structure and operation. The PoF statistical approach takes a divide-and-conquer way to reduce the complexity of reliability simulation of ultra large-scale integration (ULSI) devices. A device functional diagram is needed to divide the whole chip into functional blocks. Inside each functional block, cells are categorized and analyzed. Device operation needs to be analyzed to build a cell operation profile.
- Cell reliability simulation inputs include cell schematic and layout, Technology file, SPICE models, Stimuli file (The stimuli file should be application-oriented so that the reliability output can be directly correlated to the stresses in a real application).
- Failure mechanism models and parameters.
  - Lifetime models and parameters. With the technology information (t_{ox}, V_d, V_{th}, etc.) and acceleration test data, reliability engineers can choose or build the appropriate lifetime model for each failure mechanism described by a nth root of time dynamic change in signature parameter and its associated failure criteria. Once the lifetime models have been decided, the model parameters can be estimated from maximum likelihood estimation (MLE) analysis or other regression analysis of acceleration test data.
  - The stresses applied in testing acceleration must be representative of those present during the operation life of the product. Acceleration factors are specific to each failure mode.

o Catastrophic or otherwise random failures are sudden and complete. However, the failure may be progressive and partial through degradation or wear-out. To characterize this degradation, two families of models can be observed: continuous models and multi-state models.

- The first models are used for a change in some amplitude of a characteristic supposed to be either mechanical as for crack propagation, delamination or electrical (mobile or fixed charge generation/recovery) impacting leakage currents, V_{th} rate change, and for which modeling of the evolution trajectory is required.

- The second models are used for a series of states of varying degrees of degradation and the switching laws between these states. For example, they include members of family of Markovian processes such as piecewise-deterministic Markov processes (PDMP) [23].

o When the level of degradation is observable, monitoring can be used to acquire much more information on the wear-out phenomenon than only monitoring operating time to failure. The degradation is accelerated by stress conditions. The Standard Model of Accelerated Life [68] is justified on the assumption that only the scale factor of the degradation distribution is changed and not its form. Consequently, the acceleration factor AF is then like the one assumed for random failures but with different parameter values. If the acceptability threshold can be defined, the degradation model becomes a reliability mode.

o Continuous degradation models are usually based on the family of Lévy processes [69] with stationary independent increments (each increment depends only on the time interval), including the Gamma process, Weiner process and compound Poisson process. We can mention also the PH regression model developed by Cox [50] in which the baseline failure rate $\lambda_0(t)$ is affected by different covariates X_i facilitating or preventing the occurrence of failures (respectively either $\beta_i > 0$ or $\beta_i < 0$) to model some stress factors (temperature, biasing voltages or currents, atmosphere contents such as humidity or hydrogen, etc.):

$$\lambda(t) = \lambda_0(t) \cdot e^{\sum \beta_i \cdot X_i} \tag{3.68}$$

o Failure distributions and parameters must be determined. It is important to have the correct failure distributions to estimate device reliability. For EM, lognormal distribution is normally the first choice. Weibull distribution has been widely used to model TDDB failures. For HCD and NBTI lognormal distribution can be utilized. With given acceleration test data, the goodness-of-fit of these statistical distributions can be checked and the related distribution parameters can be estimated.

3.3.7 Applicability of Reliability Models

For a known or suspected failure mechanism, all stimuli affecting the mechanism based on anticipated application conditions and material capabilities are identified. Typical stimuli may include temperature, electric field, humidity, thermomechanical stresses, vibration, and corrosive environments. The choice of acceleration test conditions is based on material properties and application requirements; the time interval should be reasonable for different purposes as well. Acceleration stressing must be weighed against generating failures that are not pertinent to the experiment, including those due to stress equipment or materials problems, or "false failures" caused by product overstress conditions that will never occur during actual product use. Once specific conditions are defined, a matrix of acceleration tests embodying these stimuli must be developed that allows separate modeling of each individual stimulus. A minimum of two acceleration test cells is required for each identified stimulus; three or more are preferred. Several iterations may be required in cases where some stimuli produce only secondary effects. For instance, high-temperature storage and thermal cycle stressing may yield similar failures, where high temperature storage alone is the root stimulus. Test sampling, defining failure criteria (i.e. the definition of a failure depends on the failure mechanism under evaluation and the test vehicle being used) and choosing an appropriate distribution (fitting the observed data to the proper failure distribution), so that the acceleration factors are determined [29, 32, 70].

"Initial failures" happen when a latent defect is created during the device production process and is then revealed under the stress of operation. This could occur because of tiny particles in a chip or crystal defects in the gate oxide film or silicon substrate. The failure rate decreases over time since only devices with latent defects will fail, and they are gradually removed. During the initial stage of device operation, there may be devices with latent defects that fail and are then removed from the set based on the failure rate. This period is known as the initial failure period, and the failure rate is defined as a decreasing function since the number of devices with latent defects decreases as they are removed.

Most initial defects in microelectronic devices are built into the devices, primarily during the wafer process. Dust adhering to wafers and crystal defects in the gate oxide film or silicon substrate are common causes of these defects. Defective devices surviving the sorting process during manufacturing may still be shipped as passing products, but these types of devices are inherently defective from the start and often fail when stress, such as voltage or temperature, is applied for a relatively short period. Nearly all these types of devices fail over a short time period, and the failure rate becomes part of the region called the random failure part. The process of applying stresses for a short period of time to eliminate defective devices before shipping the product is known as screening or "burn-in." When validating and verifying the performance of an equipment under extreme conditions like cold starts with surge signals, utmost caution is required. Engineers involved in engineering quality assurance and design processes must avoid entering regimes with excessively high acceleration, as this could inadvertently activate some failure mechanism and reduce the Remaining Useful Life (RUL) of the equipment.

"Random failures" occur once devices with latent defects have failed and been removed. During this period, the remaining high-quality devices operate stably. Failures that occur during this time are often caused by randomly occurring excessive stress, such as power surges or software errors. Memory software errors and other phenomena caused by high-energy particles like α-rays are also classified as randomly occurring failure mechanisms. Phenomena such as electrostatic discharge (ESD), breakdown, overvoltage (surge) breakdown electrical over-stress (EOS), and latch-up can occur randomly based on the conditions of use. However, these are classified as breakdowns rather than failures since they are produced by the application of excessive stress over the device's absolute maximum ratings and are not included in the random failure rate.

"Wear-out failures" are caused by the gradual aging and wearing down of devices, leading to a rapid increase in the failure rate. To ensure long-lasting and highly reliable electronic devices, it is important to minimize the initial failure rate to prevent wear-out failures during the guaranteed lifetime of the device. These types of failures are rooted in the durability of the materials used in semiconductor devices, such as transistors, wiring, oxide films, and other elements. Wear-out failures are used to determine the useful lifespan of a device, and as time passes, the failure rate increases until all devices ultimately fail or develop characteristic defects. It is worth noting that the same failure mechanism models used to explain wear-out failures can also be applied to random failures.

4

Transition State Theory

The aim of this chapter is to enhance the understanding of failure mechanisms by comprehensively modeling their multiphysics, phenomenological, and temporal dimensions. To accomplish this, the chapter will employ the transition state theory (TST) as a framework for developing predictive reliability models that account for the acceleration of multiple failure mechanisms under various stress conditions.

The assumptions of the Physics of Failure (PoF)-based statistical approach are as follows:

- Degradation, defect creation, and accumulation. For each failure mechanism, the degradation process is traced and quantified using electrical parameters recognized to be pre-eminent electrical signature of the failure mechanism activated during time. The interim measurements are performed at some specific interval of time where the dynamic stress is shortly removed allowing for accurate measurements of the DC and/or AC signature parameters.
- Independent failure mechanisms. All the failure mechanisms are assumed to be independent. Each failure mechanism has its specific degradation characteristic inside the transistor. Time-dependent dielectric breakdown (TDDB) causes damage inside the gate oxide while hot carrier injection (HCI)/negative-bias temperature instability (NBTI) increases interface trap densities. For a P-channel metal oxide semiconductor (PMOS), HCI and NBTI have been reported to be independent. There is no confirmation of interaction in field failure from literature research (see for example [71]).
- In the Reaction–Diffusion (R–D) formulation of NBTI degradation [72], one assumes that NBTI arises due to hole-assisted breaking of Si–H bonds at the Si/SiO_2 interface as shown in Ref. [73] if the number of broken Si–O bonds are small as is the case for low drain voltage (V_D) integrated circuits (ICs) of current and future technology nodes [74]. As related by Alam et al., the fact that uniform gate stress during NBTI involves 1D diffusion of H, while the localized

Reliability Prediction for Microelectronics, First Edition. Joseph B. Bernstein, Alain A. Bensoussan, and Emmanuel Bender.
© 2024 John Wiley & Sons Ltd. Published 2024 by John Wiley & Sons Ltd.

degradation during HCI involves 2D diffusion of H. Thus, we can interpret HCI exponents within the classical R–D framework.

- HCI is a phenomenon that can lead to defects near the drain edge at the Si/SiO_2 interface and in the oxide bulk, although it has been improved for current-generation metal-oxide silicon field-effect transistors (MOSFETs). Similar to NBTI, HCI can cause a shift in device metrics and reduce performance. The damage is caused by non-equilibrium carrier and their high energetic distribution in the indirect energy conduction band when accelerated by the high electric field near the drain side of the MOSFET, which results in impact ionization and subsequent degradation. Historically, HCI has been more significant in nMOSFETs due to the higher electron mobilities (caused by lower effective mass) compared to holes, which allows them to gain higher energy from the channel electric field. HCI has a faster rate of degradation compared to NBTI, and occurs during the low-to-high transition of the gate of an nMOSFET, making the degradation worse for high switching activity or higher frequency of operation. Moreover, the recovery in HCI is negligible, which makes it more problematic for AC stress conditions.
- Negligible BTI recovery effect. BTI degradation has been observed to have a recovery effect in acceleration tests after the stress has been removed [75, 76]. Physical understanding of these phenomena is still not clear. Since BTI is a long-term reliability concern, and recovery disappears quickly when the stress is reapplied, the worst-case BTI is considered in the PoF statistical approach.
- Competing failure modes. A device is treated as a series system in which any cell failure will cause device failure. Every cell is viewed as a series system with each failure mode composing a block of the series system.

4.1 Stress-Related Failure Mechanisms

A reliability ecosystem is illustrated in Figure 4.1. Stress-related models are phenomenological parameters to establish how the stressors induce change in the activation energy of degradation mechanisms: typically, this is the case for temperature stressors as well as for other electrical (current, voltage, power, signal, frequency, on-off power cycling, surge, electrical EOS, and ESD including DC and AC variations), or mechanical (vibration, chocs, strains, and fatigue), or chemical (humidity, hydrogen, and contaminant) stressors. These parameters are managed at high operating conditions and are fundamental to determining **acceleration factors**: higher stresses accelerate degradation. To help understand the mechanisms, the TST has been developed but it has nothing to do with time and aging. It is a theory to quantify the rate of change of a system.

In addition, the change (or drift) of signature parameters (known as electrical or performance key figures) aims to account for and quantify the aging of a device or

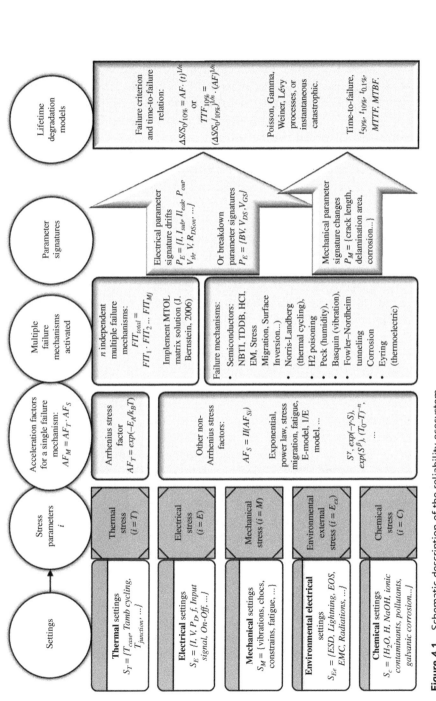

Figure 4.1 Schematic description of the reliability ecosystem.

an asset under normal or high-stress conditions of operation. They are time-related and generally based on power-law variations. They are used to determine the time to failure (TTF) of a homogeneous lot of assets, e.g. how long the devices in operation will take to lose half of their population (also related to the $t_{50\%}$ failure). Another similar factor is the mean TTF (MTTF). Note that electrical parameters can be considered either as stressors or as signatures and be careful to not confuse them. The first one is how the stress is applied to some extreme values (high temperature at high- to very high-stress current or voltage) while the second is how we measure the effect of aging under a representative low-stress-no-stress condition (room temperature at nominal bias and signal).

Many military applications today use custom commercial products and advanced components in their design. MIL HDBK 217 [1] or FIDES approaches [12] do not provide failure rate models for recently developed commercial components and typically do not address physics-based constant failure rate calculations. Degradation measurements and wear-out mechanisms must also be the basis for quantified reliability assessment at the equipment or system level. The evolution of hardware scaling continues to map efficiently on the multiple processors and software needs to be "parallelized" as explained in Ref. [77], the outstanding paper at International Reliability Physics Symposium (IRPS) 2012, Monterey, CA. In this paper, V. Huard et al. from ST Microelectronics offered an extensive discussion on this topic to highlight the reliability challenges related to technology scaling in the multicore era.

In a paper published at RAMS 2018 conference titled "Entropic Approach to Measure Damage with Applications to Fatigue Failure and Structural Reliability" [78], H. Yunm et al. suggest several entropies as candidates for proper damage measurements in fatigue process. The entropic approaches are expected to develop in order to apply them in practical reliability analyses and prognoses. The paradigm of the TST developed by E. Wigner in 1934 [79] and by M. Evans and M. Polanyi in 1938 [80] is an approach we can benefit from if adapted to the concept of a unified semiconductor reliability model. Indeed, in the early last century, the TST was applied to chemistry transformations by H. Eyring [81] and S. Glasstone et al. [82] in 1941. The TST was developed in chemistry based on Hammond's postulate [83] published in 1955 applied to physical organic chemistry.

From the 60s and during the following 40 years, several authors report complementary reliability models centered around the mobile-ion (Na+) drift within dielectric [84], to intrinsic mechanism as TDDB [85] for example, while in mid-80s, J. W. McPherson and D. A. Baglee [86, 87] report of stress-dependent activation energy and developed a generalized Eyring model to better understand thermally activated failure mechanisms (J. W. McPherson, Reliability Physics and Engineering – Time-to-Failure Modeling, 3rd ed., Springer, Chapter 9 [88]).

The Arrhenius model has been recognized for many years as the basis of all chemical and electrochemical failure mechanisms as it derives from the fundamentals of statistical mechanics. This model has often been misused in applications where multiple mechanisms simultaneously contribute to reliability. For example, activation energies cannot simply be added or averaged together, especially when the disparity of effective activation energies is significant [89]. In 2008, the Office of the Secretary of Defense (OSD) initiated a "concerted effort to elevate the importance of reliability through greater use of design-for-reliability techniques, reliability growth testing, and formal reliability growth modeling." The National Academy of Science proposed a report "Reliability Growth: Enhancing Defense System Reliability" [90] and presented 25 recommendations to overhaul the military's use of reliability prediction methods. We believe that the method detailed here, addresses many of their concerns regarding the misuse of many common practices applying reliability principles.

All PoF studies consider the activation energy as determined experimentally with respect to temperature (low versus high). Eyring models consider other stress conditions and indicators as, for example, charge detrapping for hot carrier degradation (HCD) or NBTI for PMOS devices under negative gate voltages at elevated temperature. These models are generally applicable to a given technology but also reliability figures published by manufacturers and handbooks are not transposable when technology is improved and changed. Even some end users and customers are focused on qualifying lot production instead of a process, complicating the handbook approach. There is a need to simplify the forest of existing PoF models so that it can be used in a practical manner. How can we harmonize the mathematics of the existing paradigm to achieve such a goal?

Multiple failure mechanisms and physics of degradation in semiconductors may occur in a single set of TTF data but without obvious points of inflection to help separate the mechanisms. J. McPherson, in his book, Reliability Physics and Engineering, 3rd edition, provides the basics of reliability modeling [86, 88] recalled *generally, materials/devices exist in metastable states. These states are referred to as being metastable because they are only apparently stable. Metastable states will change/degrade with time. The rate of degradation of the materials (and eventual TTF for the device) can be accelerated by an elevated stress (e.g. mechanical stress, electrical stress, electrochemical stress, etc.) and/or elevated temperature.*

The Gibbs-free energy description of material/device degradation is described in ref [87]. Considering the initial state to be a sound device before aging and the final state of a degraded device (either catastrophically failed or degraded and not compliant with the acceptable performance limit of the device). It is important to note at this point the reliability model is described by parameter drift degradation as a function of time and not as a random failure paradigm. In this diagram, the net reaction rate is a dynamic equilibrium between forward and reverse reaction, meaning the degradation could be reversible. The mathematical and PoF developments are given in Ref. [43] and depict how the equivalent activation energy is defined and dependent on the multiple stresses applied. It is shown how the level of stress (high or low) can impact the equivalent activation energy.

4.2 Non-Arrhenius Model Parameters

Reliability failure mechanisms in electronics are described from the standpoint of accelerated stress type and developed thanks to semi-empirical models. Reliability investigations reported in various books and tutorials have been synthesized as for example on Standard JEP122H. Reliability models such as electromigration (EM) [62], ohmic contact degradation [91, 92], Coffin-Manson [93], Eyring [81], humidity [94], TDDB [95], HCI [96–98], hydrogen poisoning [99, 100], thermomechanical stress [101], and NBTI [102] are generally expressed by a function of stress parameter or by a function of an electrical indicator multiplying the exponential activation energy factor.

The effect of temperature on electronic devices is often assessed by extrapolating from accelerated tests at extremely high temperatures based on the Arrhenius law. This method is known to be not necessarily accurate for prediction, particularly when stress-induced failures are driven by nonthermal dynamic electrical stresses. Application of kinetic theory, thermodynamics, and statistical mechanics have developed forms that contain exponential forms similar to Arrhenius. It is observed that the well-known Arrhenius law usually does apply, albeit with some modification, within existing models describing PoF. This is known, by example, in Black's law, Coffin-Manson or any application of Eyring's law. These include the effect of humidity or hydrogen poisoning or other effects in semiconductors [62, 81, 103, 104].

Hypotheses, Baselines, and Definitions:

✓ Relevance of non-Arrhenius models: The general Eyring model is almost the only model to include terms that have stress and temperature interactions (in other words, the effect of changing temperature varies, depending on the levels of other stresses). Most models in actual use do not include any interaction terms so that the relative change in acceleration factors when only one stress changes does not depend on the level of the other stresses.

✓ In modern devices, there is no single failure mode that is more likely to occur than any other as guaranteed by the integration of modern failure and modern simulation tools in the design process. The consequence of more advanced reliability modeling tools is a new phenomenon of device failures resulting from a combination of several competing failure mechanisms. JEDEC Standard, JESD85, gives the instruction to calculate multiple activation energy procedures for constant failure rate distributions; it is supposed that the devices failed due to several different failure mechanisms to which can be assigned the appropriate activation energies [44, 45].

✓ The key factor is that the Arrhenius relationship (proposed to model the single failure mechanism effect in accelerated tests and applied to predict the system reliability even for multiple failure mechanisms by manufacturers) is not appropriate for multiple-failure mechanism criterion.

✓ It was believed that almost all the failure mechanisms were in the wear-out part of bathtub curve. However, as technology progressed, the new era of 'constant rate processes' approximation is going to dominate the reliability of microelectronic devices and circuits. The 'constant rate processes' assumption is the result of failure mechanisms shifting from wear-out part to constant failure rate part due to deep submicron technologies as well as the shrinkage of transistor geometries.

✓ Drenick's theorem [55], as mentioned earlier in Section 3.2.1, is another way of justifying the constant failure rate approximation for a system consisting of a very large number of components, like a microelectronic chip.

✓ The central limit theorem (CLT) states that the sum of a number of independent and identically distributed random variables with finite variances will tend to a normal distribution as the number of variables grows. The CLT applies, in particular, to sums of independent and identically distributed discrete random variables. A sum of discrete random variables is still a discrete random variable. So we are confronted with a sequence of discrete random variables whose cumulative probability distribution function converges toward a cumulative probability distribution function corresponding to a continuous variable (namely that of the normal distribution).

✓ In microelectronic system reliability, one may define a system as an integrated collection of subsystems, modules, or components. In most settings involving lifetimes, the population of lifetimes is not homogeneous. That is, all the items in the population do not have the same distribution; there is usually a percentage of the lifetimes which are of a type different from the majority. For example, in most industrial populations, there is often a subpopulation of defective items. For electronic components, most of the population might be exponentially distributed, with long lives, while a small percentage often has an exponential distribution with short lives. Another way of saying this is that the failure rate of the mixture approaches the largest failure rate.

✓ Due to manufacturing and environmental variability, the strength of a component varies significantly. The component is used in an environment where it fails immediately when put into use if its strength is below some specified value.

The problem is to determine the probability that a component manufactured will fail under a given environment. It has been assumed here that the failure incidence in one component of a piece of equipment is statistically independent of the rest. If this probability is high, changing the material, the process of manufacturing, or redesigning might be the alternatives that the manufacturer and the designer might need to explore like to change the condition of biasing to reduce the stress.

Figure 4.2 shows a schematic drawing of the principle of the TST adapted from McPherson [88], which represents the amount of free energy $\Delta G^{\dagger}$ required to allow a chemical reaction to occur from an initial state (sound device – dashed line blue) to a final state (failed or degraded device). The transition state G_{TS} is the level of energy required for the reaction to occur (rate in forward balanced by rate in reverse) as a function k_F (respectively k_R) modified by various applied stressors (S_x, S_y, and T). If the chemical reaction is accelerated by a catalyst effect (induced by the stressors), the height of energy $\Delta G^{\dagger}$ is then reduced, allowing the transition Initial State $\rightarrow$ Final State to occur with less energy brought to the system at initial state.

In Figure 4.2, the Entropy $\Delta G_F^{\ddagger}(S_i, T)$ for the forward reaction and the Enthalpy $\Delta H_R^{\ddagger}(S_i, T)$ for the reverse reaction are expressed as a Taylor's series expansion around the stress condition $(S_x, S_y) = (S_a, S_b)$.

Let's define the stress parameters as:

$$S_x = x_\% \cdot S_{Br1} \text{ and } S_y = y_\% \cdot S_{Br2} \tag{4.1}$$

where S_{Br1} and S_{Br2} are respectively the breakdown and burnout experimental values of the two stress parameters considered (of course these values should probably follow a Gaussian distribution for a given population of asset, but for sake of simplicity, we consider they are fixed values). In such a case, the stresses are defined by the percentages $x_\%$ and $y_\%$, respectively, for S_x and S_y with $0 < x_\% \leq 100\%$ and $0 < y_\% \leq 100\%$.

According to Eyring law, we assume there exist two stress functions $h(S_x, T)$ and $g(S_y, T)$ defined by:

$$h(S_x, T) = \gamma_1(T) \cdot S_x \tag{4.2}$$

and

$$g(S_y, T) = \gamma_2(T) \cdot S_y \tag{4.3}$$

such as they lower the Free Gibbs energy and the Enthalpy, as shown in Figure 4.2 (b), by a symmetric amount of:

$$f(S_x, S_y, T) = h(S_x, T) + g(S_y, T) = \frac{1}{2} \cdot \gamma_1(T) \cdot S_x + \frac{1}{2} \cdot \gamma_2(T) \cdot S_y \tag{4.4}$$

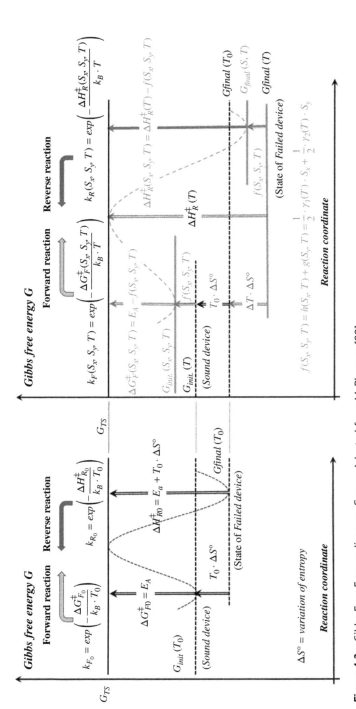

Figure 4.2 Gibbs Free Energy diagram. *Source:* Adapted from McPherson [88].

where we presuppose two temperature coefficient parameters $\gamma_1(T)$ and $\gamma_2(T)$ are defined by a linear temperature-dependence law of the form [87]:

$$\gamma_1(T) = a_0 + a_1 \cdot k_B \cdot T \tag{4.5}$$

$$\gamma_2(T) = b_0 + b_1 \cdot k_B \cdot T \tag{4.6}$$

Hence, the net reaction rate is expressed with a thermal term related to the activation energy as a function of stress conditions in a like-catalyst thermodynamic process and a term depending on the stresses applied. We define the net reaction as the balance between the forward and reverse reactions as:

$$k_{Net} = k_{Forward} - k_{Reverse} \tag{4.7}$$

with $k_{forward}$ and $k_{reverse}$ as functions of stress S_i applied:

$$k_{Forward} = exp\left(-\frac{\Delta G_F^{\ddagger}(S_i, T)}{k_B \cdot T}\right) \tag{4.8}$$

and

$$k_{Reverse} = exp\left(-\frac{\Delta H_R^{\ddagger}(S_i, T)}{k_B \cdot T}\right) \tag{4.9}$$

Finally, arranging Eqs. (2.75) to (2.77), we get the expression of net reaction rate k_{net}:

$$k_{net} = k(x_\%, y_\%) \cdot exp\left(-\frac{E_{A apparent.}(x_\%, y_\%)}{k_B \cdot T}\right) \tag{4.10}$$

with

$$E_{A apparent}(x_\%, y_\%) = E_A - \frac{a_0}{2} \cdot x_\% \cdot S_{Br1} - \frac{b_0}{2} \cdot y_\% \cdot S_{Br2} \tag{4.11}$$

$$k(x_\%, y_\%) = k_0 \cdot exp\left(\frac{a_1}{2} \cdot x_\% \cdot S_{Br1} + \frac{b_1}{2} \cdot y_\% \cdot S_{Br2}\right) \tag{4.12}$$

with $\quad k_0 = \left(1 - exp\left(-\frac{\Delta S^\circ}{k_B}\right)\right) \tag{4.13}$

All unknown parameters a_0, a_1, b_0, b_1, S_{Br1}, and S_{Br2} must be determined experimentally.

4.2.1 Hot Carrier Injection (HCI)

We demonstrate the power of using TST model to characterize the degradation of HCI in the study published in 2023 [105]. Most PoF mechanisms are modeled simply as an Eyring model [82, 106] where there are both an Arrhenius and a non-Arrhenius contribution to the acceleration factor, such that the probability of failure is due to a combination of nonthermal stress simultaneously with thermal stress. The thermal component, the Arrhenius component, is modeled as a probability that the energy exceeds some determined activation energy, E_A, whereas the energy of the electrons is in equilibrium with the lattice energy, which is the average thermal kinetic energy, $k_B T$.

$$P(T) = exp\left(-\frac{E_A}{k_B \cdot T}\right) \tag{4.14}$$

This expression accounts only for the thermal energy relative to the activation energy barrier, assuming that the barrier is an energetic difference between the quasi-stable operational state and a more stable "failed" state. The model assumes Poisson statistics of the energetic distribution and that the barrier, E_A is much greater than the average kinetic energy, $k_B T$. If the electrons are "cold" or even slightly "warm" with respect to the equilibrium lattice temperature, this approximation is sufficient. However, in the case of hot electron injection, where the electron energies are greater than the equilibrium lattice temperature, it is no longer valid to assume that the only energetic contribution is from the temperature of the lattice.

S. E. Tyaginov et al. [107] summarized the HCI degradation mechanism as the result of several combined effects. On the one hand, mechanism effect can be related to the localization on the drain side and at the oxide-semiconductor interfaces, on the other hand the effect can be due to a higher rate of degradation at low temperature, and to the increasing density of integration in new technologies. The first characteristic is related to the "hot" carriers, while the second is controlled by the fraction of quasi-equilibrium "cold" carriers. The latter is more like the NBTI effect of breaking silicon–hydrogen bonds. Thus, the NBTI effect is generally characterized by the relation to gate voltage whereas HCI is related to the conducting phase where high current density flows through the channel.

The energy-driven paradigm by Rauch and LaRosa [108] and the work of Hess [109] on defect dynamics refined the HCD model. Their work provided simple and effective approximations to the carrier energy distribution function to more accurately model HCD.

The Hess concept is the introduction of two competing mechanisms for Si–H bond-breakage, namely the single- and multiple-carrier processes. Penzin et al. [110] adapted the Hess model using TCAD device simulations by considering a phenomenological approximation. The Kufluoglu-Alam group model based on R–D framework [73] focused on the physical picture behind hot carrier degradation. The Bravaix et al. model [111] modifies the Hess and Rauch/LaRosa approach. Channel HCD models in most advanced low-power NMOS nodes point out that this degradation mode persists even at low voltages. This originates from mechanisms which have been modeled by scattering effects (electron–electron scattering, EES) and by multivibration excitation (MVE). These both contribute to increased interface trap generation due to H-release.

Hot carrier transport through a device can acquire large kinetic energies in regions of high electric fields. These hot carriers can gain sufficient energy to be injected into the gate oxide or can damage the semiconductor/gate oxide interface. The generation of defects at the interface results in fluctuations in the electrical characteristics and operating points of MOSFETs. HCI is typically a concern when the design of a device allows the presence of "high" electric fields during operation.

HCI mechanisms include four main types of HCI modes including channel hot electron (CHE) injection, substrate hot electron (SHE) injection, drain avalanche hot carrier (DAHC) injection, and secondary generated hot electron (SGHE) injection as summarized by E. Takeda [112].

A fresh look at the energetics of hot carriers moving through the channel reveals that electrons are accelerated beyond their thermal equilibrium temperature. Hence, they are "Hot" electrons. The energy barrier, E_A, is not changed. In fact, the energy distribution of the available electrons that cause the HCI damage is increased but the barrier to defect formation remains unchanged. Now, we assume that all the electron energies are shifted by the kinetic energy due to the applied drain voltage versus the source, V_{ds} where the channel electric field, $\mathcal{E}$ is $V_{ds}/L_{channel}$.

The average velocity of the electrons reaching the drain that are available to damage the oxide in the drain region (the effect of HCI) has velocity $v = \mu\,\mathcal{E}$, where μ is the carrier mobility. However, the velocity itself is generally saturated at the maximum velocity, v_s, as is well known at moderately high fields. The kinetic energy, E_k, due to this voltage acceleration is greater than the average thermal kinetic energy, $k_B T$, and shifts the probability of electrons surmounting the barrier E_A by adding to the total energy available to induce the damage. Therefore, it is easy to see that the denominator of the Poisson function should be modified by adding E_k to $k_B T$, leaving the rest of the equation deriving from the Arrhenius equation.

Now that we have established that the statistics of the damage are unchanged, the energetics must be modified by adding the field-induced kinetic energy to the thermal electron energy. Due to the nonlinearity of the energy term, $k_B T$, when $E_k > k_B T$, the kinetic energy dominates the thermal energy. It is well established in semiconductors that the electron/hole mobility is limited by collision probability along its mean-free path within the crystal. Hence, in the case of heavily doped silicon, the temperature dependence of mobility is less than the theoretical $T^{-3/2}$ law, but the mobility still increases with decreasing temperature. We have measured the conductivity of 16 nm FinFET devices by comparing the ring-oscillator

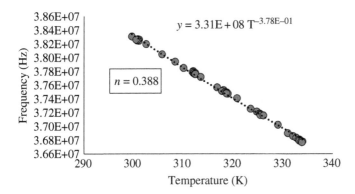

Figure 4.3 RO frequency of 16 nm FinFET versus temperature in (K).

resonant frequencies of field-programmable gate arrays (FPGAs) to find the conductivity change versus temperature as shown in Figure 4.3. We see the frequency dependence of μ goes as $T^{-0.378}$.

This measured value can be substituted for the electron mobility, μ, that represents the increased kinetic energy, E_k, due to the electric field, $\mathcal{E}$, whereas ν_s is the saturated velocity:

$$E_k = \frac{1}{2} \cdot m \cdot v_s^2 \tag{4.15}$$

The velocity saturation is temperature dependent and expressed as:

$$E_k(T) = A \cdot T^{-2n} \tag{4.16}$$

where $2n$ is the square of the mobility temperature factor found in Figure 4.3 and A is a constant that equates the temperature at which the kinetic energy and thermal lattice temperature energies are equal. This factor A can be stated if we use a normalized temperature T_0:

$$A = E_k(T_0) \cdot T_0^{2n} \tag{4.17}$$

Combining (Eq. (4.16)) to (Eq. ((4.17))), we get:

$$E_k(T) = E_{k0} \cdot \left(\frac{T_0}{T}\right)^{2n} = \frac{1}{2} \cdot m \cdot [v_s(T_0)]^2 \cdot \left(\frac{T_0}{T}\right)^{2n} \tag{4.18}$$

Thus, we now can express the denominator energy of the Arrhenius factor in expression Eq. (4.14), E_{den}, as the sum of both thermal and kinetic energies, $k_B T + E_k$.

So, the new Arrhenius model that correctly accounts for the energies of hot carriers has the form:

$$P \propto exp\left(- \frac{E_A}{k_B T + E_{k0} \cdot \left(\frac{T_0}{T}\right)^{2n}} \right) \qquad (4.19)$$

where E_{k0} is the kinetic energy at fixed temperature T_0 for each type of material (Si FinFET or GaN). For a given temperature T, the kinetic energy and thermal lattice temperature energies are equal. By inspection, when the temperature is high, the $k_B T$ term dominates but as the temperature decreases so that the energetic electrons are hot, the $\left(\frac{T_0}{T}\right)^{2n}$ term will dominate.

For FinFET technology, the mobility increases with lower temperature by a factor of $T^{-0.38}$ (from Figure 4.3). Thus, the change in this energy with temperature is related to the variation in ν_s^2, which goes as $(T^{-0.38})^2$, which is proportional to $T^{-0.76}$.

For silicon FinFET as well as older generations, we found that this n term is approximately constant, going back at least to 45 nm planar, where $n = 0.38$. Hence, $2n = 0.76$. Some other papers show [113] that for GaN HEMT devices, the temperature factor is closer to the ideal $n = 3/2$, making the hot carrier effect even more dramatic.

It has been established that the Arrhenius activation energy is simply the derivative of $log(P)$ with respect to $(1/k_B T)$ when considering the pure thermal effect, which results in the probability function having a minimum going from a positive to a negative slope with lower temperature.

4.2.2 Negative Apparent E_A

Now that we have established that the apparent negative activation energy observed for HCI damage is simply a consequence of the energetics of hot carriers as compared to the traditional Arrhenius model, we can plot the probability function for damage as a function of $1/k_B T$, and that $2n = 3$ for GaN and 0.76 for Si, using Eq. (4.16).

The left-hand side of this plot represents the high temperature, where a normal activation energy is found. This will be the result of any system where the electrons are not accelerated beyond the HCI limit. However, at lower temperatures, moving to the right, we see that the apparent activation energy appears negative. This is only a consequence of the increasing electron energy, but it has a very clear negative E_A appearance on the low-temperature side versus $(1/k_B T)$. In the hot carrier limit of low temperature where $k_B T$ is negligible, Eq. (4.19) becomes:

$$P \propto exp\left(- \frac{E_A T^{2n}}{A} \right) \qquad (4.20)$$

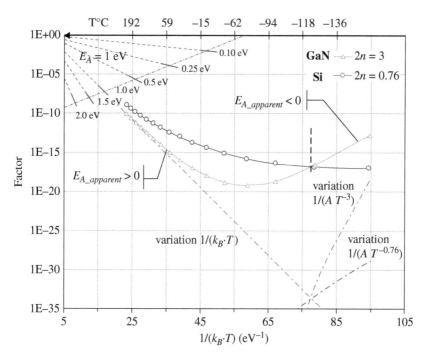

Figure 4.4 Arrhenius plot of acceleration factor versus $1/k_B T$.

We see here that as T gets higher, the probability of failure gets lower (more negative) as seen in the plot of Figure 4.4. The temperature term moves to the numerator of the exponent which gives the exact opposite effect of assuming $k_B T$ is in the denominator. The dramatic effect on voltage is also observed and is consistent with the highly nonlinear damage profile of HCI with voltage. When we consider the dynamics of electron-induced damage due to HCI, it is simple to visualize the failure mechanism due to hot carriers as an Arrhenius-type behavior with highly accelerated "hot" electrons causing damage and charge trapping at the surface of the channel.

The Arrhenius model is a well-established kinetic model that describes the rate of a chemical reaction as a function of temperature. The model is based on the TST [79, 80], which suggests that a chemical reaction proceeds through an activated complex or transition state, with an activation energy, E_A, required to overcome the energy barrier between the reactants and the transition state. However, the activation energy of a reaction may not be constant and may depend on other stress factors such as pressure and electrical stress parameters (voltage, current, or power dissipation).

To account for current stress, the Eyring model is used as a correction to the Arrhenius model. Eyring takes into consideration the influence of nonthermal stress, providing a more accurate prediction of the reaction rate. For HCI, this stressor is generally the current, I, through the channel and the stress is proportional to I^γ, where γ is the empirical power-law constant.

Figure 4.5 compares the schematic representation of the relationship between the energy band diagram of an indirect semiconductor and the associated

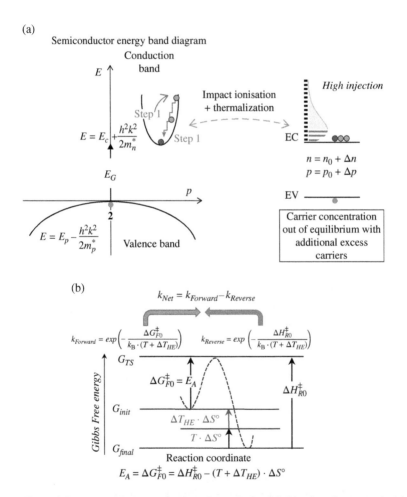

Figure 4.5 Schematic representation of the relationship between the energy band diagram of an indirect semiconductor and the associated transition state theory (TST) model. (a) Indirect semiconductor energy band diagram and impact ionization/thermalization of carriers (C. Kittel. Introduction to Solid State Physics. 8th ed. Wiley 2005. Case: High electric field and high carrier injection. (b) TST describes probability of failure of devices under various stresses (see also J. McPherson [88]).

TST model which accounts for the Gibbs free energy to model the probability of failure:

Step 1: Carriers with excess kinetic energy $(1/2\, m\, v^2)$ induced by high electric field.
Step 2: Generation of secondary electrons and positive charges by impact ionization (collision with crystal and defect Si–H bonds dandling) and thermalization.

In Figure 4.5b, the Gibbs free energy variation is shown from initial to final state when current stress is applied, with $\Delta S°$ being the variation of entropy in a chemical reaction and $\Delta H_{R_0}^{\ddagger}$, the variation of enthalpy.

As an outcome, the electric current accelerates electrons beyond their thermal equilibrium $k_B T$, adding $k_B \cdot \Delta T_{HE}$ to the probability function for damage at the interface. As depicted in Figure 4.5b, in the presence of high electric field generating impact ionization, net kinetic rate of the reaction is given by:

$$k_{Net_{HE}} = exp\left[-\frac{\Delta G_{F_0}^{\ddagger}}{k_B \cdot (T + \Delta T_{HE})}\right] - exp\left[-\frac{\Delta H_{R_0}^{\ddagger}}{k_B \cdot (T + \Delta T_{HE})}\right] \tag{4.21}$$

The probability of failure TTF_{HE} is inversely proportional to $k_{Net_{HE}}$ and proportional to the Eyring stress factor [112] as defined as the HCI model from JEDEC standard based on the peak current during stressing. So, we can set:

$$TTF_{HE} = \frac{1}{k_{Net_{HE}}} \cdot \left(\frac{I}{I_0}\right)^{-\gamma} \tag{4.22}$$

From Figure 4.5b, equation (4.10) is developed using the protocol proposed by J. McPherson, and we get:

$$TTF_{HE} = B \cdot \left(\frac{I}{I_0}\right)^{-\gamma} \cdot exp\left(\frac{E_A}{k_B \cdot (T + \Delta T_{HE})}\right) \tag{4.23}$$

where B is a time constant depending on change of entropy of the system. From this equation, it can be observed that at high temperatures, the hot carrier and impact ionization mechanisms are insignificant, leading to a negligible $k \cdot \Delta T_{HE}$ compared to $k \cdot T$. As a result, the apparent activation energy E_A is equivalent to the known Arrhenius activation energy.

Conversely, at low temperatures and high electric fields, the term $k_B \cdot \Delta T_{HE} \gg k_B \cdot T$, and knowing $\Delta T_{HE} = E_{k0} \cdot \left(\frac{T_0}{T}\right)^{2n}$ as expressed by Eq. (4.8), we get:

$$TTF_{HE} = B \cdot \left(\frac{I}{I_0}\right)^{-\gamma} exp\left[\frac{E_A}{k_B \cdot E_{k0}} \cdot \left(\frac{T}{T_0}\right)^{2n}\right] \tag{4.24}$$

By definition, the apparent activation energy $E_{A\_apparent}$ is the derivative of TTF_{HE} with respect to $1/(k_B \cdot T)$ at fixed current I and is developed as follows:

$$E_{A_{app.}}(T) = \left[\frac{\partial[\ln(TTF_{HE})]}{\partial\left(\frac{1}{k_B \cdot T}\right)}\right]_{I = constant} = C \cdot \frac{\partial T^{2n}}{\partial\frac{1}{T}} = -2n \cdot C \cdot T^{2n+1}$$

(4.25)

with $C = \dfrac{E_A}{E_{k0}} \cdot \left(\dfrac{1}{T_0}\right)^{2n} > 0$.

As observed in the plot of Figure 4.4, at low temperatures, the occurrence of hot carriers and impact ionization effect, $E_{A_{apparent}}$ as a function of $1/(k_B \cdot T)$, is negative.

4.2.3 Time-Dependent Dielectric Breakdown (TDDB)

TDDB, also known as oxide breakdown, is a failure mode of the capacitors within ICs caused by conductive defect accumulation through the dielectric (silicon dioxide), and eventually short-circuiting the device. The effect is accelerated by voltage stress and by temperature and therefore becomes worse as electronic devices decrease in size. It is one of the most important failure mechanisms in the semiconductor technology. TDDB refers to the wear-out process in gate oxide of the transistors; TDDB as well as EM and hot carrier effects are the main contributors to semiconductor device wear-out. However, the exact physical mechanisms leading to TDDB are still unclear [114–116].

A recent study was published at IRPS2022 by T. Garba-Seybou et al. [117] on a detailed analysis of off-state TDDB under nonuniform field performed in MOSFET devices from 28 nm FDSOI, 65 nm SOI to 130 nm nodes. In another paper from S. Kupke et al., [118] studied the impact of DC off-state and AC gate + off-state stress on the TDDB of ultra-short channel high-k/metal gate (HKMG) in which nMOSFETs were investigated. Under high DC off-state (drain) bias, the dielectric wear-out was found to be caused by a hot-hole injection at the drain side. The breakdown time is scaled with the gate length in accordance with a higher-impact ionization rate by an increased subthreshold leakage current for shorter channels. At identical bias, drain-only stress results in a less severe degradation in comparison to gate-only stress. However, the combination of alternating gate and off-state stress results in a lower lifetime compared to DC and AC gate-only stress. The AC gate and off-state pattern exhibit a similar degradation behavior as bipolar AC stress, attributed to continuous charge trapping and detrapping in the gate oxide.

It is generally believed that TDDB occurs as the result of defect generation by electron tunneling through gate oxide. When the defect density reaches critical density, a conducting path would form in the gate oxide, which changes the dielectric properties. Defects within the dielectric, usually called traps, are neutral at first but quickly become charged (positively or negatively) depending on their locations in regard to anode and cathode.

TDDB models for silica (SiO_2)-based dielectrics were revisited by J. McPherson in a paper published in 2012 [119] so as to better understand the ability of each model to explain quantitatively the generally accepted TDDB observations. The most frequently used TDDB models are based on field-induced degradation or current-induced degradation, or a combination of field-induced and current-induced degradation. Molecular dielectric degradation models, which lead to percolation path generation and eventual TDDB failure, tend to fall into three broad categories: field-based models, current-based models, and complementary combinations of field- and current-based models.

4.2.3.1 Thermochemical E-Model

This E-model prediction matches an important TDDB observation – nonpolar dielectrics (covalent-bonded dielectrics) do not undergo TDDB. One of the main criticisms of the E-model is its failure to account for polarity dependence where differences can arise if the anode and cathode roles are reversed. Because of the polarity-dependence deficiency of the E-model, current-based models (which can be polarity-dependent) have gained favor, even though they have their own set of deficiencies.

E-model relates the defect generation to the electrical field. The applied field interacts with the dipoles and causes the oxygen vacancies and hence the oxide breakdown. The lifetime function has the following form:

$$t_{breakdown} = t_0 \cdot exp(-\gamma \cdot E_{OX}) \cdot exp\left(\frac{E_A}{k_B \cdot T}\right) \tag{4.26}$$

where t_0 and γ are constants.

4.2.3.2 1/E Model (Anode-Hole Injection Model)

In the 1/E model for TDDB (even at low fields), damage is assumed to be due to current flow through the dielectric due to Fowler–Nordheim (F–N) conduction. Electrons, which are F–N-injected from the cathode into the conduction band of SiO_2, are accelerated toward the anode. Due to impact ionization, as the electrons are accelerated through the dielectric, some damage to the dielectric might be expected. Also, when these accelerated electrons finally reach the anode, holes can be produced which may tunnel back into the dielectric causing damage. This is the anode-hole injection (AHI) model. If the oxide layer is very thin (<5.0 nm), then direct tunneling (DT) is also possible. Since the electrons (from the cathode) and the injected holes (from the anode) are the result of F–N tunneling conduction, the TTF is expected to show an exponential dependence on the reciprocal of the electric field, $1/E$:

$$t_{breakdown} = t_0 \cdot exp\left(\frac{G}{E_{OX}}\right) \cdot exp\left(\frac{E_A}{k_B \cdot T}\right) \tag{4.27}$$

where E_{OX} is the electric field across the oxide and G and t_0 are constants.

Historically, because the $1/E$ model is a conduction model, Q_{bd} (total charge fluence to breakdown) became an often-used reliability metric. However, the usefulness of Q_{bd} as an important reliability metric diminished when film thickness became <5.0 nm because of the intrinsic leakiness of ultra-thin SiO_2 films.

The discussion proposed by J. McPherson [119] contained: "*The main criticism of the 1/E model (AHI theory) is twofold: (1) the hole generation rate at the anode is extremely small at low fields/voltages and (2) the efficiency for the injected holes to produce defects in the silica seems to be very low – intended confirmation experiments whereby holes were injected into the oxide via substrate hot-hole injection (SHH) have shown little/no impact on TDDB. In addition, since F–N tunneling is very weakly temperature dependent, no quantitative explanation exists in the present AHI theory to explain the strong temperature dependence observed during TDDB testing.*"

For thick dielectrics, the dielectric field is an important parameter controlling the breakdown process, while temperature dependence of dielectric breakdown is another key point. The E and $1/E$ models can only fit part of the electric field, as shown in (Figure 4.6); both models are not discernable for electric field above 9 MV/cm for the field acceleration values.

There are some articles trying to unify both models as well [120–122]. However, both of those models are not applicable to the ultrathin oxide layers. For ultrathin oxide layers (between 2 and 5 nm), other models are used.

4.2.3.3 Power-Law Voltage V^N-Model

"Voltage-driven model for defect generation and breakdown" is a model that applies the percolation theory to find the breakdown in SiO_2 layers and suggests the use of a voltage-driven model instead of an electric field-driven one. It says that the tunneling electrons with energy related to the applied gate voltage are the driving force for defect generation and breakdown in ultrathin dielectrics. According to this model, despite the thick dielectric layer TTF breakdown, the ultrathin dielectric time to break down does not have the Arrhenius behavior anymore [115, 123]. The power-law V^N-Model is referred to as the anode hydrogen release (AHR) model. Si–H bonds at the Si/SiO_2 anode interface are believed to be excited by single-event electron processes (coherent processes) or multievent electron processes (incoherent processes). The AHR model was originally proposed for ultra-thin dielectrics where ballistic transport dominates. Thus, if the gate voltage

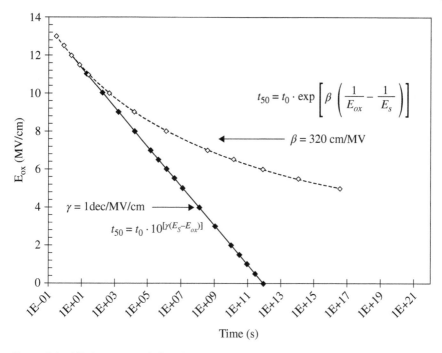

Time (s)

Figure 4.6 Lifetime extrapolations based on the linear E and 1/E models show the large discrepancies at the lower electric fields. *Source:* Bisschop [29].

is V, then the energy delivered to the anode by each electron is simply $(e \cdot V)$. For SiO_2 dielectrics, which are hyper-thin SiO_2 (<40 Å), a power-law voltage model has been proposed for TDDB of the form:

$$t_{breakdown} = B \cdot V^{-n} \qquad (4.28)$$

where B and n are constants driven from the experimental data. The power-law voltage dependence can eliminate the two unphysical results of the exponential law as well [124].

The primary weakness(es) of the AHR model – the theory makes no effort to quantitatively explain the strong temperature dependence (high activation energy) associated with TDDB and as to why the activation energy reduces with field. In addition, the AHR suffers from the dilution problem. While the amount of hydrogen at the anode Si/SiO_2 interface might be sufficient to cause TDDB in hyper-thin Si/SiO_2 films, the impact of the released hydrogen at the anode must surely be diluted as the thickness of the SiO_2 film increases.

In fact, as per Stong et al. [125], there are defect generation models that are based on the injection of carriers and the dissipation of the carrier's energy at the anode interface. This electron energy-driven approach has two main versions: the AHR model [119, 126–135] and the AHI model [136–142]. In these models, defect generation is described as a two-step process: (1) Electrons are injected through the oxide and lose their energy at the anode interface where they release some positively charged species; (2) The released species are reinjected into the oxide, travel toward the cathode, and generate defects by reaction with some unspecified precursor sites.

4.2.3.4 Exponential $E^{1/2}$-Model

Exponential $E^{1/2}$-model TDDB development [143–145] has been considered for low-k SiO_2-based interconnect dielectrics. The $E^{1/2}$ dependence is due to the Poole-Frenkel injection mechanism but at low fields, the $1/E$ dependence predominates. Experimental verification of this model was demonstrated by Lloyd et al. in [146, 147]. For high-quality SiO_2, the dominant current flow is nearly always F–N conduction and thus the damage is assumed to follow a $1/E$ model. This may not be the case for other dielectric types, or even poor-quality SiO_2 dielectrics (such as low-k interconnect dielectrics) where the conduction mechanism may be Poole-Frenkel or Schottky conduction. Thus, based on a current-induced degradation, a time-to-breakdown model of the form for low-k dielectrics is as follows:

$$t_{breakdown} = D \cdot exp\left(\frac{Q_{bh} - \lambda \cdot \sqrt{E}}{k_B \cdot T}\right) \tag{4.29}$$

where the barrier height Q_{bh} is usually only a few tenths eV and λ is referred to be the root-field acceleration parameter.

As with all the current-based models, the primary weakness of the $E^{1/2}$ model, it has great difficulty explaining, quantitatively, the strong temperature dependence (high activation energy) associated with TDDB and as to why the activation energy reduces strongly with field.

4.2.3.5 Percolation Model

In 1995, the paper presented by Degraeve et al. [148] suggested a consistent model for the intrinsic TDDB of thin oxides. The model links existing AHI and electron trap generation patterns and describes wear-out as a hole-induced generation of electron traps. The breakdown is defined as conduction via these traps from one interface to the other. The paper explains the relationship between the Weibull slope of the QBD distribution and the oxide thickness dependence. It shows that thinner oxides have a larger Weibull slope due to the statistical properties of the breakdown mechanism. The critical electron trap density (CETD) is identified as the breakdown mechanism, and the plausible breakdown mechanism is conduction via the generated traps.

The simulation results agree well with the data, showing a linear dependence of the CETD with oxide thickness. The larger Weibull slope for thinner oxides is explained as follows: in the thinnest oxide only, a few traps are needed to form the conductive breakdown path and consequently there is a large statistical spread on the average density to form such a short path. In thicker oxides, the breakdown path consists of a larger number of traps, and the spread of the trap density necessary to generate such a large path is smaller. So, it was demonstrated that the decrease of the CETD and the critical hole fluence with decreasing oxide thickness is an intrinsic statistical property of the breakdown mechanism. Overall, this paper provides a consistent model that links AHI and electron trap generation models and explains the intrinsic TDDB of thin oxides. The results suggest that the breakdown mechanism is a statistical property of the system, and the model provides a better understanding of the oxide thickness dependence of the Weibull slope and the critical hole fluence.

The statistics of gate oxide breakdown are usually described using the Weibull distribution:

$$F(t) = 1 - exp\left[-(t/\alpha)^\beta \right] \tag{4.30}$$

where α is the scale parameter and β is the shape parameter. Weibull distribution is an extreme-value distribution in $ln(x)$ and is the "weakest link"-type of problem. Here F is the cumulative failure probability, x can be either time or charge, α is the scale parameter, and β is the shape parameter.

The "weakest link" model was formulated by Suñé et al. and described oxide breakdown and defect generation via a Poisson process [149]. In this model, a capacitor is divided into a large number of small cells. It is assumed that during oxide stressing, neutral electron traps are generated at random positions on the capacitor area. The number of traps in each cell is counted, and at the moment that the number of traps in one cell reaches a critical value, breakdown will occur. Dumin [150] incorporated this model to describe failure distributions in thin oxides. The Weibull slope β is an important parameter for reliability projections. A key advance was the realization that β is a function of oxide thickness t_{ox}, becoming smaller as t_{ox} decreases. The smaller β for thinner oxide is explained as the conductive path in the thinnest oxides consisting of only a few traps and therefore has a larger statistical spread. The shape parameter's oxide thickness dependence is shown in Figure 4.6. The dependence of β on t_{ox} can be fitted to:

$$\beta = \frac{(i_{int} + t_{ox})}{a_0} \tag{4.31}$$

where a_0 is the linear defect size with a fitted value of 1.83 nm, and t_{int} is the interfacial layer thickness with a fitted value of 0.37 nm [151].

It can be found that β is approaching one as t_{ox} is near 1 nm, which means Weibull distribution will become exponential distribution. Lognormal distribution has also been used to analyze acceleration test data of dielectric breakdown. Although it may fit failure data over a limited sample set, it has been demonstrated that the Weibull distribution more accurately fits large samples of TDDB failures. An important disadvantage of lognormal distribution is that it does not predict the observed area dependence of TDDB for ultrathin gate oxides and the distribution itself does not fit into a multiple mechanism system that is linearly combined. Statistical explanations of this mechanism are detailed in Section 7.1.

4.2.4 Stress-Induced Leakage Current (SILC)

High-field stressing of SiO_2 causes low-field current leakage, called stress-induced leakage current or SILC [43].

During the initial investigation of SILC in flash EEPROM devices, multiple mechanisms were put forth to explain it, one of which was positive charge-assisted tunneling (PCAT) [152], neutral trap-assisted tunneling (TAT) [153] and thermally assisted tunneling [154]. Among various oxide stress conditions, hot-hole stress-induced oxide leakage has been found to be most serious in flash EEPROM devices [155]. Studies showed that SILC exhibits a significant transient effect [156]. A threshold voltage shift of 1.0 V resulting from a SILC transient component was reported [157].

In their paper, Chou et al. [158] summarized that SILC in ultrathin oxide metal–oxide–semiconductor devices was quantitatively modeled by the TAT mechanism. The results are compared with experimental data on samples with oxide thickness ranging from 40 to 80 Å. The model accurately describes the electric-field dependence of SILC and also predicts the increase, then decrease in SILC, with decreasing oxide thickness, which is observed experimentally. Extracted from [158], Figure 4.7a shows that damage is caused by the tunneling electron after it gains energy from the stressing electric field (ε_{stress}). The damage occurs between the tunneling distance X_T, which is a function of ε_{stress}, and the remaining thickness of the oxide. Damage occurs beyond tunneling thickness (X_T) when electrons (e^-) gain sufficient kinetic energy and suffer collisions with atoms in the oxide. The traps are distributed within the oxide band gap, shown schematically at zero electric field in Figure 4.7b. Once these traps are generated, they serve as intermediate tunneling sites, which can increase leakage current even at moderate oxide fields Figure 4.7c. The trap distribution N_T is a function of ε_{ox} since changing the applied electric field changes which traps are aligned with the cathode conduction band.

A measurement technique was implemented to investigate oxide charge trapping and detrapping in a hot carrier stressed n-MOSFET [159] by measuring a

(a)

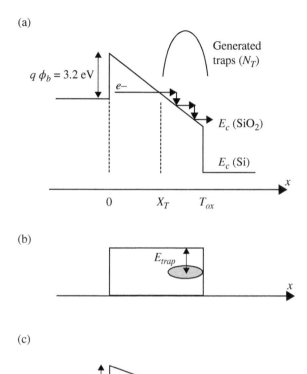

(b)

(c)

Figure 4.7 (a) Generation of traps under high-field F–N stress conditions (ε_{stress} is ≈ 11 MV/cm). (b) Schematic of trap distribution in oxide band gap at ε_{ox} ~0. (c) Trap-assisted conduction mechanism at low or moderate electric field ($\varepsilon_{ox} \approx$ 4–7 MV/cm). *Source:* Adapted from Chou [158] / American Institute of Physics.

gate-induced drain leakage (GIDL) current transient. This measurement technique is based on the concept that in a MOSFET, the Si surface field and thus GIDL current vary with oxide trapped charge. By monitoring the temporal evolution of GIDL current, the oxide charge trapping/detrapping characteristics can be obtained. An analytical model accounting for the time dependence of an oxide charge detrapping-induced GIDL current transient was derived. A specially designed measurement consisting of oxide trap creation, oxide trap filling with electrons or holes, and oxide charge detrapping was performed. Two hot carrier stress methods, CHE injection and band-to-band tunneling-induced hot-hole injection, were employed. The time dependence of the transients indicates that oxide charge detrapping is mainly achieved via field-enhanced tunneling. In

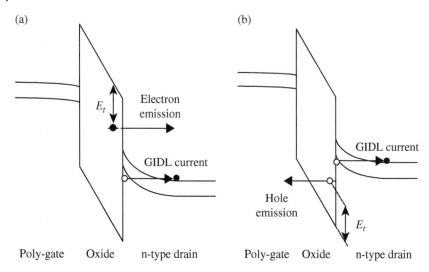

Figure 4.8 Schematic extracted from Ref. [159] showing band diagrams of oxide charge detrapping: (a) electron detrapping (b) hole detrapping. *Source:* Adapted from Wang [159]/IEEE.

addition, they used this technique to characterize oxide trap growth in the two hot carrier stress conditions. The result reveals that the hot-hole stress is about 104 times more efficient in trap generation than the hot electron stress in terms of injected charge.

The concept of the GIDL technique is illustrated in Figure 4.8 extracted from Ref. [159]. In the figure, a low-level V_{ds} was considered to minimize lateral field heating; a hot carrier-stressed n-MOSFET is biased in a GIDL condition, i.e. a negative gate bias and a positive drain bias. The band diagrams before and after oxide charge detrapping are represented by the solid line and the dashed line, respectively. Here, negative oxide charge is assumed in the figure. Due to an increased band-to-band tunneling distance after negative charge detrapping, the GIDL current is reduced. Therefore, by monitoring the temporal evolution of GIDL current, we can deduce escaping charge density and the associated charge detrapping rate.

To summarize as in JEDEC, the mechanisms having emerged as the main contributors to V_T instabilities in Flash technologies are:

- SILC, caused by TAT, resulting in leakage of charge to/from cell Floating Gate [160],
- Interface states annealing, affecting cell VT via enhancement of channel conductivity [161, 162],
- Charge detrapping, being originated by the capture of charge in oxide defects and by its subsequent neutralization [163].

The TTF of SILC-related dielectric leakage induced by program/erase cycling depends on cycle count, temperature, applied voltage, and the threshold voltage at which failure occurs:

$$t_{breakdown} = A_0 \cdot cycles^{-n} \cdot exp\left(\frac{E_{aa}}{k_B \cdot T}\right) \cdot exp[-\gamma \cdot (V_{T,Crit} - V_G)] \text{ for charge} - \text{loss}$$

$$(4.32)$$

$$t_{breakdown} = A_0 \cdot cycles^{-n} \cdot exp\left(\frac{E_{aa}}{k_B \cdot T}\right) \cdot exp[-\gamma \cdot (V_G - V_{T,Crit})] \text{ for charge} - \text{gain}$$

$$(4.33)$$

with n = cycling power-law coefficient (typically 0.4 to 0.7), E_{aa} = apparent activation energy (typically zero to 0.3 eV), cycles = number of program/erase cycles prior to the retention period, $V_{T,crit}$ = cell threshold voltage at which the cell will be sensed as having incorrect data, and V_G = applied voltage on the top or control gate of the cell during the retention stress.

SILC characterization generally steers clear of temperature acceleration, mainly due to the low E_{aa} and the possibility of oxide recovery at high temperatures. As an alternative, electric-field acceleration is commonly employed using either margin testing, which involves testing devices with a guard banded value of $V_{T,crit}$, or an applied gate voltage (V_G).

4.2.5 Negative Bias Temperature Instability (NBTI)

NBTI happens to PMOS devices under negative gate voltages at elevated temperatures. The degradation of device performance, mainly manifested as the absolute threshold voltage V_{th} increase and mobility, transconductance, and drain current I_{dsat} decrease, is a big reliability concern for today's ultrathin gate oxide devices [164]. Deal [165] named it "Drift VI" and discussed the origin of the study of oxide surface charges. Goetzberger et al. [166] investigated surface state change under combined bias and temperature stress through experiments that utilized MOS structures formed by a variety of oxidizing, annealing, and metalizing procedures. They found an interface trap density D_{it} peak in the lower half of the bandgap and p-type substrates gave higher D_{it} than n-type substrates. The higher the initial D_{it}, the higher the final stress-induced D_{it}. Jeppson et al. [167] first proposed a physical model to explain the surface trap growth of MOS devices subjected to negative bias stress. The surface trap growth was described as diffusion controlled at low fields and tunneling limited at height fields. The power-law relationship ($t^{1/4}$) was also proposed for the first time. In this section, the up-to-date research discoveries of NBTI failure mechanism, models, and related parameters will be discussed.

Most PoF mechanisms are modeled simply as an Eyring model [82, 106] where there are both an Arrhenius and a non-Arrhenius contribution to the acceleration factor, such that the probability of failure is due to an accumulation of nonthermal stress simultaneously with thermal stress. The voltage, non-Arrhenius, portion for BTI is usually modeled as an exponential function of voltage multiplied by γ [43], as

$$R(T, V) = \nu_0 \cdot exp\left[- \left(\frac{E_A}{k_B \cdot T} \right) \right] \cdot exp(\gamma \cdot V) \tag{4.34}$$

where ν_0 is a constant rate function with time.

The failure rate is normally understood as a probability function that simply describes the constant probability that an energy barrier, E_A, is exceeded by the electron potential due to temperature, T, and Voltage, V. However, this expression can only describe a linear dependence on time. In this case, the dependence is on the rate of change with voltage. This rate should describe the variation in threshold voltage over time, which would be a constant rate in time that is exponentially dependent on voltage.

As for BTI, no consensus has yet been formed on the physical mechanism governing the kinetics during DC and AC stress and for recovery after stress, in large and small area devices (J. Stathis [168]). On the contrary, two disparate viewpoints have taken hold, typically referred to as the "defect-centric" model (Grasser et al. [169, 170]) and the Reaction–Diffusion or "RD" model based on a comprehensive framework by Chakravarthi et al. in 2004 [171] for predicting interface state generation during NBTI stress.

In 2018, Patra et al. [172] proposed compact BTI models developed and verified with 14 nm FinFET and 28 nm HKMG, based on stochastic trapping/detrapping mechanisms. Their model addressed divergence between long-term and short-term stress time. Later, contrary to the accepted model that has been uniformly accepted as the cause of NBTI, the time-law that is observed for threshold voltage shift is experimentally observed to change as a power-law with time as shown, for example, in 2019 by T. Asuke et al. [173]. This group extracted a power-law model of BTI with time exponent $n = 1/6$ from smooth degradation curves. In another study, an ultrafast (10μs delay) characterization method was used by Choudhury et al. [174] to measure threshold voltage shift (ΔV_T) owing to BTI and HCD stress in N and P channel Gate-All-Around NSFETs (GAA Nano-Sheet FETs).

When we consider that the threshold voltage would be changing in the exponent, as in Eq. (4.34), the change in time should be logarithmic and not a power law. All our observations of NBTI degradation assume that the threshold voltage changes continuously with increased surface charge density, which shifts V_{th} proportionally to the charge, where ΔV_{th} is the interface charge generation, q_{it}, divided by the gate capacitance, C_{OX}. The fact is that all NBTI testing is performed with constant applied voltage while the threshold voltage slowly changes because of increased interface traps. Hence, the voltage expression in the exponent of 4.34 is the applied voltage minus ΔV_{th}. Thus, the time-resolved power-law expression is not consistent with an exponential dependence on voltage.

NBTI occurs in PMOS devices typically stressed at high temperatures (100–$200°C$) with negative gate voltages. The primary characteristic of NBTI is a decrease in negative threshold voltage for PMOS. The shift in threshold voltage over time is dependent primarily on temperature and gate voltage. The voltage shift can be readily detected by observing the frequency shift in a ring oscillator (RO) design as we will show for FinFET technology as the $\Delta F/F_0$ parameter signature where F_0 represents the initial RO frequency at time $t = 0$. Experimental data consistently show a drift in frequency as a power-law time-dependent and is generally given by:

$$X(t) = \frac{\Delta F}{F_0} = \left(\frac{t}{\tau_0}\right)^{\frac{1}{n}} \tag{4.35}$$

where $X(t)$ is a variable describing the change in frequency and where n and t_0 are, respectively, the time power-law factor and the normalization time parameter in hours for a dimensionless equation, both determined experimentally. Assuming a failure criterion of 10% change of initial frequency to define TTF, we get:

$$\left.\frac{\Delta F}{F_0}\right|_{10\%} = AF \cdot \left(\frac{t}{\tau_0}\right)^{\frac{1}{n}} \tag{4.36}$$

where AF is the accelerating factor for a given stress operation (both accounting for Arrhenius and Eyring non-Arrhenius factors). Its expression is given by:

$$AF = exp\left[\left(-\frac{E_A}{k_B}\right) \cdot \left(\frac{1}{T} - \frac{1}{T_{ref}}\right)\right] \cdot \left(\frac{V - Vth}{V_{ref}}\right)^{\gamma} \tag{4.37}$$

And so, merging Eq. (4.36) with Eq. (4.37), we get the TTF for a 10% failure criterion of relative to frequency drift:

$$TTF_{10\%} = A \cdot exp\left[\left(n \cdot \frac{E_A}{k_B}\right)\left(\frac{1}{T} - \frac{1}{T_{ref}}\right)\right] \cdot \left(\frac{V - Vth}{V_{ref}}\right)^{-n \cdot \gamma}$$

$$\text{with} \tag{4.38}$$

$$A = \tau_0 \cdot \left(\frac{\Delta F}{F_{010\%}}\right)^n$$

4.2.5.1 Time Dependence

In 2007, Islam et al. [175] presented advances in experimental techniques (on-the-fly and ultrafast techniques) to allow measurement of threshold voltage degradation due to NBTI over many decades in timescale. They considered the soft saturation of ΔVth at a long stress time. The measured time exponent was observed to be a function of time for various stress biases. They highlighted that the time exponent converges to a robust $1/n \sim 1/6$ value at a long stress time (e.g. $\Delta n \sim 0.005$/decade), the reduction being higher for higher stress V_G. This reduction arises from the decrease in oxide electric field at a constant voltage stress due to the increase in ΔVth over time.

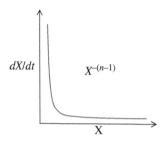

Figure 4.9 dX/dt versus X(t) for short time.

In our analysis, we consider the variation of total charge q_{it} with time to be proportional to the change of frequency considered the electrical signature of the mechanism or mode as:

$$\frac{d\,q_{it}}{dt}\bigg|_{@T,V} = \frac{d\frac{\Delta F}{F_0}}{dt} = \frac{dX(t)}{dt} = \frac{\left(\frac{t}{\tau_0}\right)^{\frac{1}{n}}}{n \cdot t} \qquad (4.39)$$

and because of 4.35, the inverse function is $t = \tau_0 \cdot [X(t)]^n$.

We get:

$$\frac{dX(t)}{dt} = \frac{X(t)^{-(n-1)}}{n \cdot \tau_0} \text{ or } \frac{dX(t)}{X(t)^{-(n-1)}} = \frac{dt}{n \cdot \tau_0} \qquad (4.40)$$

Of course, the integration of the two sides of Eq. (4.40) is consistent with the Eq. (4.35).

So, as sketched in Figure 4.9, we can now see that the time dependence of $X(t)$ will have a power-law dependence with time. However, the power law will be to a negative power, such that the zero-time change with time will be infinite since

$$\frac{dX}{dt} = \frac{t^{\left(\frac{1}{n}-1\right)}}{n \cdot \tau_0^{1/n}} \qquad (4.41)$$

which gives us a characteristic for the initial change in time for the threshold voltage and therefore the frequency, ΔF.

We see that $X(t)$ is a time-dependent change of the frequency normalized to some initial frequency, F_0. However, this initial frequency is not possible to obtain since the change in frequency happens very quickly when voltage is first applied.

4.2.5.2 1/*n*-Root Measurements

In previous studies [176–178], we demonstrated how parallel failure modeling can be used to correlate between single test elements and Weibull distribution slopes. The studies monitored the frequency degradation of 160 ROs running in parallel and sampled every minute. The degradation observed is an average of the degradation of the single stages.

Based on degradation for approximately 200 hours, *TTF* values are calculated. The *TTF* is defined as the time for the RO to decease in frequency by 10% of the original frequency. We modeled the failure statistics with Weibull distributions. Based on the CLT, with the increase of stages, the distribution slopes will increase at the same rate. The following development of N elements failing in a parallel system shows the effect analytically:

$$F(t) = \left(1 - e^{-\lambda_1 t}\right)\left(1 - e^{-\lambda_2 t}\right)...\left(1 - e^{-\lambda_N t}\right) = \left(1 - e^{-\lambda t}\right)^N$$

$$\Rightarrow R(t) = 1 - \left(1 - e^{-\lambda t}\right)^N \tag{4.42}$$

Using the approximation: $\exp(-\lambda \cdot t) \approx 1 - \lambda \cdot t$

$$R(t) \approx 1 - (\lambda \cdot t)^N \approx e^{-(\lambda \cdot t)^N} = e^{-(t/\theta)^\beta} \tag{4.43}$$

So, we see from Eqs. (4.42) and (4.43) that the beta slope will change directly with the number of stages in the ring. Although this phenomenon is always true, it must be set to the correct time rate to receive accurate results.

The ROs degrade with some n-root power rule as was detailed earlier. Here we can see how the time rule affects the Weibull slope by assuming different time rules on the 3-stage ROs. Figure 4.10 is taken from 16 nm FinFET technology. The left side shows the beta slope with a ½ root law. Moving to the right side, the n-root is increased until the most right of $n = 5$. The plots show the change in the beta slope with an increase in the n-root power law. Using Eq. (4.43) and solving for beta generates these plots where the axes are:

$$\ln(-\ln(1 - F(t))) \rightarrow {}^{"}Weibit{}^{"} = \beta \cdot \ln(TTF) \tag{4.44}$$

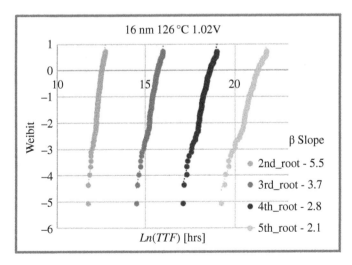

Figure 4.10 A plot showing the beta slope measurements of 16 nm data at 126°C and 1.02 V stress.

Based on a ½ root law, the beta slope received is 5.5. When n increases, the beta decreases. The most fitting plot, based on the correlation mentioned earlier is a ¼ root law where the beta is 2.8. Given the stress conditions of the test, one would expect to see BTI degradation. This is congruent with the results which show a time law with $n = 4$, characteristic of BTI.

4.2.5.3 Voltage Power Law

Once we establish the time power law, the next step is to calculate the voltage law. Since we have a direct calculation of the extrapolated TTF for every temperature measured, the data taken for many temperatures and voltages can be combined by dividing the Arrhenius factor, $AF = exp\left[\left(\frac{n \cdot E_A}{k_B}\right) \cdot \left(\frac{1}{T} - \frac{1}{T_{ref}}\right)\right]$, leaving only the voltage factor, $\left(\frac{V - V_{th}}{V_{ref} - V_{th}}\right)^{-n \cdot \gamma}$. Hence, Figure 4.11 is a plot of the normalized *TTF/AF* versus *V* and fit to a *ln(V)* model using Excel.

We see from this fit that the power law for voltage fits nicely to a factor of 10. Since we established that according to the time law, $n = 4$, we have the power-law description for voltage, $\gamma = 2.5$, which is the expression in 4.37. It could be that if $n = 5$, then $\gamma = 2$. This may reveal some underlying physics related to the non-Arrhenius portion, but we see full expression is experimentally determined for the *TTF* as a function both of time and voltage in a self-consistent compact model.

The validity of the time-dependent and effective voltage stress accounting for the V_{th} change under accumulation of charge has been set. We reconciled the laws governing voltage and time by accounting for changes in threshold voltage over time due to reduced voltage stress at the oxide-silicon interface.

The present analysis revealed that including a power law for voltage is necessary to accurately capture the time dependence as a power of $1/n$. To validate our model

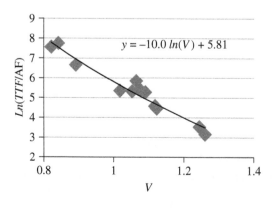

$$y = -10.0\ ln(V) + 5.81$$

Figure 4.11 *Ln(TTF/AF)* versus *V* where the *TTF* is normalized over many temperatures to measure.

in 16 nm FinFETs, we utilized Weibull slope analysis in multiple experiments. The resulting fitting plots followed a ¼ root law with a beta of 2.8, which is indicative of BTI degradation under the test conditions. The observed time law has a characteristic value of $n = 4$, consistent with BTI.

4.2.6 Electromigration (EM)

EM is widely believed to be the effect of momentum transfer from the electrons of the metal, which move according to the applied electric field, to the ions that constitute the lattice of the metal. Metal atoms (ions) travel toward the positive end of the conductor while vacancies move toward the negative end.

EM is actually not a function of current, but a function of current density. It is also accelerated by elevated temperature. Thus, EM is easily observed in Al metal lines that are subjected to high current densities at high temperatures over time.

As mentioned in JEP122 standard, EM associated with vias and contacts must be investigated separately because they show characteristics unlike single leads fed by bond pads. Vias can show different degradation rates depending on electron current flow direction (M2-to-M1 versus M1-to-M2) and the degradation rate is strongly dependent on via structure, via number, and layout. Via degradation may also have a reservoir effect.

EM in metal interconnects gives rise to two distinct effects, namely the depletion of atoms and the deposition of atoms. The former effect results in a slow reduction of connectivity and eventual interconnect failure due to voids, while the latter effect creates shortcuts due to the deposition of hillocks and whiskers. As a result, two failure mechanisms can be induced: grain boundary diffusion in aluminum (Al) leads to void formation, while surface diffusion in copper (Cu) results in hillock and whisker formation, respectively [179–181].

EM is becoming a design problem due to increased current densities related to IC down-scaling. At the critical limit, the width of the metal line becomes smaller than the grain size itself, such that all grain boundaries are now perpendicular to the current flow. Such a structure is also known as a "bamboo structure." This results in a longer path for mass transport, thereby reducing the atomic flux and EM failure rate.

Black's equation [62] is a very useful empirical equation for describing the relationship

$$t_{50\%} = t_0 \cdot J^{-n} \cdot exp\left(\frac{E_A}{k_B \cdot T}\right) \tag{4.45}$$

between the median time-to failure $t_{50\%}$, the electrical current density J, and $1 \leq n \leq 2$ and t_0 are factors determined experimentally. The symbol t_0 is a constant, which depends on a number of factors, including grain size, line structure and geometry, test conditions, current density, thermal history, etc. Black determined the value of *n* to equal 2. However, n is highly dependent on residual stress and current density and its value is highly controversial. In this approach, it is assumed that the interconnect is always at the same temperature. This is often not a realistic assumption. As EM begins, the cross-sectional area decreases. Thus, the current density increases and the temperature increases due to IR drop. This thermal runaway process reduces the MTTF value. In addition, it only assumes a single diffusion process. This assumption is fine for low-level interconnects made of Cu or Al (Cu is primarily used in modern ICs). Black's equation only describes the lowest energy EM process (bulk, grain boundary, or surface diffusion) and cannot describe multiple processes simultaneously. For other conductive materials etched in semiconductors, surface and grain boundary diffusion may occur simultaneously.

While this equation is widely used, the value for the current density exponent and its implications for EM lifetime prediction are still much debated. Budiman et al. [182] suggested in a paper published in 2010, that the higher n could be traced back to the higher level of plasticity in the lines; the closer n is to unity, the less plasticity must have influenced the EM degradation process.

It must be stressed that the definition of a long line must be that the line is substantially longer than the Blech length for the current density applied. If line length is not more than several times the Blech length, then large apparent n values can be obtained from tests.

There is also a critical lower limit for the length of the metal line that will allow EM to occur. Known as the Blech length, any metal line that has a length below this limit will not fail by EM. The effect is named after Ilan Blech of Israel's Technion, who, in the 1970s, discovered that when current was passed through gold on a substrate the upstream side moved with the current flow but the other side remained fixed. But if the distance was short enough, there would be no EM. The Blech effect [65] works in opposition to EM. Migration creates tensile stress at the upstream, cathode, end of the line, and compressive stress at the downstream, anode end. The sidewalls, being rigid, supply back pressure against this stress, slowing migration. The Blech effect is line length dependent: shorter lines offer more resistance. S. Thrasher and colleagues at Motorola found that there is a critical value of the length-current density product below which no migration occurs. Here, a mechanical stress buildup causes an atom back flow process which reduces or even compensates the effective material flow toward the anode.

Internal conductors that are in thermal contact with a substrate along their entire lengths (e.g. metallization strips, contact areas, and bonding interfaces) must be designed to ensure that no properly fabricated conductor will experience a current density above specific values at the device's maximum rated current. These values are determined based on the relevant conductor material and must

Table 4.1 Device material maximum rated current in semiconductors.

Conductor material	Maximum allowable continuous current density (RMS for pulse applications)
Aluminum (99.99% pure or doped) without glassivation	$2 \cdot 10^5 \, \text{A/cm}^2$
Aluminum (99.99% pure or doped) with glassivation	$5 \cdot 10^5 \, \text{A/cm}^2$
Gold	$6 \cdot 10^5 \, \text{A/cm}^2$
All other (unless otherwise specified)	$2 \cdot 10^5 \, \text{A/cm}^2$

account for factors like worst-case conductor composition, cross-sectional area, production tolerances on critical interface dimensions, and actual thickness at critical areas like steps in elevation or contact windows. These requirements for high-reliability application are provided, for example, in MIL-PRF-19500 standard [183] Appendix H as shown in Table 4.1.

A range of values for the EM activation energy, E_a, of Al and Al alloys is also reported. The typical value is $E_a = 0.66 \, \text{eV}$. The activation energy can vary due to mechanical stresses caused by thermal expansion. Introduction of 0.5% Cu in Al interconnects may result in $n = 2.63$ and activation energy of $E_a = 0.95 \, \text{eV}$. For multilevel Damascene Cu interconnects, an activation energy of $E_a = 0.94 \pm 0.11 \, \text{eV}$ at a 95% confidence interval (CI) and a value of the current density exponent of $n = 2.03 \pm 0.21$ (95% CI) were found [184].

Traditionally, the EM lifetime has been modeled by the lognormal distribution. Most test data appear to fit the lognormal distribution, but these data are typically for the failure time of a single conductor.

Through the testing of over 75,000 Al(Cu) connectors, Gall et al. [185] showed that the EM failure mechanism does follow the lognormal distribution. This is valid for the TTF of the first link with the assumption that the first link failure will result in device failure. The limitation is that a lognormal distribution is not scalable. A device with different numbers of links will fail with a different lognormal distribution. Thus, a measured failure distribution is valid only for the device on which it is measured. Gall et al. also showed that the Weibull (and thus the exponential) distribution is not a valid model for EM. Even though the lognormal distribution is the best fit for predicting the failure of an individual device due to EM, the exponential model is still applicable for modeling EM failure in a system of many devices where the reliability is determined by the first failure of the system.

The sensitivity of the EM lifetime can be observed by plotting the lifetime as a function of the stress parameters. For EM, the most significant stress parameters corresponding to lifetime are the temperature (T) and current density (J). It is worst to consider the two stressors' parameters are the engine to generate the failure mechanism (material property degradation), while the failure mode signature parameter to account for a failure is generally another sensitive electrical parameter related to the electrical failure mode.

In an extensive paper published in 1994 at IRPS [186] on 0.5 μm GaAs MESFET MMIC technology, we studied the reliability figures of various electronic parts integrated into test structures called technology characterization vehicle (TCV). We designed on the test structure; and 16 single independent elements as transistors, diodes, resistors, coil inductors, air-bridges, interconnects, and vias in order to submit them to a 4000-hours high-temperature bias lifetest series of experiments, respectively, at 205°C and 225°C junction temperature. Among these cells under reliability evaluation, we measured the degradation of active and passive elements to address how the EM mechanism takes place with time and stress conditions.

Prior to lifetesting, thermal analyses were performed on the devices and fixturing was developed that stressed all elements according to electrical levels which span from nominal values for MESFETs to the design rule maxima for resistors, inductors, interconnects, and capacitors. Based upon the degradation of signature parameters for each cell, we observed trends in the devices during lifetesting. Several failure criteria were set for each element on the TCV but, in general, most of the criteria set a 10% change as the degradation limit. Testing was conducted on stressed devices with interim measurement points made at ambient controlled room temperature from 4 to 4000 hours based on 14 measurements in equal log (time) intervals. We addressed the EM form interconnects (inductors, airbridges, ohmic contact resistance, NiCr resistors, etc.). For such dedicated cells, the resistance signature parameter is the resistance of the device measured at the device's ambient room temperature (no Joule heating of the device) at any interim measurement point. During stress lifetest operation, the device is powered at the desired stress current at the given stabilized stress temperature for the time specified between interim points. It should be noted that if very sensitive low-level amplitude measurement or failure criterion is selected, judgment must be used in selecting the "real" initial resistance over the first hours of test since the time-zero resistance varies due to possible conductive material micro-characteristic change and its associated interconnect interfaces.

Monitoring of the "signature parameters" was performed on pinch-of and saturation drain current of a GaAs MESFET technology. Gate sinking failure mechanism was confirmed based on electrical characteristics of a representative MESFET transistor. Figure 4.12 is an example of degradations and trends recorded on two series of GaAs MESFET transistors, life tested under the conditions given above. We can observe that the degradation of pinch-of voltage signature parameter is fitted by a square root of time law.

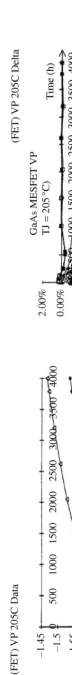

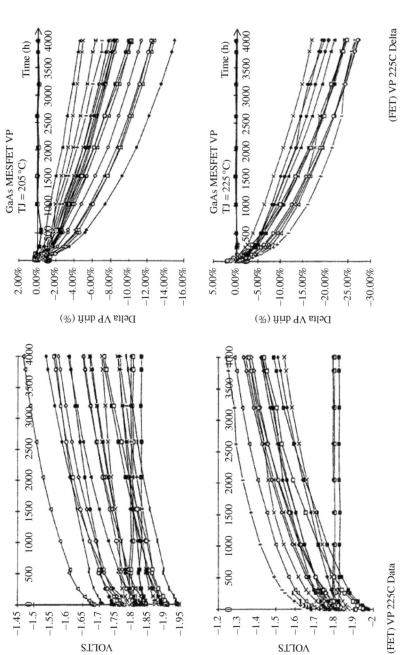

Figure 4.12 Lifetest raw data (absolute and relative drift values of pinch off voltage) for a GaAs MESFET (T_{j_test} =205°C and 225°C, respectively).
Source: Bensoussan et al. [186].

The gate sinking is considered to reduce the depletion depth d under the gate electrode after steady test lifetests.

To determine typical values for d assuming the given gate length W and doping profile, we base our analysis on the simple model from P.H. Ladbrooke [187]. This model expresses dependence of equivalent circuit elements of the intrinsic MES-FET given by:

$$V_p = -\frac{q \cdot N \cdot W^2}{2 \cdot \epsilon} + V_{B0} \tag{4.46}$$

assuming $\epsilon = 1.106 \cdot 10^{-10}$, $q = 1.6 \cdot 10^{19}C$, the built-in potential V_{B0} or barrier height at the Schottky interface (0.8V for GaAs/Au-based Schottky metal) supposed unchanged during aging (first approximation), the saturated electron drift velocity $V_{sat} = 10^5$ m/s, and Z_G the total gate width.

Using Eq. (4.46), the drift of pinch-off voltage after high-temperature lifetest is related to physical parameters such as active layer thickness and Schottky diode barrier height. Let us assume the change in the thickness of the active layer as $W - X_d$ due to sinking gate failure mechanism (e.g. diffusion of the Schottky gate interface into GaAs by the amount of X_d). So, the expression of the pinch-off voltage is:

$$V_p(t) = -\frac{q \cdot N \cdot [W - X_d(t)]^2}{2 \cdot \epsilon} + V_{B0} \tag{4.47}$$

For a given sampling lot, the mean value of the pinch-off voltage delta after aging time t is:

$$\overline{\Delta V_p(t)} \approx \frac{q \cdot N}{\epsilon} \cdot X_d(t) \cdot W \tag{4.48}$$

In our paper published at IRPS, using the simple one-dimensional diffusion model described by Fick's law, we demonstrated the sinking gate wear-out mechanism model is expressed by:

$$\overline{\Delta V_p(t, T)} \approx 5.5 \cdot \frac{q \cdot N \cdot W}{\epsilon} \cdot \sqrt{D_0} \cdot \sqrt{t} \cdot exp\left(\frac{E_A}{2 \cdot k_B T}\right) \tag{4.49}$$

$$\text{and} \quad X_d(t, T) = 5.5 \cdot \sqrt{D_0} \cdot \sqrt{t} \cdot exp\left(-\frac{E_A}{2 \cdot k_B T}\right) \tag{4.50}$$

where D_0 is the diffusion coefficient of the Schottky metal and E_A the activation energy of the failure mechanism can be extracted from aging data. If the MESFET sampling lot follows a Gaussian distribution, half of the sample population have ΔV_{pc} values greater than $\overline{\Delta V_p(t, T)}$ and the time $t_{50\%}$ (or $MTTF$) is approximated for a failure criterion ΔV_p when $\overline{\Delta V_p(t, T)} = \Delta V_{pc}$.

Rearranging Eq. (4.49), the sinking gate wear-out mechanism model is expressed by the following equation:

$$t_{50\%} \approx \left(\frac{\epsilon}{5.5 \cdot q \cdot N \cdot W}\right)^2 \cdot \frac{1}{D_0} \cdot \left(\Delta V_{pc}\right)^2 \cdot exp\left(\frac{E_A}{k_B T}\right) \tag{4.51}$$

Based on MESFET lifetesting data shown in Figure 4.12 (mean values), parameters of Eq. (4.49) can be assessed from a square root of time dependence of the mean value of pinch-off voltage drift through aging. The depth of the diffusion front gate metal into GaAs was assessed from Eq. (4.50) and is in the range of 120 Å after 4000 hours at T_j=225°C with $D_0 = 5.5 \cdot 10^{-8}$ m^{2}s^{-1} and $E_A = 1.7$eV.

4.3 Physics of Healthy

4.3.1 Definitions

It is crucial to understand the concepts of PoF, failure mechanisms, and failure modes is crucial for engineers and scientists to gain a comprehensive understanding of degradation processes, underlying mechanisms, and observable manifestations of failure in a particular system or material. By understanding these three concepts, we can develop effective strategies to prevent failures, optimize system performance, and improve reliability of engineered systems, leading to increased safety, efficiency, sustainability, cost savings, and higher customer satisfaction.

Physics of Failure (PoF): The PoF refers to the underlying physical and chemical processes that can cause degradation and failure in a system or a material: it is a cause identification concern. This concept is key to figuring out and understanding how modifications in physical or chemical properties may produce failure mechanisms. By understanding root causes by studying the PoF, engineers can identify the fundamental mechanisms that lead to failure (either sudden or by aging cumulative weaknesses). The PoF approach allows us to capture the inner workings of the symptoms of defects. This can help in identifying potential weak points and developing strategies to prevent failures from occurring and give direction for predictive modeling and simulation of degradation processes. The PoF can be complex and requires in-depth understanding of material properties, environmental factors, and other variables that influence the degradation processes. However, the PoF alone may not provide direct information about specific failure modes or their occurrence in a particular system or material, as it focuses on the underlying mechanisms.

Failure Mechanisms: Failure mechanisms are the specific processes that result in the failure of a system or material. These physical mechanisms are supported by multiple stressor interactions applied to an asset, such as temperature, load, or chemical exposure. Advantages of considering failure mechanisms include tailored mitigation strategies to describe mathematically specific stress models to allow engineers and scientists to develop targeted mitigation strategies that are tailored to the underlying causes of failure. They are based on thermodynamic laws, physics of metastable states, and can model the driving force for materials' degradation; they provide hypotheses and simple models to understand the kinetics (temperature and stress dependence) of failure mechanisms described by probabilistic PoF. Understanding failure mechanisms can help in designing more efficient assets and systems for performance and durability targets by mitigating the key mechanisms that are most likely to cause failure. Failure mechanisms alone are an incomplete coverage and a partial view of the complexity of reliability paradigm. Failure mechanisms may not capture all possible ways in which a system or material can fail, as they are usually based on specific stressors or conditions and may not be directly applicable to different scenarios without proper consideration of contextual factors.

Failure Modes: Failure modes are the expression of specific mechanical or electronic characteristics that show how a degradation manifests itself on a system or material. Some electrical parameters (leakage currents, electronic static characteristics, etc.) evolve faster than the system performance measurement and may correlate to the aging state of an asset before it fails. Based on the extreme value distribution (EVD) approach, the asset (or component) has a certain resistance to load. Failure occurs when the load induced by the operating conditions exceeds the resistance. The failure mode parameter is a kind of measure of the stress state that a system can withstand (capacity) in the face of stress. The degradation of the system performance is correlated to the degradation of some parameters called "hypersensitive." For example, the leakage current in a microelectronic component is the result of an accumulation of defects in certain active areas that can lead to catastrophic final failure beyond a certain threshold. The ability of the component to withstand more accumulated loads is reduced as it ages. Of course, as the end, a failure is observed when the performance or functionality is lost with respect to the target specification limit. So, failure mode can be a feature often characterized by observable and measurable precursor signatures (sometimes hidden or not measured), such as their changes evolve faster than the performance does. The kinetics of this "signature parameter" change provides a key visualization of the time-related pre-degradation (pre-aging) of the asset before the complete loss of performance and functionality.

When considering failure modes in isolation, there are some advantages and disadvantages to keep in mind. One of the advantages is the practicality and simplicity they offer. Failure modes provide a straightforward and efficient way to categorize, understand, and anticipate the observable signs of failure, which can be helpful during testing, inspection, and troubleshooting. They can also be used to assess the reliability of a system or material, as they represent the actual outcomes of failure events in a specific context.

If one were to rely solely on failure modes, without considering the underlying mechanisms, he would find some drawbacks. One of these drawbacks is that the manifestation may not reveal the underlying mechanisms or root causes of failure, as they focus primarily on the observable expression of the cause. This means that they may not provide the full insight needed to address the problem at its source. Additionally, relying solely on failure modes can result in a reactive approach to reliability, where failures are addressed after they occur rather than proactively preventing them, which may be a less effective strategy in the long run.

In conclusion, none of these levels, considering *Physics of Failure*, *Failure Mechanisms*, and *Failure Modes*, may be addressed separately. The key to understanding reliability lies in the interaction between these three concepts: this is what we call *Physics of Healthy*. By considering the Physics of Healthy as a whole, engineers can gain a comprehensive understanding of the degradation processes, underlying mechanisms, and observable manifestations of failure in a particular system or material. This holistic approach allows for more effective strategies to prevent failures, optimize system performance, and improve reliability of engineered systems. Figure 4.13 details the three concept interactions in Physics of Healthy of semiconductors from (a) related trap-assisted PoF (TAPF) and mobile carrier PoF (MCPF) phenomenological transport mechanisms, (b) model-based stressors inducing Failure Mechanisms, to (c) Failure Modes expressed by signature parameters to describe aging embodied by degradation rate figures.

4.3.2 Entropy and Generalization

The Eyring-Arrhenius law determines the lifetime τ for a material or a device experiencing combined action of an elevated temperature and external stress. The general expression of τ is given by:

$$\tau = \frac{1}{\lambda} = \tau_0 \cdot f(S) \cdot e^{\left(E_A/k_B T\right)} \tag{4.52}$$

where τ_0 is the time constant, λ is the failure rate during the steady-state operation of the device, and $f(S)$ is the Eyring stress factor as a function of stress S:

$$f(S) = S^\gamma \text{ or } f(S) = e^{\gamma \cdot S} \tag{4.53}$$

By way of a simple example, we may show a case when the objects of interest are subjected to an elevated temperature and, in addition, are loaded by stressors S as for example a voltage V. Let us assume that the failure rate λ_I, which characterizes the propensity of the material or the device to failure, is determined by the level of

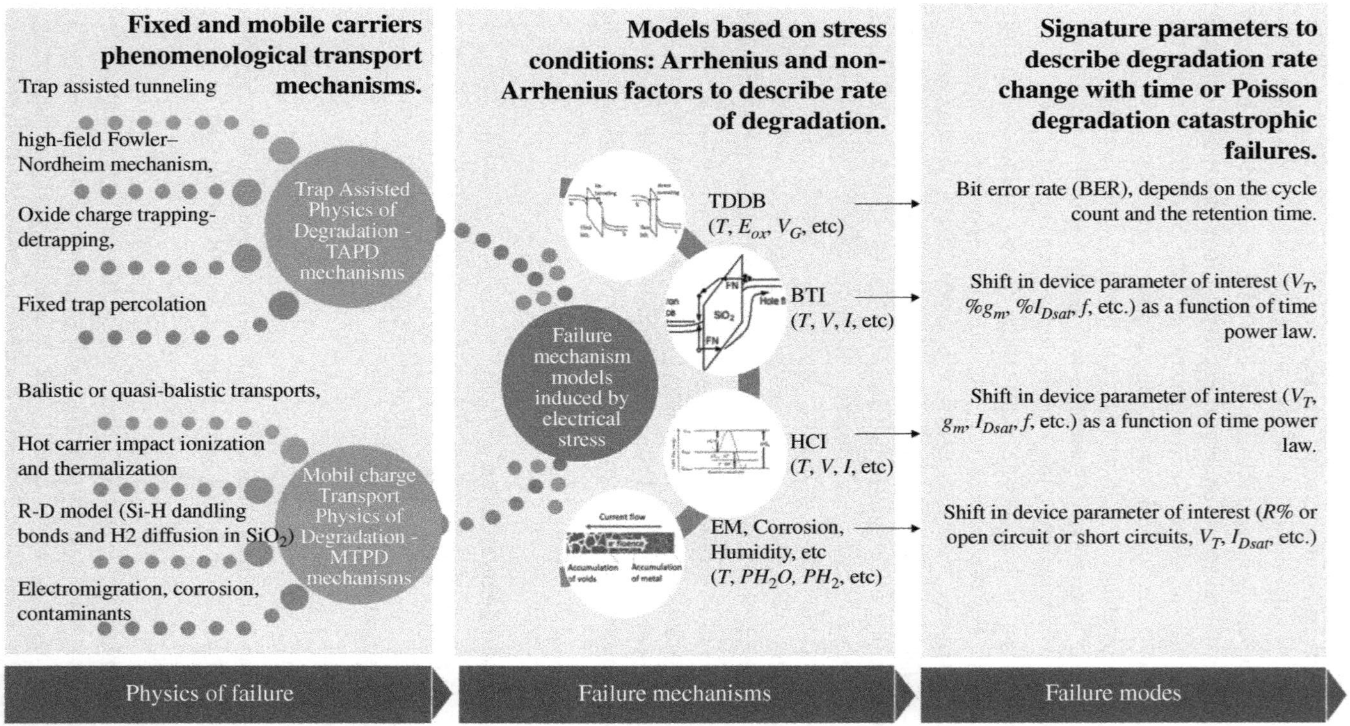

Figure 4.13 PoH – How the three concepts interact: PoF → Failure mechanisms → Failure modes.

the failure signature parameter I_S. As discussed in Chapter 5, we considered the drift in frequency as a power-law time-dependent to be the signature parameter of the failure mechanism. This kind of signature parameter is indeed the failure mode used as the degradation (failure) indicator. Data experiments of various failure mechanisms are showing similar I_S time variation expressed by:

$$\frac{\Delta I_S}{I_{S0}} = \left(\frac{t}{\tau_0}\right)^{\frac{1}{n}} \tag{4.54}$$

assuming I_{S0} is the reference signature measurement done after a short period of time starting the applied stress and Δ_{IS} the drift observed for this parameter at a time t with n the time power factor.

If we assume a failure criterion of 10% change of the initial signature parameter at $t = \tau$ to be the TTF, we get:

$$\frac{\Delta I_S}{I_{S0}}\bigg|_{10\%} = AF \cdot \left(\frac{\tau}{\tau_0}\right)^{\frac{1}{n}} \tag{4.55}$$

considering the accelerating factor AF for a given stress operation with respect to the accelerated reference stress conditions inferred from Eqs. (4.52) and (4.53):

$$AF = exp\left[\left(-\frac{E_A}{k_B}\right) \cdot \left(\frac{1}{T} - \frac{1}{T_{ref}}\right)\right] \cdot \left(\frac{S}{S_{ref}}\right)^{\gamma}$$

$$\text{or} \quad AF = exp\left[\left(-\frac{E_A}{k_B}\right) \cdot \left(\frac{1}{T} - \frac{1}{T_{ref}}\right)\right] \cdot e^{\gamma \cdot (S - S_{ref})} \tag{4.56}$$

Compiling Eqs. (4.55) to (4.56), we can write:

$$\tau = \tau_0 \cdot \left(\frac{\Delta I_S}{I_{S0}}\bigg|_{10\%}\right)^n \cdot exp\left[\left(n \cdot \frac{E_A}{k_B}\right) \cdot \left(\frac{1}{T} - \frac{1}{T_{ref}}\right)\right] \cdot \left(\frac{S}{S_{ref}}\right)^{\gamma \cdot n} \tag{4.57}$$

$$\text{or} \quad \tau = \tau_0 \cdot \left(\frac{\Delta I_S}{I_{S0}}\bigg|_{10\%}\right)^n \cdot exp\left[\left(n \cdot \frac{E_A}{k_B}\right) \cdot \left(\frac{1}{T} - \frac{1}{T_{ref}}\right)\right] \cdot e^{\gamma \cdot n \cdot S} \tag{4.58}$$

The lifetime τ is viewed as the MTTF related to the Weibull law of reliability used to evaluate the probability of nonfailure during the steady-state operation of the system:

$$P(t) = exp\left[-(\lambda \cdot t)^{\beta}\right] = exp\left[-\left(\frac{t}{\tau}\right)^{\beta}\right] \tag{4.59}$$

and reduces to the exponential law for $\beta = 1$, when failures are rare and occur randomly.

So, it becomes:

$$P(t) = \exp\left\{ -\frac{t}{\tau_0} \cdot \left(\frac{\Delta I_S}{I_{S0}}\bigg|_{10\%}\right)^{-n} \cdot \exp\left[\left(-n \cdot \frac{E_A}{k_B}\right) \cdot \left(\frac{1}{T} - \frac{1}{T_{ref}}\right)\right] \cdot \exp(-\gamma \cdot n \cdot S) \right\}$$

(4.60)

The MTTF corresponds to the moment of time when the entropy of the law 4.58 reaches its maximum value [188, 189]. The scale of uncertainty was first provided by C. Shannon in 1948.

$$H = -P \cdot \log P$$

(4.61)

For a continuous random variable X, characterized by the probability density function $f(x)$, the entropy was introduced in 1872 by L. Boltzmann in the kinetic of gases, as follows:

$$H = -\int_{-\infty}^{\infty} f(x) \cdot [\ln f(x)] \cdot dx$$

(4.62)

The entropy of Shannon and Boltzmann is based on a similar idea. In thermodynamics, this concept is a quantitative measure of the disorder in a physical system. Indeed, the distribution 4.58 for $\beta = 1$ can be derived as follows:

$$\frac{\partial P}{\partial \lambda} = -\frac{H(P)}{\lambda},$$

$$\frac{\partial P}{\partial t} = -\frac{H(P)}{\lambda \cdot t},$$

$$\frac{\partial P}{\partial E_A} = n \cdot \frac{H(P)}{k_B \cdot T},$$

$$\frac{\partial P}{\partial S} = -H(P) \cdot n \cdot \gamma$$

(4.63)

These formulas in Eq. (4.63) reveal several important observations following E. Suhir [189]. First, as the failure rate increases, the probability of nonfailure decreases proportionally to the entropy of this probability and is inversely proportional to the failure rate. Second, the probability of nonfailure also decreases with time, in proportion to the entropy of this probability and inversely proportional to the duration of operation. Third, an increase in stress-free activation energy results in a proportional decrease in the probability of nonfailure, which is inversely proportional to the absolute temperature. Finally, the stress sensitivity factor can be determined by calculating the ratio of the derivative of the probability of nonfailure with respect to the applied stress to the derivative of this probability with respect to the stress-free activation energy.

We can observe that the decrease in the probability of nonfailure with an increase in the failure rate is proportional to the level of the entropy of this probability and is inversely proportional to the level of the failure rate; that the decrease in the probability of nonfailure with time is proportional to the entropy of this probability and inversely proportional to the time in operation; that the decrease in the probability of nonfailure with an increase in the stress-free activation energy is proportional to the entropy of this probability and is proportional to the time factor n and is inversely proportional to the absolute temperature; and that the stress sensitivity factor can be found as the ratio of the derivative of the probability of nonfailure with respect to the applied stress is proportional to the level of the entropy of this probability and is proportional to the stress factor γ and time factor n.

5

Multiple Failure Mechanism in Reliability Prediction

Silicon device reliability engineering assists chip fabrication in all stages of development. Early designs are simulated with failure modeling based on closely related older technologies. The prediction tools model the chip's end-of-life (*EOL*) based on known failure mechanisms, such as time-dependent dielectric breakdown (*TDDB*), negative bias temperature instability (*NBTI*), electromigration (*EM*), and hot carrier injection (*HCI*). Although the tools are beneficial in profiling each mechanism individually, they are insufficient to find their relative impact based on the models alone. This is only possible by measuring the pre-factors of the models from physical devices [190–193].

In the intermediate stages of the process, custom test circuits of the new technologies are used to characterize the mechanisms. In most cases, the devices under test (*DUTs*) are designed to locate a specific mechanism. This testing procedure also lacks analysis of all the mechanisms in a single *DUT*.

Packaged products are released after undergoing qualification testing to verify that they reach the design requirements. This final stage of testing is crucial in that it is only examination of the fully packaged product. The high-temperature operating life (*HTOL*) qualification test is usually the test that is performed [37]. The *HTOL* method is designed to predict the failure rate of devices by administering accelerated conditions to a large number of packaged devices. The standard is to use 77 devices from 3 lots and to test them at 125°C for 1000 h. In most cases, the test is completed without any device failures. The *HTOL* test has two points that shed doubt on its ability to accurately qualify the technology. First, it lacks sufficient statistical data. Companies are pressured to present zero failures in their results [38, 39]. The results are pass/fail. When there are no failures, no statistics are received. Second, the standard for *HTOL* is to perform accelerated tests assuming a single acceleration factor (*AF*). This negates known physics of failure models

Reliability Prediction for Microelectronics, First Edition. Joseph B. Bernstein, Alain A. Bensoussan, and Emmanuel Bender.

which show that there are multiple failure modes in the device at any given time [33]. Such a method will not locate the dominant failure mechanism in the device.

HTOL is based on the outdated Joint Electron Device Engineering Council (JEDEC) standard that has not been updated for many years. The major drawback of this method is that it is not based on a model that predicts failures in the field. Nonetheless, the electronics industry continues to provide data from *HTOL* tests as their main proof of manufacturing qualification. These results, which do not constitute a failure prediction of the devices, are then fit to an average *AF*, which is the product of a thermal factor and a voltage factor. The result is a reported failure rate as described by the standard failure in time (*FIT*) model, which is the number of expected failures per billion part-hours of operation. *FIT* is still an important metric for failure rate in today's technology. However, it does not account for the fact that multiple failure mechanisms simply cannot be averaged for either thermal or voltage *AF*s.

One of the major limitations of modern electronic systems qualification, including advanced microchips and components, is providing reliability specifications that match the variety of user applications. The standard *HTOL* qualification that is based on a single high-voltage and high-temperature burn-in that does not reflect actual failure mechanisms that would lead to a failure in the field. Rather, the manufacturer is expected to meet the system's reliability criteria without any real knowledge of the possible failure causes or the relative importance of any individual mechanism. More than this, because of the nonlinear nature of individual mechanisms, it is impossible to reveal the failure mechanism that will be dominant at operating conditions using *HTOL*. Therefore, reliance on the *HTOL* testing results is essentially sweeping the potential cause of failure under the rug while generating an overly optimistic picture of the actual reliability.

Two problems exist with the current *HTOL* approach, as recognized by JEDEC in publication JEP122G: (i) multiple failure mechanisms compete for dominance in our modern electronic devices and (ii) each mechanism has a vastly different voltage and temperature *AF*, depending on the device operation. This more recent JEDEC publication recommends explicitly that multiple mechanisms should be addressed in a sum-of-failure-rates approach. They agree that a single-point *HTOL* test with zero failures can, by no means, account for a multiplicity of competing mechanisms.

The multiple-temperature operational life (*MTOL*) testing method [194, 195], an alternative qualification test, is designed to reveal and classify all substantial failure mechanisms in packaged devices. *MTOL* resolves both shortcomings mentioned regarding *HTOL*.

The *MTOL* testing approach enables the acceleration of different mechanisms in the same set of accelerated tests. Figure 5.1 presents the flow of the *MTOL* testing method.

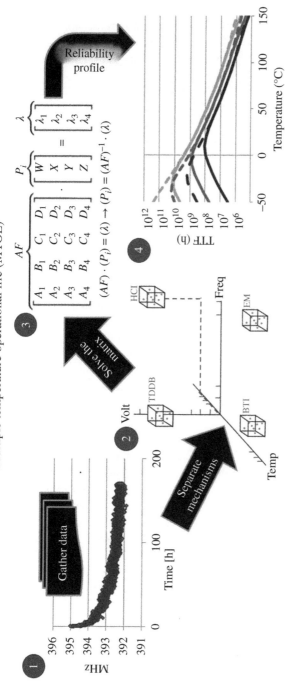

Figure 5.1 The flow of the *MTOL* testing method.

1) Data acquisition from multiple stress modes.
2) Isolation of individual mechanisms to find the failure rates (λs).
3) Solution of the λs in a matrix with the mechanism's pre-factors.
4) Generation of a failure profile with the relative weights of the mechanisms.

The first stage of the process is the acquisition of the data. Devices programmed with a test system, which will be explained in detail in Section 5.1, are tested in multiple combinations of voltages and temperatures (high and subzero). Current is also a stressor. Current stress is actualized in digital circuits by transitioning cells [rising and falling of the voltage between drain and source (*VDS*)]. The system is designed with a large range of frequencies which translates into current stress. Depending on the stress conditions, individual failure mechanisms are revealed. Degradation data showing frequency dependency indicate a current stimulated failure mechanism such as *HCI* and *EM* [196]. *BTI*, on the other hand, is not frequency dependent. It develops from static voltage stress [43, 192, 197]. *HCI* and *EM* are discernable with the variation of temperature stress. *EM* is dominant in high temperatures due to its relatively large (1.2–1.4 eV) positive activation energy (E_A). *HCI* has an apparent negative E_A. This makes it well discernable in subzero temperatures. This constitutes the second stage of the process which is the separation of the failure mechanisms based on the stress conditions. Multiple tests facilitate the finding of a clear failure rate trend for each mechanism. The failure rate is labeled with lambda (λ) specific for each failure mechanism. We assume a constant failure rate based on the concept that a single cell with fail randomly. Therefore, we can use a simple first-order Poisson distribution to describe the rate of failure as follows:

$$R(t) = exp(-\lambda \cdot t) \tag{5.1}$$

Given the presence of multiple failure mechanisms in the chip simultaneously, we can expand the concept of random failure phenomenon to be accurate for each separate mechanism. That justifies the characterization of failure mechanisms (FM) as λ's. So, in the second stage, each *FM* is separated by testing the devices with stress modes that isolate the distinct mechanisms.

After the temperature, voltage, and current *AF*'s are obtained individually from the data, proportionate weights of the *FM*'s are solved in a matrix of extrapolated time-to-failure (*TTF*) values as is depicted in the third stage. The outcome of the process is the creation of a full reliability portrait [195] of the technology as is shown in the final fourth stage.

The traditional single-model *HTOL* gives an unrealistically low value for the expected *FIT* and customers invariably find that their application shows a much higher reported failure rate than what was provided by the supplier using the traditional approach. Our *MTOL* matrix methodology (Figure 5.2) will give a more

M-HTOL system matrix approach:

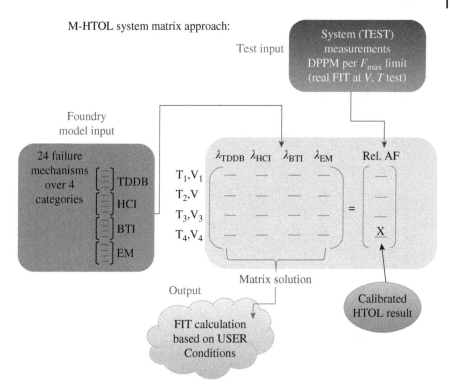

Figure 5.2 Matrix methodology for reliability prediction.

accurate prediction for the expected field failure rate that is based on actual test data and on the reliability models provided by the foundry to the chip designer. This way the designer has a much more dependable picture for device reliability and the customer will be satisfied that his design will match their customer's expectations for performance life.

5.1 *MTOL* Testing System

The field-programmable gate array (*FPGA*) configuration designed for the *MTOL* testing consists of three main parts: (i) accelerated element (*FPGA* chip); (ii) measurement system (binary counter); and (iii) control and communication interface to a PC (see Figure 5.3). The first generation of test systems consisted of two development boards containing either a Xilinx XC6SLX9 Spartan 6 *FPGA* (45 nm) built on a Mojo® board or a Zynq-7000 *FPGA* (28 nm) built on a Zybo® board.

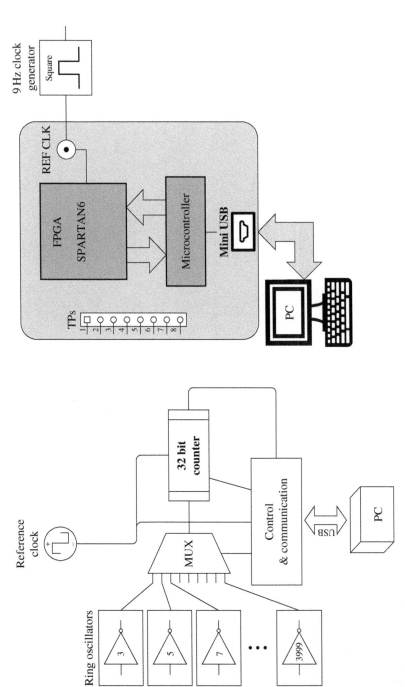

Figure 5.3 Testing setup consisting of ring oscillators, counter, and communication.

The Spartan 6 uses Taiwan Semiconductor Manufacturing Co. (*TSMC*)'s 45 nm low-power process for 1.2 V core voltage and the Zynq-7000 *FPGA* is from high-K/metal gate (*HKMG*) high-performance/low-power (*HPL*) process that is optimized for 1.0 V core voltage. The devices contain over 9000 logic cells (look-up tables – *LUTs*) and the programs were designed to cover the full scope of the device's components. The devices included inputs for reference signals, testing points, and a micro-USB interface. In order to allow various voltage levels, an external DC power supply delivered voltage to the *FPGA* cores after overriding the internal voltage controls in the board. We monitored the device temperature both internally and externally using an infrared (*IR*) camera and internal temperature measurements.

5.1.1 Accelerated Element, Control System, and Counter

The accelerated element ran with several different frequencies, allowing independent measurements of the degradation effects over time as a function of frequency. In order to create a measurable accelerated system, ring oscillators (*ROs*) consisting of inverter chains were constructed (shown in Figure 5.4). The frequency of each *RO* is given by: $1/(2 \cdot N \cdot T_p)$, where *N* is the number of inverters and T_p is the time propagation delay per inverter. Each inverter chain was implemented as a complete logical cell using predefined Xilinx primitives and thus each *RO* was made up of the basic components of the *FPGA*. When degradation occurred in the *FPGA*, a decrease in performance and frequency of the *RO* could be observed. For optimal testing and chip coverage, different-sized *ROs* were selected, ranging from three inverters, giving the maximum frequency possible in accordance with the intrinsic delays of the *FPGA* employed (400–700 MHz), and up to 4001-inverter oscillators, giving a much lower frequency (around 200 kHz). The system implemented on the chip starts operating immediately when the *FPGA* core voltage is connected. This allows seeing the frequency dependence of the failure mechanisms without any recovery effect.

The control system includes a programmable multiplexer (*MUX*), which switches between the various *R* sampled outputs. Each cycle transfers a different *RO* for measurement and a communication controller that connects to a PC. The

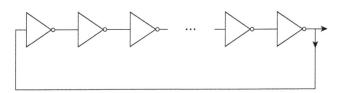

Figure 5.4 Ring oscillator is made of 2*N* + 1 Inverters connected in a chain.

control system communicates with the computer via a USB connection with a simple serial protocol. The data received from the board is saved on a file in the computer in .csv format to be editable using a standard data-processing program such as Excel.

Frequency measurements were carried out from a crystal on the board. For high-resolution measurements, a 32-bit counter was chosen. Assuming the highest available frequencies to be around 700 MHz, a Reference Clock signal of 9 Hz was fed into the *FPGA*. The reference signal was generated outside of the *FPGA* (and the accelerated environment) allowing as accurate measurements as possible. The reference signal is also returned to the exterior in order to verify that the measurement remains accurate throughout the experiments.

The test conditions, i.e. voltage and temperature levels are defined in a test plan for imposing an acceleration of individual mechanisms to dominate others at each applied condition. For example, testing in subzero temperatures and high voltages will exaggerate *HCI* for high frequency [196], but very low degradation due to *BTI* [197]. For each test, the *FPGA* board was placed in a temperature-controlled oven, similar to those used in *HTOL* testing, with an appropriate voltage connected at the *FPGA* core from an external supply. The board was connected to a computer via USB cables. The tests were performed for 100–200 h, while the device was operating at accelerated voltage conditions – the frequencies of 140 *RO*s of different sizes were sampled every 5 min.

The measured data was stored in a database from which one could draw statistical information about the degradation in the device performance. The particular testing conditions were chosen to isolate each failure mechanism allowing examination of the specific effect of that mechanism on the system and thus define its unique physical characteristics. A close inspection of the results in comparison to one another yielded more precise parameters for the *AF* equations and allowed adjusting them to fit all *DUT*s.

Finally, after completing the tests, some of the experiments with different frequency, voltage, and temperature conditions were chosen to construct the *MTOL* matrix. The calculated *FIT* was then plotted as a function of temperature and frequency for every operating voltage.

Since the *FIT* calculation is, in essence, an average of a large number of devices (107 transistors), assuming the Poisson model, we can postulate that the standard deviation is equal to *TTF*.

5.1.2 Separating Failure Mechanisms

The common intrinsic failure mechanisms affecting electronic devices are *HCI*, *BTI*, *EM*, and *TDDB*. In our tests, no signature of *TDDB* was observed. This result is not surprising considering that in other accelerated test results on comparable technologies, *TDDB* is only observed in voltages higher than 1.6 V [192]. The standard models for failure mechanisms in semiconductor devices are classified by JEDEC Solid State Technology Association and listed in publication JEP-122G [43]. The failure mechanisms can be separated due to the difference in physical nature of each individual mechanism.

5.1.3 E_A and γ Extrapolation

Our tests for various mechanisms included exposing the core of the *FPGA* to accelerating voltages above nominal. 45 nm defines the nominal voltage at 1.2 V and for 28 nm, 1.0 V. Our method of separating mechanisms allowed the evaluation of actual activation energies for the three failure mechanisms detailed earlier. We plotted the degradation in frequency and attributed it to one of the three failure mechanisms as will be explained further on.

The results of our experiments give both E_A and γ for the three mechanisms we studied at temperatures ranging from −50 to 150°C. The Eyring model [72] is utilized here to describe the *FIT* for all of the failure mechanisms. This model, using a simple constant rate (1-parameter) Poisson function, conserves strict linearity. Any additional terms would be nonlinear and invalidate the matrix.

This linear, constant failure rate, approximation is born out of the observation that the statistical variability across the chip converges to the constant rate, Poisson process, allowing the linear matrix calculations to hold. The specific constant failure rate of each failure mechanism, calculated as *FIT*, follows these equations:

$$FIT_{BTI} = exp(\gamma_{BTI} \cdot V_G) \cdot exp\left(-\frac{E_{a_{BTI}}}{k_B \cdot T}\right) \tag{5.2}$$

$$FIT_{HCI} = f \cdot V^{\gamma_{HCI}} \cdot exp\left(-\frac{E_{a_{HCI}}}{k_B \cdot T}\right) \tag{5.3}$$

$$FIT_{EM} = f \cdot V_D{}^{\gamma_{EM}} \cdot exp\left(-\frac{E_{a_{EM}}}{k_B \cdot T}\right) \tag{5.4}$$

The degradation slope, α, is measured as the degradation from initial frequency as an exponential decay, approximated by taking the difference in frequency, divided by initial frequency over time. In our experiments, we found that when the decay was dominated by *BTI*, the decay was proportional to the 4th root of time, while *HCI* and *EM*, being diffusion-related mechanisms, have decay that is proportional to the square root of time [198], as seen in Figure 5.5, exhibiting degradation slope, α (voltage 2.4 V and high temperature 150°C showing a strong *BTI* degradation at low frequency).

This result is consistent with literature and with the JEDEC document that lists the failure mechanisms, JEP-122G. For each oscillator, the ring frequency was

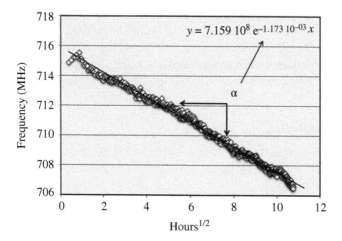

Figure 5.5 Typical graph showing frequency versus square root of time.

measured and plotted against the square root of time in 45 nm devices in the cases where the data was taken from *HCI* and *EM*-induced degradation. The slope, α, was then converted to a *FIT* for each test as determined by extrapolating the degradation slope to 10% degradation from its initial value. One *FIT* is defined as 1 failure in 10^9 part-hours. Each set is plotted as an exponential decay dependent on the square root of time (for failure mechanism that n-root power law of 2) as shown by example in Figure 5.5. This slope is then used to find the *TTF* as seen in the following development of *FIT*:

$$\alpha_{slope} = \frac{\Delta f}{f_0 \cdot \Delta \sqrt{t}} \tag{5.5}$$

$$TTF = \left(\frac{10\%}{\alpha}\right)^2 \tag{5.6}$$

$$FIT = \frac{10^9}{TTF} \tag{5.7}$$

$$FIT = 10^9 \, (10 \cdot \alpha)^2 \tag{5.8}$$

This makes our *FIT* easy to calculate since *FIT* is defined as:

$$FIT = \frac{10^9}{MTTF} \tag{5.9}$$

where *MTTF* is the mean time to fail in hours.

The *TTF* for each point was then calculated as the square of the inverse slope times the failure criterion, which is 10% degradation. Hence, the *FIT* for each slope is simply determined as $(10 \cdot \alpha)^2$. The average *FIT* is the metric to determine the reliability since that corresponds to the *MTTF* in Eq. (5.9). This *FIT* value is plotted as a function of the frequency to determine the failure mechanisms and to fit the model parameters for each mechanism. Two typical degradation plots taken from 45 nm tests are shown in Figure 5.6. *FIT*s, determined by the slopes, are plotted against frequency in two different experiments. The data demonstrates the clear advantage of *RO*-generated frequencies in a single chip [199].

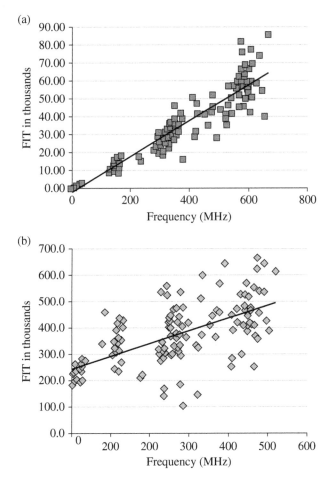

Figure 5.6 Failure rate, *FIT*/1000, versus frequency in MHz for (a) *HCI*, stressed at −35°C with 2.0 V core voltage. (b) *BTI*, stressed at 145°C with 2.4 V at the core.

Examples of *FIT* to frequency plots are shown in Figure 5.6. We see the *HCI* effect at a temperature of −35°C and voltage of 2.0 V in (a), showing *FIT* that is directly proportional to frequency [151]. In (b), a chip was stressed at high voltage (2.4 V) and high temperature (145°C) showing a strong *BTI* degradation at low frequency and a much shallower slope. The slope is due to *EM* caused by an increase in effective resistance and therefore a frequency-dependent effect [200]. Such curves were made for each experiment, incorporating all the oscillators across the chip spanning the range of frequencies, also reflecting the averaging effect of the longer chains. Hence, the variability is much lower than at higher frequencies, demonstrating that the averaging of many variations results in a consistent mean degradation. This strengthens credibility of modeling the failure mechanisms with Eqs. (5.2)–(5.4) since the data shows a clear Poisson statistical nature. The slope of *FIT* versus frequency is then related at low temperatures as occurring only from *HCI* while at higher voltages and temperatures, it can be due to *BTI* [179] and *EM*. *BTI* is only responsible for low-frequency degradation [185].

The dependencies of each mechanism must be determined experimentally. The activation energy relates to the temperature factor (*TF*) while the voltage acceleration factors (*VF*) are determined from Eqs. (5.2)–(5.4). The results of *FIT* versus *V*, *T*, and *F* are plotted as follows. *HCI* voltage constant γ was found by plotting *FIT*/*TF* versus *V* as seen in Figure 5.7a and Figure 5.7b shows the *FIT*/*VF* versus $1/k_B T$ looking only at temperatures below 5°C in order to determine the activation energy, E_A. Since both plots depend on each other, the two are performed simultaneously, where E_A is used to determine *TF*, where:

$$TF = exp\left(-\frac{E_A}{k_B T}\right) \tag{5.10}$$

$$V_F = V^\gamma \text{ for } HCI \text{ and } EM \text{ and } V_F = exp(\gamma \cdot V) \text{ for } BTI \tag{5.11}$$

Hence, we were able to find the correct activation energy simultaneously with its corresponding voltage factor. Our procedure was followed for all three mechanisms for the 45 nm as well as the 28 nm devices. In the 28 nm device, there was no apparent effect from *HCI* or *EM*. That is to say that no slope was found versus frequency. This is in contrast with the 45 nm devices, showing frequency-related effects at both low temperatures due to *HCI* and a minor *EM* effect at high temperatures [102].

The plots for 45 nm technology of E_A and γ for *EM*, *BTI*, and *HCI* are shown, respectively, in Figures 5.7–5.9.

The 28 nm technology of E_A and γ *BTI* curves are shown in Figure 5.10. The E_A and γ constants for both technologies are summarized in Table 5.1.

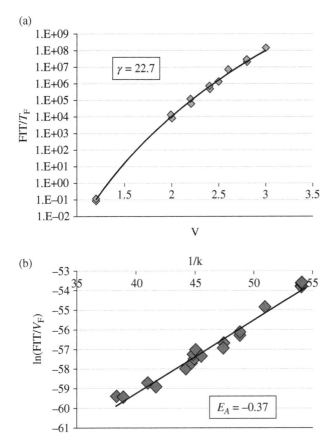

Figure 5.7 (a) γ plot for *HCI* at 45 nm. (b) Activation energy plot for *HCI* at 45 nm.

5.2 *MTOL* Matrix: A Use Case Application

Once we confidently identified the mechanisms by separating their effects based on voltage, temperature, and frequency, we can plot the expected constant failure rate, measured in $FIT = 10^9/MTTF$.

The *MTTF* is the mean extrapolated *TTF* for all the oscillators that were ordered by internal, asynchronous, ring frequency. Since the oscillators were all independent from each other, the measured frequency change over time is what allows extrapolating to 10% degradation in ring frequency, which is our definition of *TTF*. Hence, by averaging all the oscillators for each frequency, we have a very good approximation for *FIT* based on the mean *TTF*.

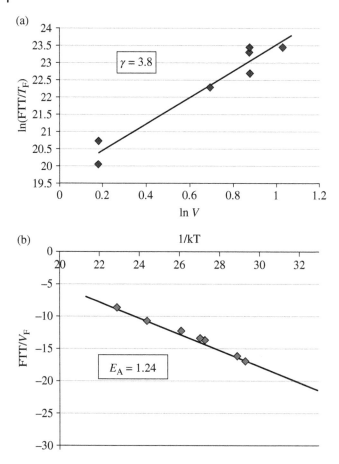

Figure 5.8 (a) Gamma plot for *EM* at 45 nm. (b) Activation energy plot for *EM* at 45 nm.

After a full characterization of the physics of failure models relating to all three mechanisms for both 45 and 28 nm *FPGAs*, we can build the Matrix Model by choosing three points, one from each mechanism, and then solve Eqs. (5.2)–(5.4) against the measured *FIT* for each condition. In the 45 nm device experiments, we chose the following data shown in Table 5.2. The relative factors that solve the matrix are shown in Table 5.3.

The procedure for finding the results of the matrix is described in previous papers [194, 201, 202]. This matrix is then used to construct the full reliability profile, whereby *FIT* is calculated versus temperature for several conditions, as shown in Figure 5.11.

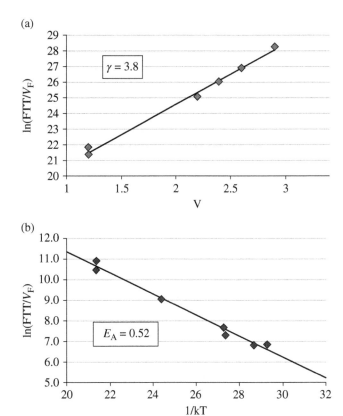

Figure 5.9 (a) Gamma plot for *BTI* at 45 nm. (b) Activation energy plot for *BTI* at 45 nm.

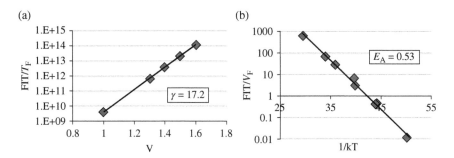

Figure 5.10 (a) Gamma plot for *HCI* at 28 nm. (b) Activation energy plot for *HCI* at 28 nm.

Table 5.1 Summary of E_A and γ from the curve fits.

	E_A @ 45 nm (eV)	γ @ 45 nm (V^{-1})	E_A @ 28 nm (eV)	γ @ 28 nm (V^{-1})
HCI	−0.4	22.7		
BTI	0.52	3.8	0.53	17.2
EM	1.24	3.8		

Table 5.2 T, V, F, and matrix versus measurement *FIT*s.

T (°C)	V	F (GHz)	HCI	BTI	EM	FIT
153	1.2	1	1.71E + 6	6.73E−05	4.27E−15	3672
−35	2.5	0.5	4.7E + 16	1.30E−07	8.96E−26	2.37E + 7
154	1.2	0	0	6.93E−05	0	2420

Table 5.3 Relative weighting factors that solve the matrix.

HCI	5.03873E−10
BTI	34761994.46
EM	3.11618E+17

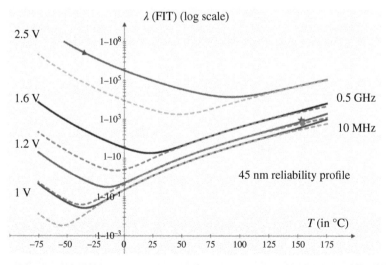

Figure 5.11 Reliability curves for 45 nm technology showing *FIT* versus temperature for voltages above and below nominal (1.2 V) and frequencies from 10 MHz (dashed line) to 0.5 GHz (solid line).

Hence, a family of 2-*D* reliability curves can result as seen in Figure 5.11 where dots are some experimental data. These curves represent the failure rate in *FIT* as a function of temperature for two representative frequencies (0.001 and 2 GHz) and three voltages: nominal, 0.2 V higher, and 0.2 V lower in all three plots.

The full system, including all mechanisms, can be calculated as the sum of each contribution from the solved equation for any expected operating voltage, temperature, and frequency:

$$FIT_M = P_{BTI} \cdot FIT_{BTI} + P_{HCI} \cdot \cdot FIT_{HCI} + P_{EM} \cdot FIT_{EM} \tag{5.12}$$

Likewise, Figure 5.12 is the same profile in 3D for 45 nm technologies showing experimental data dots, calculated lines, and 3D surface for various voltages, frequencies, and junction temperatures (Matrix model). In the 2D orientation, the voltages are represented with separate curves (1–2.5 V). The 3D orientation is added to portray a calculated continuous *FIT* trend with the stress conditions (using 5.2–5.4).

Figure 5.13 shows the *FIT* trend for 28 nm technology in the same format as the 45 nm data in Figure 5.11. The most notable difference between 45 and 28 nm is the lack of frequency effect at both low and high temperatures, leaving only one, dominant, failure mechanism at 28 nm. The consequence seems to be that there will be significantly improved reliability at low temperatures using 28 nm technology.

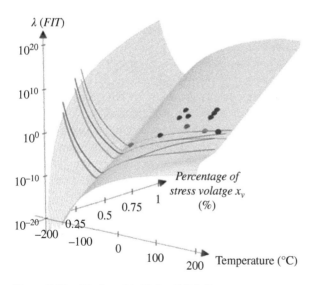

Figure 5.12 3D plot of *Ln(λ)* for *CMOS* 45 nm process versus voltage percentage and junction temperature.

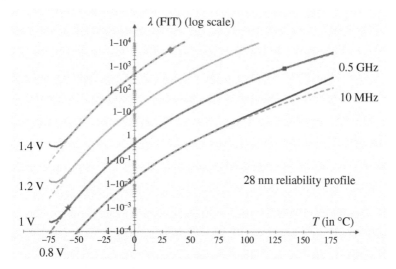

Figure 5.13 Reliability curves for 28 nm technology showing *FIT* versus temperature for voltages above and below nominal (1 V) and frequencies from 10 MHz (dashed line) to 0.5 GHz (solid line) (dots are some experimental data).

Furthermore, there is no effect on frequency related to reliability. Hence, there is no purpose for frequency de-rating using this technology. It is also clear that cooling the device is the major challenge and most important part of increasing the reliability of devices made using this technology [75]. Another observation is that the voltage acceleration is much greater at 28 nm, γ being over 17, as compared to 3.8 for 45 nm technology. This means that the core voltage is much more sensitive. A much greater reliability advantage would be gained by lowering the voltage. The temperature effect is relatively consistent for both technologies.

One speculation as to why there are no hot carrier effects in the 28 nm technology is that the mean free path of electrons transported through the gate is larger than the gate length below a certain temperature. This would suggest that the electrons are transported by ballistic means below that temperature due to the properly strained channels. Hence, electrons are not able to accelerate to the point of causing damage due to *HCI* [203].

Data of normalized *RO* frequency (F/F_0) versus temperature supports this claim. The 28 nm devices have a distinct transition at a particular temperature (around 60°C), whereas the 45 nm devices have not shown any transition along the entire range of temperatures. At higher temperatures, both 45 and 28 nm devices have similar slopes. This correlates with published data attributing the transistor channel conductance to the effect of "short channel ballistic conductance" at lower temperatures [204].

One striking benefit gained by the *MTOL* method detailed above is the summation of a full data set for application on 45 and 28 nm devices in the field. We show here how a minimal number of experiments can produce a full reliability evaluation curve including parameter extractions over a wide range of use-conditions. We see from this work how the reduction in geometry impacts the interaction of multiple known failure mechanisms. This methodology gives a baseline for assessing predictive reliability figures as requested by health monitoring and other predictive strategies for field-failure prediction.

5.2.1 Effective Activation Energy Characteristics (Eyring-M-STORM Model)

An important consequence of plotting our data as failure rate versus temperature allows us to determine the effective activation energy as a function of temperature and stressor parameters, voltage, and frequency. The principle illustrates the assumption that the failure rate λ is exponentially dependent on the activation energy divided by the absolute temperature, T. According to the Eyring-*BAZ* model [205] which is in fact similar to *COX* model [50], we find λ for a material or a device experiencing the combined action of an elevated temperature and external stress S as given by the following equation:

$$\lambda = \lambda_0 \cdot exp\left[-\frac{(E_A - \gamma \cdot S)}{k_B \cdot T}\right] = \lambda_0 \cdot exp\left[-\frac{E_{a_{eff}}}{k_B \cdot T}\right] \tag{5.13}$$

The equivalent key reliability physics parameters are the kinetic values (γ, $E_{A,effective}$) and these can be determined from actual FIT_M data based on equations also detailed by J. W. McPherson in his book "Reliability Physics and Engineering – Time-to-Failure Modeling" [88]:

$$E_{a_{eff}} = k_B \cdot \left|\frac{\partial(ln(\lambda))}{\partial\left(\frac{1}{T}\right)}\right|_S = \left|\frac{1}{FIT_M} \cdot \frac{\partial(FIT_M)}{\partial\left(\frac{1}{k_B \cdot T}\right)}\right|_S \tag{5.14}$$

$$\gamma = -\left|\frac{\partial(\lambda)}{\partial(S)}\right|_T = \left|\frac{kT}{FIT_M} \cdot \frac{\partial(FIT_M)}{\partial(S)}\right|_T \tag{5.15}$$

There are two options to show how the activation energy is related to the multiple stress conditions (V, f, and T): either to show $E_{a\_eff} = f(\% \ of \ V_{max}, \ temperature)$ or $E_{a\_eff} = f(\% \ of \ V_{max}, \ frequencies)$.

Figure 5.14 shows the effective activation energy for 45 nm as a function of junction temperature (for various voltages and frequencies calculated from the model using Eqs. (5.14) and (5.15). Figure 5.14a is the plot in two dimensions where the curves are defined by applied core voltage from 0.85 to 1.6 V and Figure 5.14b is a 3D depiction of the same data.

In a similar nature, Figure 5.15 displays the effective activation energy for 45 nm as a function of voltages and frequencies (calculated for various junction temperatures) calculated from the model also using 5.14 and 5.15. Figure 5.15a is the plot in

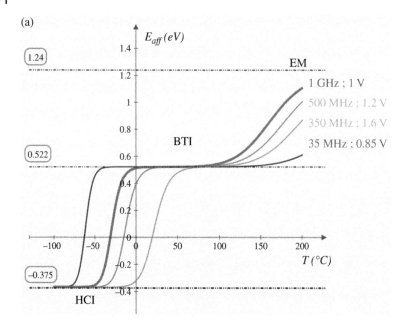

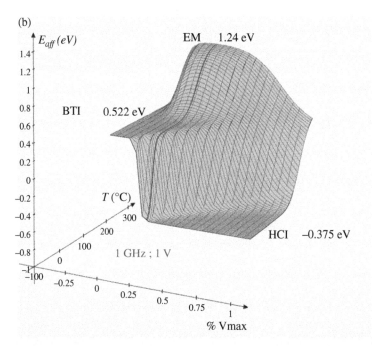

Figure 5.14 Effective Activation Energy E_{aeff} (T; V) evolution for *CMOS* 45 nm process versus percentage of maximum allowed voltages, maximum allowed frequencies, and junction temperatures. (a) 2D plot E_{aeff} (T) for various stress frequencies and voltages. (b) 3D plot E_{aeff} (T; V) at f = 1 GHz and voltage range from 0 to V_{max} = 3.75 V.

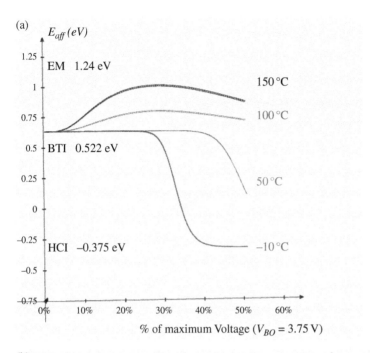

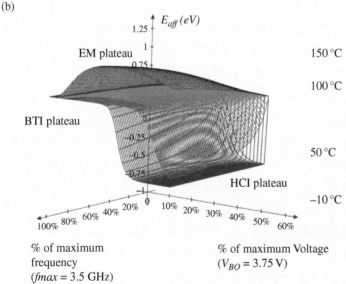

Figure 5.15 Effective activation energy E_{aeff} (f; V) evolution for *CMOS* 45 nm process versus percentage of maximum allowed voltages, maximum allowed frequency, and junction temperatures. (a) 2D plot E_{aeff} (% V_{max}) at f = 1 GHz and for various junction stress temperatures. (b) 3D plot E_{aeff} (f; V) for f = range from 1 MHz to 1 GHz and voltage range from 0 to V_{max} = 3.75 V.

two dimensions where the curves are defined by core temperature from −10°C to 150°C and Figure 5.15b is a 3D representation of the same data.

5.3 Comparison of DSM Technologies (45, 28, and 20 nm)

The test data that we elaborated upon in the previous sections was created solely from Xilinx *FPGA* devices. All the information pertinent to the fabrication of the *TSMC* process from 45 to 20 nm is available on the website, TSMC.com. The 45 nm technology is listed on the website as 40 nm, but it is known to be 45 nm technology by Xilinx. 28 nm is the first generation that the TSMC foundry started to use HKMG process." A cross-section of the PMOS device is shown in Figure 5.16.

The PMOS includes SiGe Source/Drain contacts to add compressive stress for holes, decreasing the hole's effective mass and increasing its conductivity. N-type metal oxide semiconductor (NMOS) has the same gate but, instead of SiGe, the NMOS device has a thick Si_3N_4 capping layer, causing tensile stress in the channel, and increasing the electron mobility. This NMOS cross-section is shown in Figure 5.17. The contact etch-stop layer (CESL) is seen in NMOS but not in PMOS. The metal gate and high-K dielectric are clearly seen. The use of CESL has been in production since the 90 nm node. So, this will be the same for all the technologies analyzed here.

Figure 5.16 TSMC 28 nm high-K/metal gate (HKMG) process for PMOS transistor. *Source:* Chipworks.

Figure 5.17 TSMC 28 nm high-K/metal gate (HKMG) process for NMOS transistor. *Source:* Chipworks.

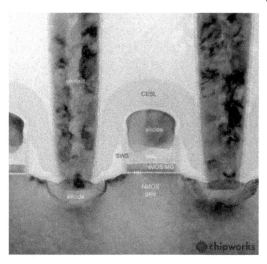

The significance of stress engineering is that it provides for higher mobility both in the electrons in NMOS as well as holes in the PMOS. This is seen graphically in Figure 5.18. Hence, we see that the only real differences in terms of structure for 45 and 28 nm processes are the metal gate and the channel length. Both processes use high-K dielectric whereas 45 nm technology still uses the polysilicon gate.

As we noted in Section 5.2, there was a significant reliability difference between 45 and 28 nm regarding *HCI* and *EM* whereas *BTI* seemed to be about the same (aside from the higher voltage *AF*). This is probably best explained by the shorter channel alone. It seems that the stress engineering techniques successfully increase the electron and hole mobilities, such that when the temperature is cooled, the electrons may have relatively no collisions with the oxide interface,

(a) NMOS

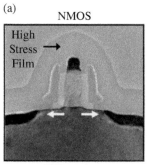

SiN cap layer
Tensile channel strain

(b) PMOS

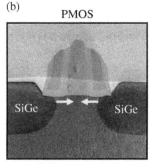

SiGe source-drain
Compressive channel strain

Figure 5.18 90 nm technology showing high-stress contact layer in NMOS and corresponding tensile stress as well as SiGe S/D in PMOS showing compression. *Source:* Chipworks.

hence completely avoiding the *HCI* trapping. Similarly, the current would be limited due to the inability of the electrons to move quickly enough before being absorbed in the drain. Hence, the current is automatically limited and there is also no *EM* resulting due to the inherent current limit of the shorter channel.

5.3.1 *BTI*'s High Voltage Constant

MTOL results for short-channel transistors show a voltage constant considerably higher than that of 45 nm transistors. This is due to the change in physical properties of the new technologies which use *HKMG* materials. Downsizing to 20 and 16 nm or smaller transistors promises the same since they are composed of the same materials. Samsung manuscripts on 14 nm *Fin-FET* devices strengthen this claim [206]. Figure 5.19 shows a graph of V_{th} shift degradations between 1.6 and 1.7 V stresses. The voltage constant is found with the help of the power law comparing the acceleration of 1.6–1.7 V as follows:

$$\gamma = \frac{\log AF}{\log\left(\dfrac{V_{HS}}{V_{LS}}\right)} \rightarrow \frac{\log 3}{\log\left(\dfrac{1.7}{1.6}\right)} = 18.1 \tag{5.16}$$

This value is consistent with the results of our 28 nm evaluation, causing us to believe that the same voltage and thermal *AF*s for *BTI* will be the same for 20, 16, and even 10 nm technologies.

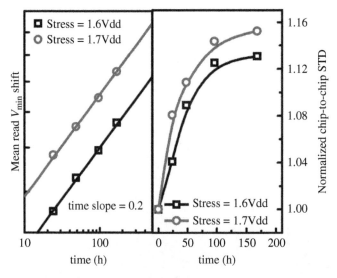

Figure 5.19 Aging time dependence of (left) mean read V_{min} shifts, (right) chip-to-chip read V_{min} variation. *Source:* Liu [206]/IEEE.

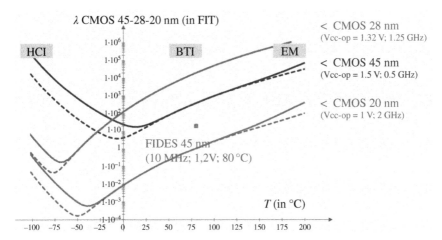

Figure 5.20   Reliability curves for 45, 28, and 20 nm technologies.

To support the previous discussion, the series of *MTOL* tests implemented on *CMOS* technologies are summarized in Figure 5.20 where the 3 *CMOS* processes 45, 28, and 20 nm are compared, showing *FIT* versus temperature for nominal voltages and nominal frequencies (solid lines) and 1/10 nominal frequencies (dashed lines). *FIDES* equivalent calculation at $f = 10$ MHz, Vcc-op = 1.2 V, and $T_j = 80°$C.

In conclusion, we demonstrate an innovative and practical way to use the various physics of failure equations together with accelerated testing for reliability prediction of devices exhibiting multiple failure mechanisms. We presented an integrated accelerating and measuring platform to be implemented inside *FPGA* chips, making the *MTOL* testing more accurate, allowing these tests at the chip and perhaps at the system level, rather than only at the transistor level. The calibration of physics models with highly accelerated testing of complete commercial devices allows for actual reliability prediction. The *MTOL* matrix can provide information about the proportional effect of each failure mechanism allowing extrapolation of the expected reliability of the device under various conditions.

This practical platform can be implemented on almost any *FPGA* device and technology to enable making *FIT* calculations and reliability predictions. The results of this approach provide the basis for improvements in performance and reliability given any design or application. This method can be extended to other processes and new technologies, and can include more failure mechanisms, thus producing a more complete view of the system's reliability. Research areas include thermal, mechanical, and electrical interactions of failure mechanisms of ultra-thin gate dielectrics, nonvolatile memory, advanced metallization, and power devices. He also works extensively with the semiconductor industry on projects relating to failure analysis, defect avoidance, programmable interconnect used in field programmable analog arrays, and repair in microelectronic circuits and packaging.

5.4   16 nm *FinFET* Reliability Profile Using the *MTOL* Method

FinFET and other 3D gate devices have long flooded the semiconductor industry. With their growth in the field, the importance of understanding their reliability characteristics increases greatly. A consensus is that *FinFETs* are susceptible to the same failure mechanisms as planar silicon technology with an adjustment due to geometry and the resulting directed heat flow down the Fin.

Numerous studies show detailed effects of *BTI*, *EM*, *HCI*, and *TDDB* in FinFET technologies. Reliability simulation models show that these mechanisms remain the most notable concerns [207]. Reduction of channel width in the fins causes a trade-off between an increase in current density and a decrease in short channel effects [208]. In addition to the other failure mechanisms, self-heating effect (*SHE*) appears in *FinFETs* due to the heat dissipation challenging the 3D gate architecture [209]. The characteristics of the *SHE* and how it interacts with other mechanisms is still a topic of concern [210].

A notable absence in most recent reliability studies is the ability to assess the relative impact of failure mechanisms in marketable devices. Most studies use either simulations or custom circuits to assess individual mechanisms. Although reliability simulations are an integral part of design, they have some notable disadvantages. Chip-level reliability prediction only focuses on the chip's end of life, while the known wear-out mechanisms are dominant; however, these prediction tools do not predict the random, post-burn-in failure rate that would be seen in the field [211, 212]. Beyond that, the tests are commonly performed with exaggerated stress conditions. Such studies, using unsynchronized testing conditions, fail to locate the impact of the different mechanisms relative to each other. For instance, many test studies show evidence of *TDDB* starting to occur in voltage stresses far beyond normal testing conditions [213].

In custom-fab-designed testing circuits, each mechanism is tested in isolation from the other mechanisms. As a result, the relative impact of each mechanism is lost before reaching the hands of the vendors. Instead, validation testing is used on finished products using the *HTOL* testing method. *HTOL* testing assumes a single failure mechanism that occurs completely randomly with a constant failure rate, which of course, is not the reality. In our system, the testing was carried out on fully fabricated *FPGA* products. This allowed us to find the relative weight of the mechanisms and to reveal a direct correlation between *SHE* and *BTI*. We also see the relative impact of all mentioned failure mechanisms.

The *MTOL* testing method was utilized in a study to provide a comprehensive reliability assessment for the *FPGA* [214]. Previous studies performed on 45 nm and 28 nm planar technologies demonstrate the separation of mechanisms *BTI*, *HCI*, *EM*, and *TDDB* from real data [194]. Proportionate weights of the mechanisms result from solving a matrix of extrapolated failure rates from known accelerated operational conditions [195]. Ultimately, the results showed a clear transition between having three working mechanisms: *BTI*, *EM,* and *HCI* at 45 nm to a single dominant mechanism: *BTI*, at 28 nm. Results from our 16 nm testing conclude by showing a single dominant failure mechanism, *BTI* with a *SHE* adjunct. This study fortifies the conclusion that, from 28 nm forward, *BTI* is the dominant failure mechanism that will lead to failure under realistic operational conditions.

In this study, the *MTOL* testing method was used to analyze the Xilinx Ultra96 *FPGA* device [215] fabricated on *TSMC* 16 nm technology [216]. The test setup is detailed and numbered in Figure 5.21. The setup included an Ultra96 *FPGA* board containing an MPSoC development taking advantage of the different resources of the board (#1). The *RO* logic, created with *VHDL*, was sectioned into ip blocks each

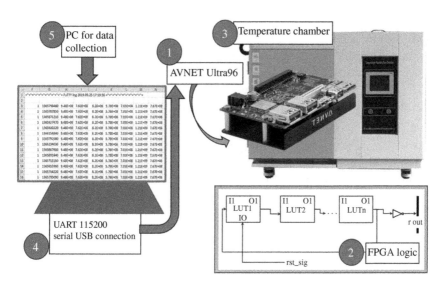

Figure 5.21 Diagram of the test setup for *MTOL* testing using the Ultra96 board.

containing 20 rings (#2). Two orientations were used in the tests: one included 15 ip blocks which, by running 300 simultaneous *RO*s, the core heat of the system was raised to as much as 90 degrees above the ambient temperature. This orientation allowed tests with relatively low voltages and high core temperatures.

The second setup was slimmer where only 80 rings were programmed. This allowed us to perform tests with higher voltage and low temperature. Tests with voltage levels of 1.25 V retained a temperature displacement of the ambient to core of no more than 40 degrees. The rings in the blocks were configured with different numbers of stages to cover a wide span of frequencies. With a frequency range from 10 MHz to over 1 GHz, the systems were sensitive enough to identify even minor frequency effects. This served to decipher between degradation caused by mechanisms strongly stimulated by current stress such as *HCI* [217, 218] and *EM* [179, 180] frequency dependent, and those with very low frequency dependency such as *BTI* [164, 165]. In this study, *BTI* is acknowledged generally since the measurement mechanism used did not allow the differentiation between *pBTI* and *nBTI*.

Table 5.4 shows the allocation of the rings. For logistical reasons, the low-frequency rings in the two orientations are different; Test setup #1 includes 300 rings producing heat in low-voltage tests. Test setup #2, with 80 rings, allows for high voltage conditions without surpassing the upper limits of the board. One 10 MHz and the other 30 MHz. Using the two orientations, we were able to perform tests from −50°C to 125°C and voltages up to 1.27 V. The *AXI* master-slave element harmonizes the blocks into a single component. For ease of use, the effort of control is performed using the hardware's *ARM* module programmed in C-based languages. The temperature of the surroundings is controlled using a standard lab temperature chamber [219] for heating and cooling (#3). A UART serial connection (#4) is possible using the external JTAG pod board [220] attached to the Ultra96. The data was collected and interpreted in a PC (#5).

Table 5.4 Resource allocation of the ring oscillators in the two orientations.

Test setup	No. of rings	No. of inverters per ring	Average frequency (MHz)
#1	20	101	32
	40	11	315
	80	5	530
	160	3	970
#2	20	333	11
	20	33	112
	40	3	970

The Ultra96 includes an automatic thermal shutdown mechanism when the core temperature exceeds 125°C. Therefore, 125°C was the upper limit of the testing. The core voltage was controlled using a PMIC, the Texas Instrument: TPS650864 [221]. The nominal voltage of the board is 0.85 V. Although the PMIC can be adjusted to 1.67 V, the strain of the logic caused the internal temperature to overheat long before reaching the highest voltage. The highest voltage conditions were tested in subzero Celsius temperatures to a maximum of 1.27 V.

Frequency degradation was used to identify and separate the failure mechanisms. In Figure 5.22, the degradation data of a test with a core temperature of 107°C and core voltage of 1.1 V is plotted to a power law with three frequency regions representing three different RO sizes. This plot allows us to keep track of the initial frequency and the change in frequency over time due to the stress. The results reveal a power law *n*-factor of 0.25. Other studies show n-factor values in *FinFETs* for *HCI* varying from 0.22 to 0.41 [221] and *BTI* between 0.19 and 0.29 [222, 223]. In 28 nm and older planar technologies, *HCI* has a half root law generated by a simple diffusion process [224, 225] and *BTI* is 0.25. Our test data shows consistency with a BTI process. The *TTF* is calculated individually for each ring using the n-factor slope extrapolated 10% of the total frequency reduction. 10% degradation was the failure criterion chosen for this study considering that a 10% performance degradation from the original design is not acceptable for most applications. Using this procedure, we receive a large collection of *TTF* values in each stress mode. Figure 5.23 displays a *TTF* to ring frequency plot of test data with

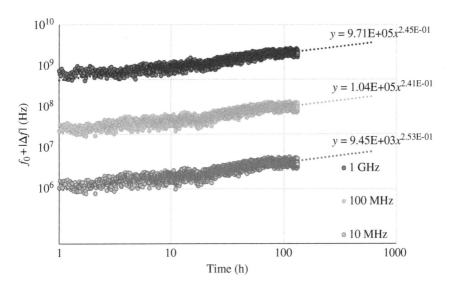

Figure 5.22 Power law plot of three different rings with frequencies of 10 MHz, 100 MHz, and 1 GHz, respectively. The core stress conditions used are 1.1 V and 107°C.

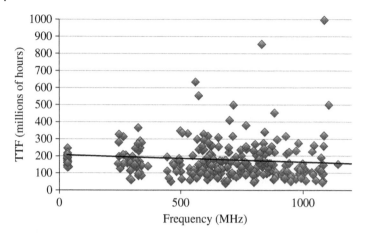

Figure 5.23 Plot showing the calculated *TTF* of each ring to frequency.

core temperature of 117°C and core voltage of 1.06 V. The high frequencies (around 1 GHz) have a *TTF* on average of about 75% of that of the low frequencies.

The data fans out in the higher frequency rings. This pattern has powerful implications in statistical failure studies. Although the subject is entitled to a separate study, for our purposes, it is enough to know that even the small rings with the most dispersed *TTF* values retain a high level of determinism. The level of randomness is indicated by the slope, β, from the group of values formulated into a Weibull distribution. Cumulative Weibull distribution of the small rings in the test data is shown in Figure 5.24. In the small rings, the slope of the Weibull plot is about 3.

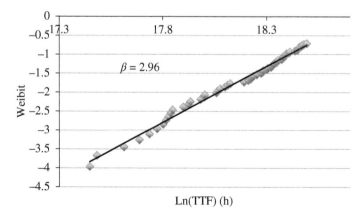

Figure 5.24 Weibull plot of early failures (1st 60 failures) which shows a β slope of approximately 3.

The mean time-to-fail (*MTTF*) value or θ is the zero point of the Weibull plot. Since we have actual *TTF* measurements with their Weibull characteristics, each test reveals the reliability distribution function of the device for that specific stress condition.

With a β slope of 3, the probability of failure over time for this voltage, frequency, and temperature is:

$$R(t) = exp\left(-\frac{t}{\theta}\right)^{\beta} = exp\left(-\frac{t}{1.4 \cdot 10^8 \, h}\right)^3 \tag{5.17}$$

Next, we detail the *EOL* assessment for an ensemble of microelectronic devices in a chip for the full range of temperature, voltage, and frequency modes.

One can also observe a mild decrease in *TTF* values relative to the increase of frequency for each *RO* circuit. In this example, we see about a 25% decrease in *TTF* at 1 GHz compared to very low frequency of 10 MHz. Interestingly, we found similar slopes in data with different applied stress conditions. Normally, a frequency effect suggests the existence of either *EM* or *HCI* [224, 225]. If this was true here, the severity of the effect should correspond to the applied stress. For *EM*, as the temperature and voltage are increased, the slope on the frequency should become steeper [200]. Similarly, in *HCI* which has a negative activation energy (E_A), decreased temperature and increased voltage should worsen the frequency effect [199]. The above suggests that *EM* and *HCI* are not the cause of the observed decline in lifetime with frequency. *BTI*, being a voltage- and temperature-stimulated mechanism, is not known to have a frequency effect. *BTI* with a minor interaction with a different mechanism, such as *SHE*, could explain the data. Since *SHE* is caused by lower heat dissipation of the fins, the high frequency of switching will exacerbate the effect [33]. Therefore, *BTI* with a contribution of *SHE* will produce a uniform, slight raise in *TTF* with frequency. *SHE* causes the cells of the higher frequency rings to heat up with the rapid switching of normal voltage swings. The relationship between change in temperature and frequency should be expressed using the formula:

$$\Delta T_{SHE} = C_L \cdot R_{th} \cdot V^2 \cdot f \tag{5.18}$$

α is a constant indicating the *SHE* contribution. "*f*" is frequency and "*V*" is voltage.

The square in the voltage corresponds to the power consumption of an inverter in transition which is ½ CV^2, where C is the input capacitance of each node. The Eyring model [198] is utilized to relate the *TTF* of the failure mechanism to the E_A and γ values. Equation (5.19) formulates the *TTF* of the single dominant mechanism, *BTI* with the influence of *SHE*.

$$TTF \sim [exp(-\gamma \cdot V)] \cdot exp\left[\frac{E_A}{k_B \cdot (T + \Delta T_{SHE})}\right] \tag{5.19}$$

ΔT is the frequency-related temperature increase defined in Eq. (5.18). The *TTF* of each data sample is measured as the degradation from initial frequency as an exponential decay. As mentioned above, the *n*-factor is found for each test. We used the Excel™ slope function to calculate slope ε. Inside the slope function:

The Y-values are the log of the full array of the 1-min frequency samples.
The X-values are the array of times raised to the power law n-factor.

The formula is as follows:

$$\varepsilon = slope\left(\frac{\ln\left(f_0 : f_f\right)}{\left(t_0^n : t_f^n\right)}\right) \tag{5.20}$$

f_0/t_0 is the initial frequency/time and f_f/t_f is the final frequency/time. "*n*" is the *n*-factor. With the slope ε, the *TTF* is extrapolated (given that 10% degradation is a failure) as follows:

$$TTF = \left(0.1/\varepsilon\right)^{1/n} \tag{5.21}$$

The power of "$1/n$" is used to convert back to time from the exponential decay.

Figure 5.25 shows the collection of tests arranged to find the slope of the activation energy generated from temperatures from $-30°C$ to $120°C$ and voltages from 0.85 to 1.27 V. The *SHE* contribution, found using an *RMS* function, is: $\alpha = 3.85 \cdot 10^{-9}$.

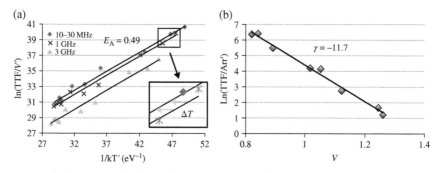

Figure 5.25 *TTF* trends are displayed revealing an activation energy of 0.49 (in average). (a) The 10–30 MHz and 1 GHz slopes are generated from *FPGA* data and the 3 GHz slope is extrapolated. V' is exp$(-\gamma \cdot V)$. T' is $T + \Delta T$. (b) Plot of the *TTF* data showing the voltage constant γ.

Table 5.5 Effect of frequency on the device temperature and *TTF*.

Frequency	Average junc. temp (°C)	TTF related to 0 Hz (%)
10–30 MHz	62.3	96.0
1 GHz	66.2	33.9
3 GHz	74.2	3.7

There are three sets of data points in Figure 5.25a for three frequency regions, all having an activation energy of 0.49 *eV*. 0.49 *eV* is well within the *BTI* range. Since E_A slope stays the same with increased frequency, the decrease in *TTF* is likely caused by increased temperature, i.e. enhanced *BTI*.

The voltage constant γ, shown in Figure 5.25b, is found to be 11.7. This voltage relationship is fitting for *BTI*. The average decrease in *TTF* with operation frequency is presented in Table 5.5. By inspecting the values of the table and the lines in Figure 5.25, one can see that the 1 GHz line has a *TTF* reduced to 30% compared to static mode. When extrapolating to 3 GHz, the same trend is received with the *TTF* reduced to 3%.

In the case of totally random system of failures, the *TTF* results like those mentioned above are used to find the random failure rate of the system. Since our system is more deterministic, we are able to pinpoint an actual EOL.

Figure 5.26 displays several Weibull plots of tests of different stress conditions. The outcome of all the data is compiled to find the reliability of the complete system expressed in Eq. (5.22) by substituting the *TTF* from Eq. (5.19):

$$R_{T,V,f}(t) = exp\left(-\frac{t}{TTF_{T,V,f}}\right)^3 \tag{5.22}$$

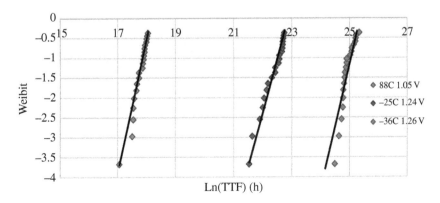

Figure 5.26 Weibull plots for three different stress conditions.

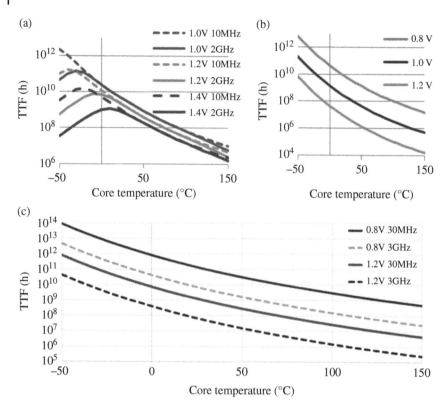

Figure 5.27 A display of a set of reliability profile curves for 45 nm (a), 28 nm (b), and 16 nm (c) technologies.

For example, to project a reliability/confidence level of 99.99% for 100,000 h or 10 yr and an operation speed of 3 GHz, Eq. (5.22) produces a maximum voltage of 0.8 V and temperature of 40°C.

Finally, we can see the reliability profile of 16 nm *FinFET* technology compared to the profiles of 45 and 28 nm in Figure 5.27. Figure 5.27a,b summarizes the *TTF* trends developed in previous sections for those technologies. The distinct contrast in these plots to the plots above for 45 nm and 28 nm is that they are *TTF* profiles instead of *FIT* profiles. The proportion will, naturally, be the same because the *TTF* of an element is the reciprocal of its failure rate as is described in Eq. (5.9). The *TTF* of the 16 nm devices, shown in Figure 5.27c is displayed over the temperature range of −50 to 150°C (125–150°C is extrapolated from the data). The plot shows that both the increase in frequency and the increase in voltage cause the same reliability hazard which is the increase in core temperature. Likewise, the raise in temperature causes a more severe *BTI* process. The key to ensuring longevity of the devices is to keep them cool.

5.4.1 Thermal Dissipation Concerns of 16 nm Technologies

The system-level concerns of *SHE* are assessed based on a standard power dissipation model published in Proceeding of IEEE 32nd International Conference on Microelectronics (Miel), Niš, Serbia, [226]. Core temperature change directly corresponds to the dynamic power dissipation caused by the logical activity as shown in the following equation:

$$T_{increase} = P_{dynamic} \cdot R_{th} = \left(N \cdot f \cdot C_L \cdot V^2\right) \cdot R_{th} \tag{5.23}$$

where N is the total number of logic elements, f is frequency, C_L is the load capacitance, and R_{th} is the thermal resistance. The testing was carried out by tracking the temperature changes of different *FPGA* programs created using a variable sum of *ROs*. The temperature was gauged with the Xilinx Ultrascale+ built-in Sysmon internal temperature sensor [227]. The board was powered using a Regal DP832 power supply [228] to track the current consumed and, therefore, the total board power consumption. Plotting the power versus temperature reveals the thermal resistance which is 14.45 Ω. The board is retained at an ambient temperature of 25°C with a thermal chamber.

Figure 5.28 displays the effects of power dissipation on the system. First, ROs with a fixed number of stages, three inverters running at about 1 GHz, were used. Figure 5.28a shows the core temperature increase for three different voltages with different amounts of 3-stage inverter chains programmed into the *FPGA*. The static chip operation temperature, found by extrapolating to zero rings, increases the core temperature to 41.8°C. The voltage-squared trend is shown clearly in Figure 5.28b. The data presented reveals a load capacitance of about 1.49 pF per cell.

Since the frequency of *ROs* inversely corresponds to the number of stages in the ring, we were able to compare the temperature increase for added logic to that of increased frequency. Tests were carried out using a constant voltage of 1 V with changes in the number of inverters in the rings. 3-stage rings were compared to 33-stage rings to create a difference of one order of magnitude (1 GHz and 100 MHz, respectively). There is a 37.5% temperature offset from increasing the frequency by one order of magnitude as shown in Figure 5.28c. Equation (5.23) can be enhanced to consider two separate load capacitances: one for simple power dissipation and the second for dissipation due to self-heating:

$$T_{increase} = N \cdot f \cdot (C_{L,PD} + C_{L,SHE}) \cdot V^2 \cdot R_{th}$$

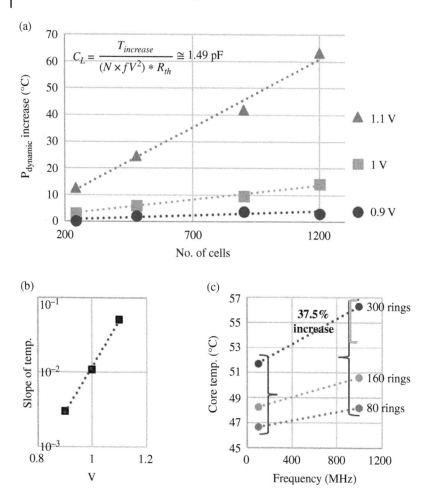

Figure 5.28 Variation of core temperature. (a) Increase of core temperature due to dynamic power dissipation with an ambient temperature of 25°C. (b) Temperature gradient increasing with voltage squared. (c) Temperature offset with frequency with an increase of 37.5% from 100 MHz to 1 GHz.

$$T_{SHE} = N \cdot f \cdot C_{L,SHE} \cdot V^2 \cdot R_{th} \tag{5.24}$$

Taking the total load capacitance mentioned above, the standard power dissipation load capacitance: $C_{L,PD} = 930\,fF$ and the load capacitance caused by self-heating: $C_{L,SHE} = 558\,fF$. Temperature changes from 33-stage to 333-stage rings retained to same ratio showing a linear increase with frequency. Theoretically, frequency and logical elements should produce an analogous result as seen in 28 nm technology. The nature of *SHE*, heating the logical elements during transition times, is fitting to explain this offset. As mentioned in the previous section concerning device reliability, we calculated $3.85 \cdot 10^{-9}$ $[F \cdot \Omega]$ for load capacitance and thermal resistance. If we estimate that the value of the load capacitance alone is the same as that found in the power dissipation study, we receive a thermal resistance for the *FinFET* devices of about 7 kΩ. Considering the extreme contrast in dimensioning between the base area of the channel of a device and the area of the chip, the contrast in the thermal resistance is easily explained.

Since these results are received using a relatively small number of cells at speeds up to 1 GHz, complex, high-speed applications should be designed with caution.

5.5   16 nm Microchip Health Monitoring (*MHM*) from *MTOL* Reliability

We have demonstrated in the section above how *BTI* with *SHE* is the prime consideration for degradation failure in 16 nm devices. Therefore, chip monitoring in 16 nm technologies can be accomplished using a single-parameter control set, allowing the development of a monitoring solution to be relatively simple. Our studies show the benefits of using *RO*-type circuits to find the failure behavior of the devices tested. Likewise, such orientations are very suitable for a health monitoring system design. By implementing the *FLL* circuit in place of the *RO* circuit, the resources required to create a monitor are reduced considerably. The proposed *FLL* circuit differs from *FLL* circuits implemented in other studies [229]. Previous FLL implementations are designed to keep the frequency constant by correcting the internal voltage. In the case of health-monitoring systems, it is imperative to measure the frequency of degradation to sense the health of the microchip. Our *FLL* circuit innovation, which measures frequency changes over time to monitor the performance of the system, has not been introduced in previous studies. Therefore, the *FLL* circuit proposed in this work is the seamless solution for anticipating early microchip failures.

In the following section, the procedure for extracting *TTF* figures from *RO* test systems is detailed. In the discussion, we demonstrate how the process requires the gathering of data from a large number of circuits. This poses a serious design challenge, as processing of a large amount of frequency data will be very costly to microchip space and power resources.

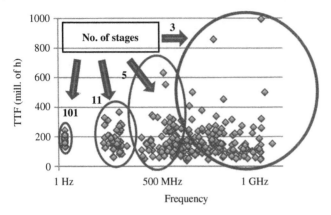

Figure 5.29 An example of a *TTF*-to-frequency plot.

5.5.1 Weibull Distribution Tapering by Increasing Devices

Prior to the *FLL* circuit design, frequency dependency was checked using variable-length *RO*s, as mentioned previously. Acquisition of accurate results required large amounts of data. Frequency of each ring circuit is dictated by the number of stages included in the ring. More stages create a longer propagation delay, decreasing the frequency. Figure 5.29 displays an example of a *TTF* value to ring frequency plot of the test data. The frequency varies with number of stages in the rings (listed in the figure). The *TTF* values become more dispersed with increase of frequency.

The data fans out in the higher frequency rings. The rings with many stages have a much tighter distribution than those with few stages. Since the *TTF* values in higher frequency rings are very dispersed, 160 3-stage ring circuits were programmed into the devices to receive a precise average. For 11-stage rings, similar accuracy is received from averaging 20 circuits.

The investigation of the dispersion phenomenon starts with calculating the *TTF* values of the data. Figure 5.30a shows the frequency degradation data of a single 101-stage ring circuit stress tested for about 160 h. Extrapolation is realized by transforming the degradation curve into a straight line. The degradation processes of failure mechanisms are not linear. Since in most materials, damage develops due to a diffusion process, the deterioration will advance with time raised to some fraction. For example, *BTI* arises due to hole-assisted breaking of Si–H bonds at the Si/SiO_2 interface [230]. Different failure mechanisms have different time scales of degradation. Based on empirical studies, *HCI* degradation can be transposed to a square root time scale [167] and *BTI* to a fourth root

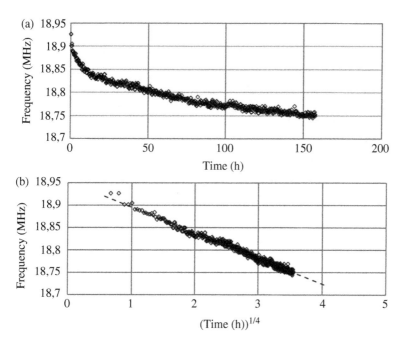

Figure 5.30 Plot of frequency degradation of an *RO* composed of 101 stages in hours (linear time scale at the top, 1/4th power law time scale at the bottom). The data as-is cannot be extrapolated to a *TTF* value without being converted to the correct time law.

[231]. From the stress conditions of the test in Figure 5.30a, the power in the test is determined to be a 4th root law as shown in Figure 5.30b. The result is a uniform slope over all the data. Analysis of the degradation data is actualized by extrapolating the frequency decrease to the point where the device is nonfunctional. Our definition of chip failure is a 10% depletion in performance. That level of deterioration will cause the device to be inoperable according to most standards.

5.5.2 The *FLL* Measurement Circuit

The reliability profiles generated with the *MTOL* method before this study used *RO* testing systems. Many other reliability testing methods use *RO*s as their degradation indicator [225, 232, 233]. In this study, we present a highly accurate solution for chip performance monitoring over multiple frequencies. The motivation for changing the measurement circuit from standard *RO*s to the new *FLL* circuit is the *RO*s lack the ability to control the ring frequency unless the number of ring stages is changed. The only way to generate high frequency is by implementing rings with few stages. Consequently, the precision of the *TTF* values received for these circuits will be poorer. This forces the designer to create a cumbersome amount of ring data to achieve a good average of *TTF* values, and thus a precise measurement. The disadvantages of using the *RO* solution are significantly increased in a health monitoring system. The *TTF* values must be calculated on the monitored microchip. To process the large data structures of *TTF* values, the microchip must perform heavy and resource-costly computations. A health monitor is only a successful solution if it is resource-efficient and transparent.

Another configuration, the *PLL* circuit is used to monitor performance degradation in *FPGAs* [234]. The signal is forked at the beginning of the circuit. One route has an inverter chain and the other a free path. The measurement indicator is the shift in phase (see Figure 5.31). This allows testing of inverter chains to be any length desired. The downside of using the *PLL* circuit is that the phase drift is hardly discernible from the noise in the signal. We base this conclusion on results of *PLL* testing models performed on previous technologies. For this reason, the PLL circuit was not implemented on the technology tested. In contrast, frequency is a convenient parameter to measure microchip performance. Hence, we preferred to design a health-monitoring system that uses frequency as its indicator.

In light of the above, a frequency-monitored circuit where the number of stages and their level of frequency are controlled separately would be the optimal circuit for a performance degradation monitor circuit. This can take the pros of both the *RO* and *PLL* circuits. The implementation of this circuit design resolves the problem of *TTF* value dispersion in high frequencies. This solution facilitates the development of circuits with a large chain of inverters that can be stressed at high frequencies.

As was demonstrated in the section above, such circuits produce very exact results. The *FLL* circuit offers stress frequency control without compromising on precision. The circuit operates in two modes: a stress mode and a measurement mode. For the stress mode, an external clock (*ext_clk*) delivers a predetermined frequency through the inverter stages that remain in an open chain. Since the inverters remain in this stress mode for relatively long period of time compared to the measurement time, the test can be considered in situ or constantly stressed throughout the duration of the test. For the measuring stage, the circuit transitions

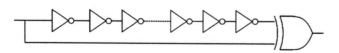

Figure 5.31 Diagram of a phase-locked loop circuit.

into an *RO* (ring mode) for a short period. The frequency is sampled to observe the degradation trend. Between transitions, the circuit is reset (*rst*).

Figure 5.32a details the logic layout programmed into the *FPGA*. A 4-input lookup table (LUT4) is connected to a chain of single-input LUTs (LUT1). We chose to use 151 LUT1s to receive a good average. The transitions are initiated by a ring enable (ring_en) switch.

Figure 5.32b shows a detailed wire diagram of the *FLL* circuit design. The design has two MUX layers that are connected to inverters. The logic of the two MUXs is using the Xilinx generate command in VHDL with the INIT in mode: X"5410". The duration of the frequency stress is 10 min between measurement samples. This provides ratio of about 200 times more stress-on compared to stress-off.

Figure 5.32c illustrates the time allocation of the different modes of the *FLL*.

The *FLL* circuit was initiated on 16 nm *FPGAs* and *MTOL* tests were performed. The testing setup included four stress frequency modes: 31, 125, 250, and 500 MHz with 10 rings instantiated for each frequency mode. Figure 5.33 is an example plot of the *TTF*-to-frequency. The *TTF* values have a tight distribution which decreases with frequency increase. The results fit in line with the results performed using standard *RO*s. In Figure 5.34, the results of five tests using different stress conditions are displayed. The *TTF* values in each frequency node are averaged. The trend of *TTF* decrease with frequency is conserved for all the tests. The benefit of the *FLL* circuit is clearly seen by showing the contrast of the *FLL* results of Figure 5.34 to the *RO* results in Figure 5.30. In Figure 5.30, the high-frequency

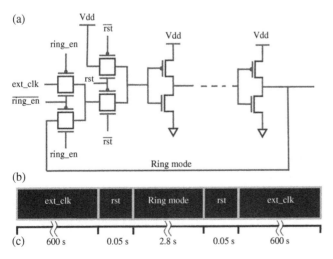

Figure 5.32 Diagram of the *FLL* circuit: (a) Block diagram of LUTs. (b) The wire layout showing the connections of the circuit. (c) Timeline of the *FLL* operation cycle.

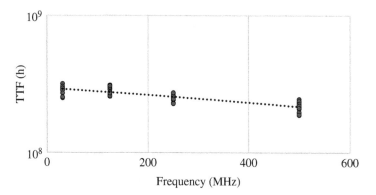

Figure 5.33 Three *TTF*-to-frequency plots averaged over the frequency. The decrease in *TTF* resembles the results received using *ROs*.

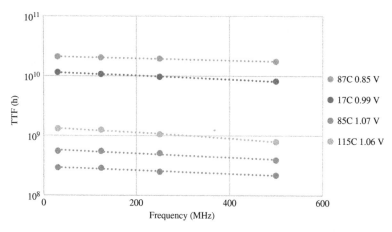

Figure 5.34 Five *TTF*-to-frequency plots averaged over the frequency. The decrease in *TTF* resembles the results received using *ROs*.

rings produce extremely dispersed *TTF* values and the results throughout all the frequencies retain tight distributions.

5.5.3 Degradation Data Correction with Temperature Compensation

A method was developed to increase the accuracy by means of compensation of frequency deviations caused by temperature change. The data was collected from *MTOL* tests with different ambient temperatures. The data was collected from multiple tests to find the frequency changes in different base frequencies. Although the tests used a temperature-regulated chamber, there were temperature fluctuations up to 5°. The frequency variations due to temperature change were recorded. The results were normalized to show a linear change with a slope of $1.2 \cdot 10^{-3}$. Using the slope found in the data, we developed an expression for correcting the degradation pattern in any frequency as detailed in the following equation:

$$f = f_0(1 + m \cdot \Delta T) \tag{5.25}$$

where f_0 is the frequency measured, m is the slope of the frequency change due to the temperature, and the formula below represents the offset of the temperature:

$$\Delta T = 60°C - T_0 \tag{5.26}$$

meaning that 60°C is the expected average chip temperature and T_0 is the temperature measured.

The technique was used in tests in several temperature and frequency modes. Figure 5.35 shows an example of data corrected using the temperature compensation formula. The original data (Figure 5.35a) has an undiscernible signal due to the distortion of the temperature fluctuations. The second plot (Figure 5.35b) is the corrected data showing a clear degradation signal. Since monitors are designed to operate in regular use conditions which are prone to non-negligible temperature fluctuations, the thermal compensation procedure is critical for the use of the monitor.

5.5.4 Accurate Lifetime Calculations Using Early Failure

We found that temperature compensation alone did not sufficiently decrease the data noise from the internal monitor. It was apparent that more effort would be necessary to bring the SNR to an acceptable level. In previous studies, Weibull distributions were used to show that the amount of dispersion of the TTF values directly corresponds to number of stages in the ring circuits [176]. When the TTF values are set to a ring frequency plot, the data in low frequencies conserves a tight pattern and fans out in the higher frequency rings. We used the Weibull distribution to analyze the transition of TTF pattern from the low frequencies to the high frequencies.

We analyzed the data from the low-stress tests, which are detailed in a previous study [210]. The results that we received seem to contradict the trend described above. Figure 5.36 displays a set of four Weibull distributions of 3-stage rings tested with a nominal core voltage of 0.85 V. The single distinguishing factor between the stress conditions of the tests presented are the ambient temperatures. Each is changed by 20°C from the other plots: 10°C, 30°C, 50°C, and 70°C. The plots are marked

(a)

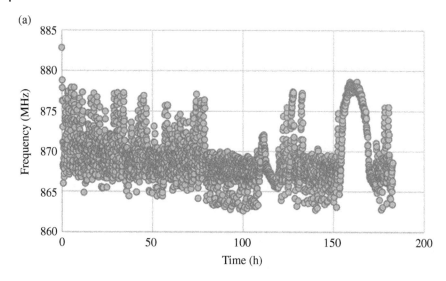

(b)

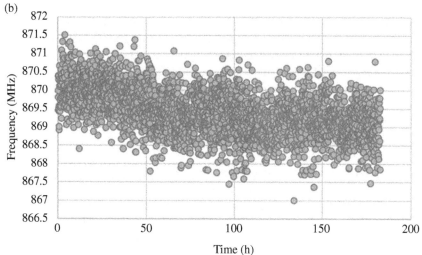

Figure 5.35 An example of degradation data before compensation (a) and after compensation (b). The stress conditions are 45°C ambient temperature and a core voltage of 1.1 V.

with a straight dotted line to mark a β slope of 3. The 70°C data shows a characteristic Weibull distribution with the expected β slope of 3. As the temperature of the test decreases, the data veers more and more from the expected Weibull distribution trend. Since the tests are under low stress, the outcome is easily explained. The *TTF* is figured from the slope of frequency degradation. When

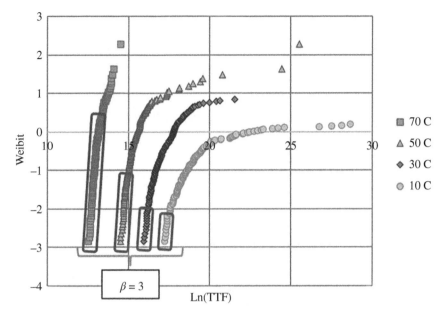

Figure 5.36 Weibull distributions of 3-stage rings are displayed for the following four different external stress temperatures: 10°C, 30°C, 50°C, and 70°C. The section that conserves a characteristic β slope of three for 3-stage rings is encircled.

the slope is too slight, the degradation signal is indiscernible from the noise of the system. Therefore, the data veering from the expected Weibull trend does not contribute to the failure trend.

It is very important to point out that the early failure times (values starting from the bottom left side of the plots) are consistently congruent with the expected Weibull trend. We found that the Weibull slopes of the early failure data are comparable to those calculated from high-stress tests. This validates the theory that the early failures are an authentic part of the signal. Basing a failure trend from the early failures is acceptable since the Weibull distribution is a weakest-link-type distribution. The early failures are a prediction of the later failures [88]. Therefore, it is sufficient to observe the early failures to see the trend of the system. The $MTTF$ value or θ is the zero point of the Weibull plot. Since we have actual TTF measurements with their Weibull characteristics, each test reveals the reliability distribution function of the device for that specific stress condition.

We have demonstrated that, as the stress of the test is lowered, the noise in the results is more visible in the data. Correction of the data by excluding longer failure times is less effective with the decrease in stress. In any case, the results obtained by filtering out nonfitting parts of the plot show a high level of accuracy.

The outcome shows frequency dependence as it is seen that the *TTF* is shorter in the higher frequencies. The results of the tests remain well within the range of values received in the high temperature, voltage, and frequency stress tests run in previous studies [235].

5.5.5 Algorithm to Calculate the *TTF* of Early Failures

We have determined from the thermal management study elaborated above that the lifetime estimate of 16 nm *FinFET* technology can be characterized using a single *TTF* model as defined in Eq. (5.19). The data extracted from *MTOL* tests based on this model were used as a control set for comparison to *TTF* measurements of the devices in real time. We implemented a microchip health monitor system into Ultra96 MPSoC development boards based on the design plan detailed in C language. The *FPGA* of the system included 140 ROs, all with three stages developed using *VHDL*. The system calculated the *TTF* of the boards over aging windows of 16 h. The *TTF* calculation was preformed using the following formula:

$$TTF = \left(\frac{0.1}{\propto / \left(\Delta t^{1/n} \right)} \right)^n \tag{5.27}$$

where "*t*" is time in hours, and "*n*" is the time power law for the degradation of *BTI*.

We found that the n is about 4. α is the relative change of the frequency in the test defined by:

$$\alpha = \frac{\Delta f}{f_0} \tag{5.28}$$

"Δf" is the displacement of ring circuit frequency due to degradation in the test. "f_0" is the initial frequency of the circuit.

Initially, we attempted to achieve accurate *TTF* measurements by averaging the *TTF* values of the 140 rings. This is the procedure used to obtain accuracy in the tests with high stress. In this case, where devices are operating at regular use conditions, the results lacked stability and were unsatisfactory. We concluded that the measurements received were not recording the degradation of the system, but rather system noise. We then developed an algorithm that implemented the noise reduction techniques explained in Section 5.5.4. The flow of the algorithm is presented in Figure 5.37. The first two actions detailed in the first level of the flow are acquisitions of the initial and final sample frequencies of all the rings and are stored in arrays F_0 and F_1. The data extracted is adjusted using the temperature compensation formulas (5.25) and (5.26) and is stored in arrays F0_comp and F1_comp. Next, the normalized displacement of each individual ring is calculated.

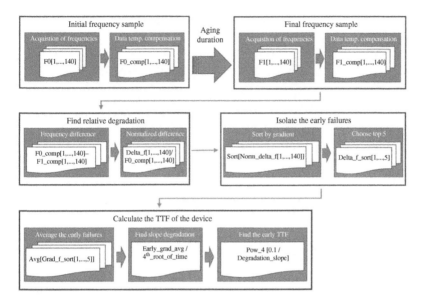

Figure 5.37 A diagram of the flow for implementing temperature compensation and early failure filtering into an *MHM* system. The techniques are performed on 140 3-stage *ROs*.

The difference in frequencies is found by storing F0_comp – F1_comp into Delta_f and then normalizing it by dividing by F0_comp and sorting the results in Norm_delta_f. The normalized differences are sorted using the "sort" function from largest to smallest. The solution is in array Delta_f_sort. We chose the first five values from the array to obtain the early failures. These values were averaged and stored in the scalar Early_grad_avg. This value was divided by the time value which in our case was simply 2 h (since the 4th root of 16 is 2) to obtain the slope of degradation. The result is used to calculate the *TTF* of the system.

Figure 5.38 shows results of the *MHM* system calculation over about 600 h. The first plot is the measurements of the system using the original approach which is simply averaging the *TTF* values. One can see that the results are very erratic. In contrast, the second plot shows the results of measurements incorporating the noise reduction techniques. The results remain stable throughout the test. The variance of the noise-reduced data is about 0.2 orders of magnitude, and the nonoptimized data is about six orders of magnitude.

The system with noise reduction is an excellent self-gauge of expected lifetime of a chip. The results of this study are significant for testing individual chips to mitigate thermal hazard occurrences. The benefit of the method is far more powerful in monitoring the activity of a large system with hundreds or thousands of chips such as in data centers. This system can be expanded to provide the information

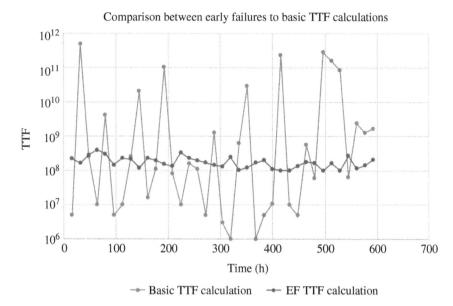

Figure 5.38 A comparison of measurement data between *MHM* testing with noise reduction (in dark gray) and without noise reduction (in Light gray).

needed to choose how to allocate resources based on a study performed previously [236]. Tasks that require heavy resources can be allocated to strong servers and low-resource tasks to weaker servers.

5.5.6 The Microchip Health Monitor

The *MHM* system is formulated using the concepts detailed in the previous sections. According to reliability trends in the latest technologies, stress due to low temperatures is no longer a factor since there is an acute reduction in *HCI*. Since in high temperatures, there is only one dominant mechanism, *BTI*, separation of failure mechanisms is not necessary. The *MHM* system has the lab data tested using the *MTOL* testing method stored in a database. Figure 5.39 displays the flow of the monitor. The following parameters are measured every 16 h: frequency of the 140 5 MHz FLL circuits, 140 500 MHz *FLL* circuits, and internal voltage and temperature. The *TTF* values are calculated using Eqs. (5.20) and (5.21) and averaged. The averaged *TTF* results called *TTF*1 and *TTF*2 are compared to the *TTF* values for the two frequencies stored in the database. Precision of the *FLL* circuits allows minimal sensitivity for the measured *TTF* values. If there is a decrease in *TTF* of a full order of magnitude, the monitored microchip with generate a

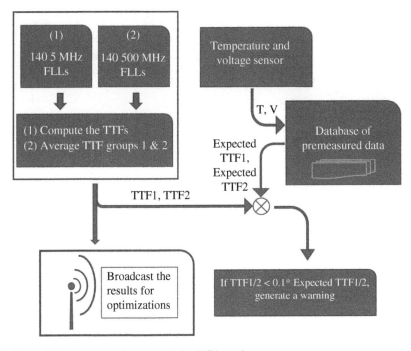

Figure 5.39 An operation flow of the *MCH* monitor.

warning flag. Given that the *MCM* system is implemented in a large scale of devices, the results will be broadcast to a central hub to optimize the formula of the database data in the future. Having such a system in advanced infrastructures can allow cautioning of harmful chip degradation before damage is caused.

6

System Reliability

Complex deep sub-micron (DSM) components are expected to operate under harsh conditions depending on their mission profiles. The reliability is required to be described by adequate models to estimate the failure probabilities and appropriate distributions. The newest technologies for modern components have very small gate lengths (nano-scale dimensions) employing new materials and innovative arrangements (nanosheet, and nanotube technologies). Literature has acknowledged complementary metal-oxide semiconductor (CMOS) bulk reliability distribution parameters accounting for electrical aging mechanisms, such as high-temperature degradation, so-called negative or positive thermal instability (NBTI or PBTI), hot carrier injection (HCI), and the hard or soft breakdown of dielectric gates. These mechanisms can cause significant reductions in performance and software errors that are very important to quantify under nominal operating conditions. Their impacts on the lifetime of the single sensitive active area of a device are a key feature to be understood and controlled.

It is important to note that the majority of these mechanisms are characterized by progressive parametric drift of signature parameters that ends in catastrophic device failure or loss of performance. It has also been observed that parametric degradation, inducing random collapses and other wear-out failures, are similar in their physical description, but they seem to be activated over large variations in time frames (exception for handling cause of failure as electro-static discharge [ESD], for example). For this reason, the evaluation of the probability of failure beyond the equivalent (necessarily short) duration of the test is misleading.

Reliability Prediction for Microelectronics, First Edition. Joseph B. Bernstein, Alain A. Bensoussan, and Emmanuel Bender.
© 2024 John Wiley & Sons Ltd. Published 2024 by John Wiley & Sons Ltd.

6.1 Definitions[1]

The instantaneous failure rate (IFR), also named the hazard rate $\lambda(t)$, is the ratio of the number of failures during the time period Δt, for the devices that were healthy at the beginning of testing (operation) to the time period Δt.

$$\lambda(t) = \frac{f(t)}{1 - F(t)} \tag{6.1}$$

The cumulative probability distribution function $F(t)$ for the probability of failure, also called cumulative distribution function (CDF), is related to the probability density distribution function $f(t)$ (or PDF) as

$$F(t) = \int_0^t f(x) \cdot dx \tag{6.2}$$

The CDF for the population is called a life distribution and is also known as the "unreliability" function. CDF has two useful interpretations:

1) $F(t)$ is the probability that a random unit drawn from the population fails by t hours.
2) $F(t)$ is the fraction of all units in the population that fail by t hours.

The reliability function $R(t)$, the probability of non-failure, is defined as
$$R(t) = 1 - F(t) \tag{6.3}$$

Figure 6.1 shows the schematic representation of failure distribution functions. *The reliability function $R(t)$ may be thought of in either of two ways:*

1) *As the probability that a random unit drawn from the population is still operating after t hours.*
2) *As the fraction of all units in the population that will survive at least t hours.*

If n identical units are operating and F(t) describes the population they come from, then n·F(t) is the expected (or average) number of failures up to time t, and n·R(t) is the number expected to still be operating.
We will consider two rules in this paragraph as follows:

1) *Multiplication rule: The probability that several independent events will all occur is the product of the individual event probabilities.*

---

1 Quotes from Ref. [37] in italic format.

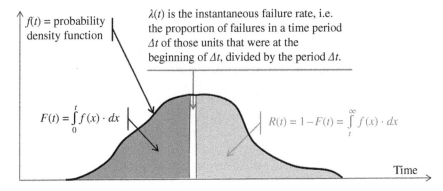

$f(t)$ = probability density function

$\lambda(t)$ is the instantaneous failure rate, i.e. the proportion of failures in a time period Δt of those units that were at the beginning of Δt, divided by the period Δt.

$$F(t) = \int_0^t f(x) \cdot dx$$

$$R(t) = 1 - F(t) = \int_t^\infty f(x) \cdot dx$$

Time

Figure 6.1 Instantaneous failure rate (IFR), cumulative distribution function (CDF), probability density function (PDF), and probability of non-failure or reliability function $R(t)$.

2) *Complement rule: The probability that an event does not occur is 1 minus the probability of the event.*

The independent events are the failure or survival of each of n randomly chosen, independently operating units. The probability of all of them still operating after t hours is defined by the multiplication rule and forms the product of n·R(t) terms. In other words, the probability that n independent identical units, each with reliability R(t), all survive past t hours is $[R(t)]^n$.

The probability that at least one of the n units fails is one minus the probability that all survive. Applying the complement rule, the probability that at least one of n independent identical units fails by time t is given by:

$$1 - [R(t)]^n = 1 - [1 - F(t)]^n \tag{6.4}$$

If the life distribution for each of these components is F(t), then the probability that the system does not have a failure by time t is $[R(t)]^n$. If the system fails when the first of its components fails, and we denote the life distribution function for a population of these systems by Fs(t), then the complement rule gives us:

$$F_s(t) = 1 - [R(t)]^n \tag{6.5}$$

Numerical application: Assume a ring oscillator contains 15 identical elements using a FinFET technology, all operating independently and each critical to the operation of the ring oscillator. If each element has a reliability estimated at 0.995 for the working period, what is the probability the ring oscillator fails while under working period?

Applying Eq. 6.5, we get 7.24%.

6.2 Series Systems

This section considers the formula for the reliability of a series system, parallel system, or a mix series/parallel system as fully detailed in Ref. [37] in *Applied Reliability*, third edition book by Tobias and Trindade.

The most commonly used model for system reliability assumes that the system is made up of n independent components, all of which must operate in order for the system to function properly. But this series model is applied specifically when a single integrated circuit with several independent failure modes is analogous to a system with several independent elementary constituents. The failure mechanisms are competing with each other in the sense that the first to reach a failure state causes the component to fail: the open question is still to consider what a failure state is. Is it for catastrophic failure or related to a failure criteria or performance? As argued by Tobias, "the more general competing risk model where the failure processes for different mechanisms are not independent can be very complicated since one must know how the random times of failure for different mechanisms are correlated."

To describe component failure rate characteristics, we must take into account multiple failure mechanisms superimposed as dependent or independent, being sudden (cataleptic failures) or progressive. What failures result in an open path, what is a short circuit path, and consequently, what do they account for in series or parallel system modeling?

Assuming the first hypothesis (open circuit) and the ith element have a CDF $F_i(t)$, the probability the series system fails at time t is the probability that one or more of the n independent components have failed at time t. In terms of total probability, we consider the product of the n CDFs:

$$R_s(t) = Pr(E) = Pr(E_1, E_2, ..., E_n) = Pr(E_1) \cdot Pr(E_2), ..., Pr(E_n)$$

$$= \prod_{i=1}^{n} R_i(t) \tag{6.6}$$

with

$$R_i(t) = exp(-\lambda_i \cdot t) \tag{6.7}$$

where λ_i is the failure rate of ith component. Hence,

$$\lambda_s(t) = \sum_{i=1}^{n} \lambda_i \tag{6.8}$$

If the components have the same reliability R_{s0}, meaning $\lambda_i = \lambda_0$ for $i = 1$ to n, this becomes:

$$R_s(t) = R_{s0}^n = [exp(-\lambda_0 \cdot t)]^n = exp(-n \cdot \lambda_0 \cdot t) \tag{6.9}$$

LEMMA #01: A series system composed of n identical and independent elements, each described by a Poisson distribution (λ_0) reliability model, can be approximated by a general equivalent Poisson distribution with parameter $n \cdot \lambda_0$.

6.2.1 Parallel Systems

A system that operates with n elements in parallel survives until the last of its elements fails. Consider here that a failure mode is defined only as an open circuit. Thus, a system with n elements in parallel fails only when all the n elements fail.

Assuming the hypothesis "fail in open circuit," and the ith element has a CDF $F_i(t)$, the probability that the parallel system fails at time t is the probability that all the components have failed at time t.

6.2.2 Poisson Distribution Function

If the CDF is defined by a Poisson or an exponential distribution function, the probability that the system fails, e.g. if at least one of the n elements fails, is the reliability or probability of success at time t and is expressed as follows:

$$R_p(t) = 1 - \prod_{i=1}^{n} (1 - exp(-\lambda_i \cdot t)) \tag{6.10}$$

If $\lambda_i = \lambda$ for any $i = 1$ to n, then:

$$R_p(t) = 1 - [1 - exp(-\lambda \cdot t)]^n \tag{6.11}$$

If we consider the lifetime, t, to be within a portion of the plateau of the bathtub curve $\lambda \cdot t \ll 1$, then the first-order development of the exponential around 1, gives an approximation of the reliability at time t by:

$$exp(-\lambda \cdot t) \approx 1 - \lambda \cdot t \text{ or } \lambda \cdot t \approx 1 - exp(-\lambda \cdot t) \tag{6.12}$$

and merging Eq. (6.11) with Eq. (6.12) gives:

$$R_{p_{approx.}}(t) \approx 1 - (\lambda \cdot t)^n \approx exp[-(\lambda \cdot t)^n] = exp\left[-(t/\theta)^\beta\right] \tag{6.13}$$

Similar to a Weibull distribution, the parameter n is also the β parameter. We can quantify the error due to this approximation by:

$$RE(in\%) = \frac{R_p(\lambda \cdot t) - R_{p_{approx.}}(\lambda \cdot t)}{R_p(\lambda \cdot t)} \cdot 100 \tag{6.14}$$

(a)

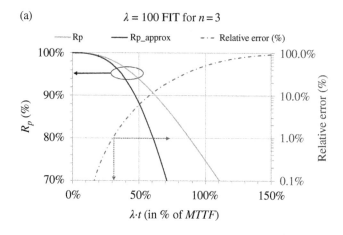

(b)

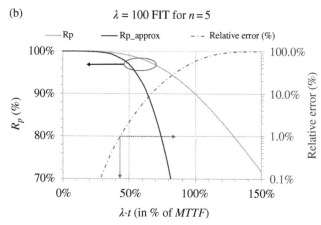

Figure 6.2 (a) Plot of Eqs. (6.11) and (6.13) and relative error for λ = 100 FIT and n = 3 and (b) plot of Eqs. (6.11) and (6.13), and relative error for λ = 100 FIT and n = 5.

Figure 6.2 shows plots of Eqs. (6.11)–(6.14) versus parameter $\lambda \cdot t$ in unit of mean time to fail (MTTF). We can observe that for $\beta = n = 3$ and 5 (respectively, figures a and b), the relative error generated by the approximation is less than 1% for $\lambda \cdot t \leq 0.3$.

Thus, we see that the error is less than 1% for any time t from initial and up to 30% of $MTTF$ (for $n = \beta = 3$, respectively, 5, $MTTF = 1/\lambda = 10^7$ h or $\lambda = 100$ FIT). It is noted that this relative error decreases for higher values of n at a given $\lambda \cdot t$.

As a major fact, such parallel systems, based on n independent and identical elements having the same Poisson failure rate constant λ (e.g. with unique equivalent $MTTF_i = MTTF = 1/\lambda$ for $i = 1$ to n), show an equivalent Weibull distribution for short times $t \leq 30\% \cdot MTTF$ range for an error lower than 1%.

LEMMA #02: A parallel system of n identical and independent elements having failure mode as an open circuit, each described by a Poisson distribution (λ) reliability model, can be approximated by a general equivalent Weibull distribution with parameters λ and $\beta = n$ with an error lower than 1% for a delay operation lower than 30%·MTTF for the example shown.

6.2.3 Weibull Distribution Function

The reliability or probability of success at time t of a random variable T with a Weibull distribution defined by parameters α_i and β_i is given by:

$$R_{i_W}(t, \alpha_i, \beta_i) = exp\left[-\left(t/\alpha_i \right)^{\beta_i} \right] \tag{6.15}$$

For a parallel system of n Weibull elements, and assuming independence, the probability that the system functions is computed as $1 -$ Prob[system fails to function]. Consequently, the reliability of such a system is given by:

$$R_{P_W}(t, \alpha_i, \beta_i) = 1 - \prod_{i=1}^{n}\left\{ 1 - exp\left[-\left(t/\alpha_i \right)^{\beta_i} \right] \right\} \tag{6.16}$$

Since identical independent elements exist, we get $\alpha_i = \alpha$ and $\beta_i = \beta$ for any $i = 1$ to n, so:

$$R_{P_W}(t, \alpha, \beta) = 1 - \left\{ 1 - exp\left[-(t/\alpha)^{\beta} \right] \right\}^{n} \tag{6.17}$$

Considering the life mission t is defined by a portion of the plateau of the bathtub, using the same approach as we do for Eq. (6.12), we can approximate for short time t:

$$R_{P_{W_{approx}}}(t, \alpha, \beta) \approx exp\left[-(t/\alpha)^{\beta \cdot n} \right] \text{ for } \lambda \cdot t \ll 1 \tag{6.18}$$

Therefore, the error due to this approximation is defined by:

$$RE_W(in\%) = \frac{R_{P_W}(t, \alpha, \beta) - R_{P_{W_{approx}}}(t, \alpha, \beta)}{R_{PW}(t, \alpha, \beta)} \cdot 100 \tag{6.19}$$

Figure 6.3 displays plot of Eqs. (6.17) and (6.18) versus parameter $\lambda \cdot t$.

As a major fact and from a generalization point of view, such parallel systems based on n independent and identical elements having the same constant Weibull failure rate parameters (α, β) are shown to be modeled by an equivalent Weibull distribution with reliability parameter (α, $\beta \cdot n$). The error induced by such approximation is estimated for short times $t \leq 55\%$·MTTF, ranging for an error lower than 1%.

It is noted that this relative error is decreasing for high values of β and n for a given $\lambda \cdot t$.

(a)

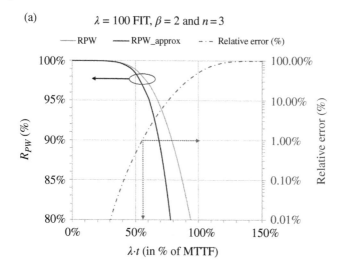

(b)

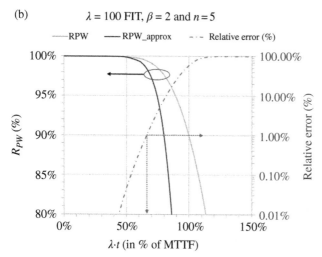

Figure 6.3 (a) Plot of Eqs. (6.17) and (6.18) relative error for λ = 100 FIT, β = 2 and n = 3 and (b) plot of Eqs. (6.17) and (6.18) and relative error for λ = 100 FIT, β = 2 and n = 5.

LEMMA #03: A parallel system constituted of n identical and independent elements, each described by a reliability Weibull distribution model (α, β), can be approximated by a general equivalent Weibull distribution with parameter (α, $\beta \cdot n$) with an error lower than 1% for time operations lower than 55%·MTTF.

6.2.4 Complex Systems

Physical configurations in series or parallel do not necessarily indicate the same logic relations in terms of reliability. An integrated circuit (DSM) composed of look-up tables (LUTs) connected in logical paths contains billions of interconnected elementary structures. From a reliability perspective, the LUT includes multiple transistors in series-parallel configurations, but a LUT is said to have failed if one or more transistors failed. So, the transistors in a LUT are considered in series from a reliability perspective.

There are systems that require more than one component to succeed in order for the entire system to operate. In addition, the performance of a product is usually measured by multiple characteristics. In many applications, there is one critical characteristic that describes the dominant degradation process. This one is an open door to characterize product reliability. The failure of a product can be defined in terms of performance characteristics, also called indicators, crossing a specific threshold. Figure 6.4 depicts the relation between the degradation path of an indicator or a signature parameter, pseudo-life *TTF*, and life distribution.

The maximum likelihood method is used for estimation of distribution parameters. Assuming general behavior of a non-monotonic drift degradation, due to possible temporary remission phases, we can model them by a Weiner process. In such occurrence, definition of a degradation acceptability threshold criteria makes it possible to switch from a degradation model to a reliability model. One can note that there is no first-passage distribution of a threshold in the case of Weiner process. Nevertheless, reliability can be estimated by Monte Carlo simulation as well as to assess the remaining useful life.

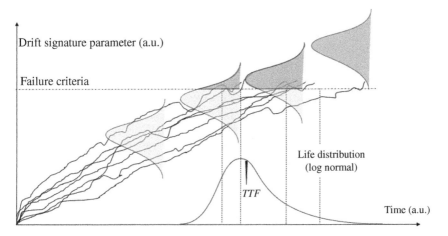

Figure 6.4 Relation of degradation path, pseudo-life *TTF*, and life distributions.

Once the estimates of reliability parameters are quantified, we can use a Monte Carlo simulation to generate a large number of degradation paths (known as the bootstrapping method proposed by Efron and Tibshirani [237]). The probability of failure $F(t)$ is approximated by the percentage of simulated degradation paths crossing a specified threshold after a stress time is applied.

The signature degradation measured is the cumulative effect of multiple physics of failure mechanisms and can be seen either as a series configuration or as a parallel configuration failure mechanism combination.

Let us again consider a ring oscillator structure (RINGO) as a series of cascaded odd-numbered NAND gates as represented in Figure 6.5. On the left side of the figure, the NAND gate structure is detailed with respect to existing failure mechanisms activated by temperature (ranked low to high). The failure rate λ equivalent ($\lambda_{equ.}$) is shown as a function of λs and accelerating factors related to stress conditions for each mechanism.

We must realize that several failure mechanisms can be activated simultaneously. Thus, we need to understand how these mechanisms can interact and lead to failure of the RINGO. Should these failure mechanisms be considered in series or in parallel? To answer this question, we must go back to the basics.

A series system is a configuration such that, if any one of the system components fails, the entire system fails. Conceptually, a series system is one that is as weak as its weakest link.

A parallel system is a configuration such that, if not all of the system components fail, the entire system works. Conceptually, in a parallel configuration, the total system reliability is higher than the reliability of any single system component.

Nevertheless, considering parametric failure mechanism hypothesis, the following schematic should be proposed as shown in Figure 6.6. In such a case, a single component with four independent parametric failure mechanisms (HCI, BIT, EM, or TDDB) and occurring simultaneously, is analogous to a series system. Each failure mechanism "competes" with the others to cause a failure. Failure rates are additive, mechanism by mechanism, and so drifts are cumulated. In a RINGO structure made of DSM technology (either FinFET of FD-SOI or others), a failure is observed when the frequency drift reaches the failure criteria assigned. In such a case, the reliability model is similar to a parallel system and can be related to the occurrence of either catastrophic or parametric failures affecting each logical NAND element representing LUTs. Therefore, in a RINGO structure, a failure can be originated by single or multiple NAND gate failures and can yield to:

1) Either a short path delay causing a RINGO frequency drift being possibly able to reach the failure criteria (several gates can be affected simultaneously before the failure criteria are caught); or
2) An open path of a single NAND gate, then a RINGO frequency catastrophic failure.

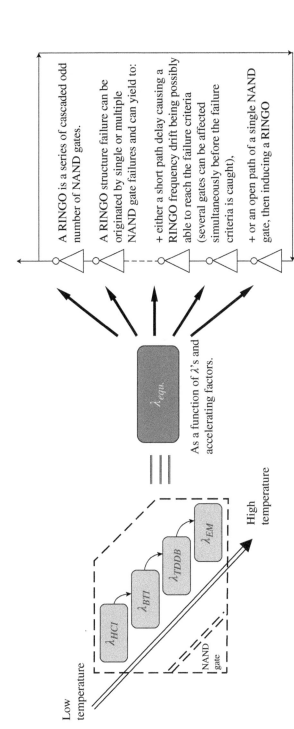

Figure 6.5 Series reliability configuration for DSM RINGO catastrophic failure mode modeling (four failure mechanisms and m = RINGO size defined as number of NAND interconnected).

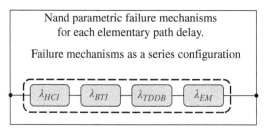

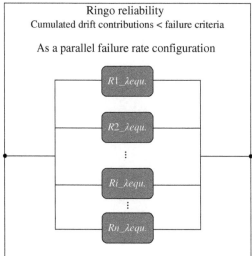

Reliability of a series-parallel system is defined in Figure 6.6, based on *a **Poisson distribution function***, and is applicable as:

$$R_{p-s_P}\left(t, \alpha_{i,j}, \beta_{i,j}\right) = 1 - \prod_{j=1}^{n}\left[1 - \prod_{i=1}^{4} exp(-\lambda_i \cdot t)\right] \tag{6.20}$$

considering *n* NAND elements in parallel each with four types of failure mechanisms in series.

Then, we get:

$$R_{p-s_P}(t) = 1 - \prod_{j=1}^{n}\left[1 - exp\left(-t \cdot \sum_{i=1}^{4} \lambda_i\right)\right] \tag{6.21}$$

Say all *n* NAND elements are similar in terms of reliability point of view, we can write:

$$R_{p-s_P}(t) = 1 - \left[1 - exp\left(-t \cdot \sum_{i=1}^{4} \lambda_i\right)\right]^n \tag{6.22}$$

If $t \cdot \sum_{i=1}^{4} \lambda_i \ll 1$, e.g. assuming true up to few percent's MTTF, we can approximate the reliability at time t by:

$$\left[1 - exp\left(-t \cdot \sum_{i=1}^{4} \lambda_i\right)\right]^n \approx \left(t \cdot \sum_{i=1}^{4} \lambda_i\right)^n \tag{6.23}$$

and $R_{p-sp}(t) = 1 - \left[1 - e^{-t \cdot \sum_{i=1}^{4} \lambda_i}\right]^n$ (6.22) reduces to:

$$R_{p-sp_{approx.}}(t) \approx 1 - \left(t \cdot \sum_{i=1}^{4} \lambda_i\right)^n \tag{6.24}$$

Then to get:

$$R_{p-sp_{approx.}}(t) \approx 1 - \left(t \cdot \sum_{i=1}^{4} \lambda_i\right)^n \approx exp - \left(t \cdot \sum_{i=1}^{4} \lambda_i\right)^n = exp - (t/\theta)^\beta \tag{6.25}$$

as a Weibull general equivalent representation with the following parameters:

$$\frac{1}{\theta} = \sum_{i=1}^{4} \lambda_i \text{ and } \beta = n \tag{6.26}$$

With β is indeed the number of NANDs incorporated in a RINGO test structure.

The error due to the approximation taken in $\left(1 - e^{-\sum_{i=1}^{4} \lambda_i \cdot t}\right)^n \approx \left(t \cdot \sum_{i=1}^{4} \lambda_i\right)^n$ (6.23) is defined by:

$$RE_{P-Sp}(in\%) = \frac{R_{p-sp} - R_{p-sp_{approx.}}}{R_{p-sp}} \cdot 100 \tag{6.27}$$

Assuming all λ_i are the same (but indeed, this is not true as the correct math theory and experiment approach will be addressed in Chapter 3 with the Matrix multiple-temperature operational life [MTOL] concept), Figure 6.7 shows plot of Eqs. (6.22) and (6.25) versus $\lambda \cdot t$ variable. Hence, we see the calculated relative error is assessed to be lower than 1% for time operation lower than 20%·MTTF when considering a 21 NAND RINGO with 4 failure mechanisms characterized by a failure rate of 10 FIT each.

LEMMA #04: In a series-parallel system composed of n identical and independent elements, each described by a reliability Poisson distribution (λ), its reliability model can be approximated by a general equivalent Weibull distribution with parameter (θ, $\beta = n$ as the number of NAND gates) with an acceptable less than 1% error and for time delay up to 20% of MTTF.

6.3 Weibull Analysis of Data

Further details are obtained by studying the Weibull distribution of each group of ring sizes: 3, 5, and 11 inverters. The Weibull distribution [238], named after the Swedish Professor Waloddi Weibull, is perhaps the most commonly used

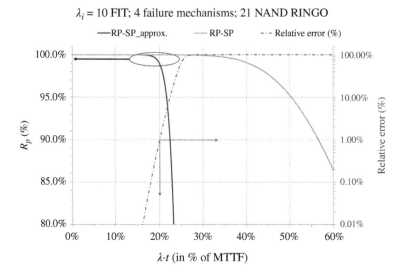

$\lambda_i = 10$ FIT; 4 failure mechanisms; 21 NAND RINGO

Figure 6.7 Plot of Eqs. (6.22), (6.25), and relative error for a Poisson distribution with λ_i = 10 FIT, four failure mechanisms and number of NAND gates $n = \beta$ = 21.

distribution for lifetime data analysis. While being straightforward compared to other distribution formulas, the Weibull distribution is also versatile enough for analyzing diverse types of aging phenomena. We found a direct correlation between the number of stages in the rings and the Weibull distribution slope for the TTF of that group of rings. The level of randomness of the TTF values is indicated by the slope, β, from the Weibull reliability probability distribution, as detailed in equation of [185]:

$$R(t) = exp - (t/\theta)^\beta \qquad (6.28)$$

where θ is the characteristic failure time. β is the slope of the distribution also referred to as the shape parameter. When the β slope is about 1, the system exhibits a failure distribution that is almost completely random. Distributions where the β slope is higher than 1 illustrate a more deterministic failure characteristic. As the shape parameter increases, the failure distribution approaches a single failure time. This transition is clearly differentiated in the frequently used "bathtub curve" displayed in Figure 6.8 [239]. The accented point on the right side of the curve is the transition from a constant failure rate to the end-of-life time of the device. Our assumption, based on the orientation of the data displayed in Figure 6.9, is that Weibull distributions categorized by ring size will reveal a correlation between β slope and the number of stages in a ring. The procedure for plotting a Weibull distribution is provided in the steps below. Figure 6.1 is

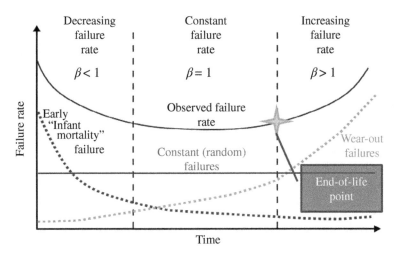

Figure 6.8 The bathtub reliability curve is used to describe device failure-rate characteristics for nearly all devices. The relationship of Weibull plot slopes to the bathtub curve is set by the value of β.

Figure 6.9 Weibull distribution created using 3-stage ring oscillators on 45 nm technology with stress conditions of 35 °C oven temperature and 1.2 V core voltage.

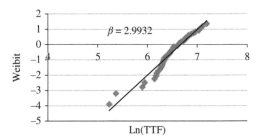

rearranged to isolate β. More appropriate names are used for the $R(t)$ and θ parameters creating the following formula:

$$\ln(-\ln(1 - R(t))) = \beta \cdot \ln(\theta) \rightarrow \ln\left(-\ln\left(1 - \frac{failure\#}{\#of\ rings}\right)\right)$$

$$= \beta \cdot \ln(TTF) \tag{6.29}$$

The ring data for each ring size is categorized by TTF value starting from the shortest failure time. Calculation of the TTF value, found on the right side of the equation, is performed in two steps:

1) The slope of the degradation curve is calculated using the following Excel™ formula

$$normalized\ slope\ (NS) = -slope(\ln(ring\ frequncy), time\ [4th\ root]) \quad (6.30)$$

2) The formula for calculating degradation down to 90% (TTF), assuming the n-root law of 0.25, is:

$$TTF = \left(0.1/_{NS}\right)^4 \quad (6.31)$$

The Weibull slopes are presented by plotting the "Weibit," which is based on the number of failures as follows:

$$Weibit = \ln\left(-\ln\left(1 - \frac{failure\#}{\#of\ rings}\right)\right) \quad (6.32)$$

The x-axis is plotted to: $\ln(TTF)$ and the y-axis is the Weibit. The slope of the plot is β as presented in Figure 6.9. A collection of Weibull distributions of 3, 5, and 11 stage rings is displayed in Figure 6.10. A clear one-to-one correlation between the number of stages in the rings and the β slope appears. As will be explained later in an analytical study, these results produce a good practical example of the central limit theorem (CLT). Based on Drenick's deduction, one can expect a completely random failure rate for each stage [55]. In any case, there is a large difference between the TTF distribution of small rings to large rings. The explanation is that in the small rings, the output signal is an average of few stages, resulting in highly diverse TTF values. Larger rings produce a TTF value averaged over more stages, producing a tightly bound distribution of TTF values.

The same conclusion is reached by inspecting the ring circuits analytically with reliability models. The Weibull function takes only the extreme value approach. In other words, only TTF values much smaller than the mean time to fail (MTTF) are considered. This allows the use of a constant failure rate model. The reliability function for a single element, $R(t)$ and the failure function, $F(t)$ are listed in Eq. (6.32). This equation is built from the first-order Poisson function. As the conditions of the system are time-independent, the failure rate λ is constant. Thus, we have:

$$R(t) = exp(-\lambda \cdot t), F(t) = 1 - exp(-\lambda \cdot t) \quad (6.33)$$

Equation (6.32) describes a single element system. Ring oscillators include multiple elements. It is imperative to clarify what behavior best describes how the elements contribute to the failure of the system. The seemingly most obvious fit for a failure system model for microelectronic devices is the series system model [240]. For example, field programmable gate array (FPGA) devices, consisting of a matrix of logical elements called lookup tables (LUTs), will only operate if all the LUTs are functional. Therefore, just like the strength of a chain is as strong as its weakest

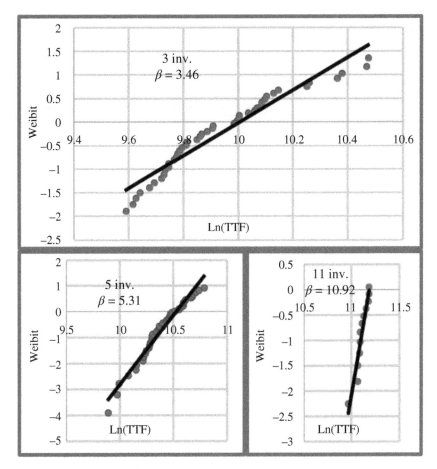

Figure 6.10 Weibull plots of the three different sizes of ring oscillators that demonstrate the direct correlation between number of inverters and Weibull slope β.

link, the reliability of an FPGA is only as robust as its worst LUT [29]. The diagrams in Figure 6.11 give a graphic representation of the parallel system model (respectively, series series) in reliability terms.

The total reliability is described as follows:

$$R(t) = [exp(-\lambda_1 \cdot t)] \cdot [exp(-\lambda_2 \cdot t)] \cdots [exp(-\lambda_i \cdot t)] \sim exp(-i \cdot \lambda_1 \cdot t) \qquad (6.34)$$

To suggest that the rate of failure in practice can be developed from this model is contradictory to evidence in the field. One would be forced to say that the *TTF* values decrease proportionally with the number of transistors or gates in the device. The series model suggests that the failure rate should increase as a function

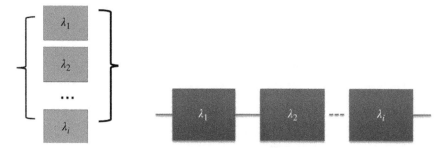

Figure 6.11 Parallel or Series weakest link failure system model.

of the number of devices in the system. Since device numbers are increasing exponentially over time, chip failure rates should also be increasing at a comparable rate. Based on in-field data, this is not the case. Therefore, the serial system model alone does not properly reflect the failure characteristics of a full microprocessor.

In a parallel system, the system only fails after all the components have failed. Assuming that the failure probability of each component is random, the probability for a single element is a Poisson process with a failure rate of λ_i. One can ask this question: What is the justification for describing the microelectronic device failure behavior as a parallel system. It seems obvious that each element in the chip is prone to fail, thus causing the whole device to fail. In other words, it can be defined as a classical series system. This perspective is misleading because it suggests that elements in a device are prone to catastrophic or complete failures. This is not commonly observed in test data. Rather, performance degrades disproportionately for each different element. Since the different elements in a logical path influence the response time of the logical path, they average together into a comprehensive failure rate λ value. Consequently, the interaction between the stages in a ring can best be described as a parallel failure system. This is because the stages are averaged together to generate the TTF of the ring. The failure probability of the parallel system is defined as:

$$F(t) = [1 - exp(-\lambda_1 \cdot t)] \cdot [1 - exp(-\lambda_2 \cdot t)] \cdots [1 - exp(-\lambda_i \cdot t)] \tag{6.35}$$

In the case where the variance in the rate (λ_i) of these processes is negligible, the equation collapses to the following for the functions for $F(t)$ and $R(t)$:

$$F(t) = [1 - exp(-\lambda \cdot t)]^N \Longrightarrow R(t) = 1 - [1 - exp(-\lambda \cdot t)]^N \tag{6.36}$$

where N is the number of stages in a single ring. Each ring is described as a system of multiple elements. The λ is always much smaller than 1 assuming an early failure model: ($t \ll 1/\lambda$) [24]. We can therefore make the following approximation:

$$exp(-\lambda \cdot t) \approx 1 - \lambda t \tag{6.37}$$

So, based on the failure function in Eq. (6.36), the reliability probability function is:

$$R(t) \approx 1 - (\lambda t)^N \approx exp\left[-(\lambda \cdot t)^N\right] = exp\left[-(t/\theta)^\beta\right] \tag{6.38}$$

The failure probability function is:

$$F(t) = 1 - exp\left[-(\lambda \cdot t)^N\right] \tag{6.39}$$

In Figure 6.12a, Eq. 6.39 is plotted with N values of 3, 5, and 11. Note that the function flattens out at some point after θ. This is not an issue of concern since the system uses an extreme value approach as mentioned previously. The derivative of Eq. (6.39) produces the failure distribution over time:

$$F'(t) = N \cdot \lambda \cdot (\lambda t)^{2N-1} \cdot exp\left[-(\lambda \cdot t)^N\right] \tag{6.40}$$

Figure 6.12b displays the result of Eq. (6.40) for N values of 3, 5, 11, and 101. As the number of stages becomes larger, the gradient steepens. We have demonstrated both empirically and analytically that the shape of the failure distribution of a system of rings directly correlates to the number of stages programmed into the rings. Additionally, by averaging the degradation of many stages in an RO, a very precise failure time or end-of-life (EOL) is obtained.

6.4 Weibull Analysis to Correlate Process Variations and BTI Degradation

In Chapter 5, we demonstrated how 16 nm technology shows BTI as the dominant mechanism. There is an additional element that compounds FinFET wear at higher frequencies, the self-heating effect (SHE). We see that in all tested technologies, a strong signature of BTI is observed. Therefore, it is natural to assume that the presence of process variations (PVs) will directly affect the BTI influence in all technologies.

PV classification in silicon device manufacturing has been covered by numerous studies from the early development of device structures. Traditional PV investigations distinguish between wafer-level variations and die-level variations [241]. One cause of PV can be roughly defined as random fluctuations of dopant atoms affecting the device's electrical parameters and is identified from fixed deviations in the device from nominal behavior [242].

In older technologies, the ratio of PVs to the nominal dimensions of devices was small enough that it did not greatly affect circuit performance. As the nodes of

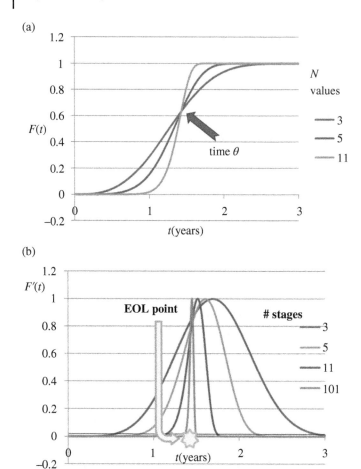

Figure 6.12 (a) Plots of the failure distribution function (Eq. (6.39)) pivoting around the θ point and (b) the derivative of the failure distribution function (Eq. (6.40)) revealing the EOL point.

technology are scaling down, the PV influence on the delays of logic gates increases significantly [243]. One can say that atoms are not scalable. Nominal dimensions of modern nano-scale devices are just a few atomic cells thick.

Since the severity of BTI is strongly related to the T_{ox} and V_{th} parameters, variances due to PV will generate a dispersion in the degradation [244]. The observations of this study suggest that PV may be a BTI concern, particularly in scaled devices beyond most estimates.

We demonstrated in Section 6.3, both experimentally and analytically, how the length of the RO chain directly coincides with the slope β of Weibull plots taken from the *TTF* measurements of the rings [24]. The Weibull CDF is the following formula:

$$F(t) = 1 - exp\left[-\left(t/TTF\right)^{\beta}\right] \tag{6.41}$$

where β is the slope of the distribution, also referred to as the shape parameter.

When the β slope is about 1, the system exhibits a failure distribution that is almost completely random. Distributions where the β slope is higher than 1 illustrate a more deterministic failure characteristic. As the shape parameter increases, the failure distribution approaches a single failure time. The Weibull distribution is plotted using special axes to give a linearized representation as follows:

$$\ln(-\ln(1 - F(t))) \rightarrow \text{``Weibit''} = \beta \cdot \ln(TTF) \tag{6.42}$$

Figure 6.13 shows the direct correlation between ring size and Weibull slope. The upper part of the figure has a plot of the *TTF* values taken from a single MTOL test example as a function of frequency. The viewer can easily perceive that the distribution of *TTF* values in the low frequencies has a small variance (upper left side), whereas the variance of the distribution in the high frequencies is very large

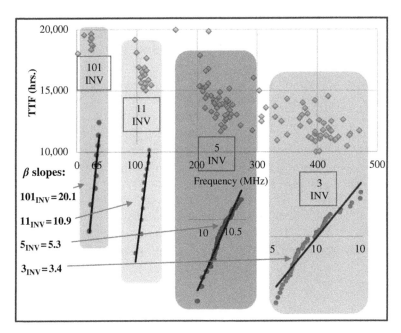

Figure 6.13 A *TTF* to frequency plot sectioned into four sectors defined by ring size 3, 5, 11, and 101 inverters. The corresponding Weibull plots for the four sectors are displayed below the *TTF* values for each ring group with their β slope values.

(upper right side). This is also verified by presenting the data in Weibull distributions. The results are labeled with their respective ring sizes. Each sector was formulated into a Weibull plot with the axes in Eq. (6.42). The results are represented in the lower part of the figure. There are two important points to learn about the results of this inspection. First, the β slopes increase with the number of stages in the rings. Second, the increase of the β slope fits nearly one-to-one with the number of stages in the rings of each sector apart from the 101-stage rings.

The correlation displayed above is the expected result as was elaborated upon in Ref. [176]. The outcome of the Weibull study detailed earlier demonstrates how a testing system with an ensemble of multi-sized rings is used to extract vital failure data from the device under test (DUT) [177]. We see an example of the advantages of the MTOL testing method over other qualification tests. The fact that large rings (101 stages in our case) stray from the β slope trend presented still needs clarification.

Several tests per technology were aggregated to produce the Weibull plots. Precision was increased with the addition of data. The objective of the experiment was to find the β slope in each technology that ceases to steepen. This will be considered the cut-off point for additional precision by adding stages to the test system ring oscillators. The cause of the β cut-off slope is statistical noise in the system. Due to the noise, the test results will show a convolution of the signal, the Weibull failure distribution, and the noise, some normal noise distribution. As a result, one would assume that the mean of the distribution would not be altered by the noise but rather only by the variance.

The interaction between the distributions can be visualized in terms of angular deviation from the direct north direction or positive $90°$. At this angle, there is no variance described as a β of infinity. A Weibull distribution with a β of 1 will have an angular increase of $45°$ or positive $135°$. Using the angular perspective is helpful because it gives a simple representation of how much impact the noise variance has on the signal variance. If we take, for example, a failure distribution with $\beta = 1$ and a noise distribution with $\beta = 20$, the deviation caused by the noise will only be about $2.9°$ or 0.1 deviation of the β. However, if the noise $\beta = 6$, the deviation will be 1.1 which makes the failure signal unintelligible.

Analytically, the result of having two distributions interfering can be described in simple terms as the parallel summation of the two slopes as follows:

$$\beta \parallel \varphi = \left(\frac{1}{\beta} + \frac{1}{\varphi}\right)^{-1} \tag{6.43}$$

where β is the slope of the failure distribution and γ is the slope of the noise distribution. Therefore, the accurate function to describe the data is:

$$F(t) = 1 - exp\left[-(t/\theta)^{\beta \parallel \varphi}\right] \tag{6.44}$$

Our study is aimed at finding the origin of this disturbance. If the precision cut-off point is the same for all the technologies, one can infer that the noise of the test system is the limiter. If the cut-off is lower as the technology becomes more scaled, the disturbance is caused by variations in the process.

A synopsis of the results for the different technologies including 45, 28, and 16 nm is presented in Figure 6.14: Weibull distribution plots of 3, 5, 11, and 101 stage

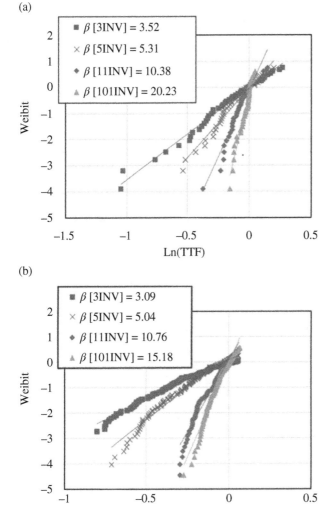

Figure 6.14 Weibull distribution plots of 3, 5, 11, and 101 stage rings for (a) 45 nm, (b) 28 nm, and (c) 16 nm where the mean is normalized to Weibit = 0.

(c)

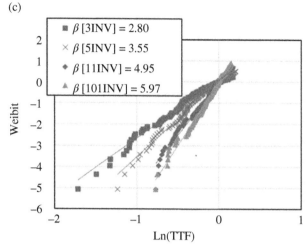

- $\beta\,[3\mathrm{INV}] = 2.80$
- $\beta\,[5\mathrm{INV}] = 3.55$
- $\beta\,[11\mathrm{INV}] = 4.95$
- $\beta\,[101\mathrm{INV}] = 5.97$

Figure 6.14 (Continued)

rings 45 nm (a), 28 nm (b), and 16 nm (c) where the mean is normalized to Weibit $= 0$. The data used to generate the plots is collected from several tests. The distributions of the different ring-size groups are normalized to a mean of zero on the Weibit to add clarity in comparing the slopes. The results show that the Weibull slope β is upper bound to a slope unique to the technology measured.

The 45 and 28 nm results shown in (a) and (b) both have Weibull slopes for the 3, 5, and 11 stage rings that are with one whole number of the ring size. The β slope in the 45 nm devices for 101 rings is slightly over 20. The 101-ring Weibull distribution for 28 nm devices drops to a β of about 15. This shows that the point where adding stages to the rings will no longer add precision to the TTF distribution is about 20 stages in 45 nm and 15 stages in 28 nm.

In the 16 nm results in (c), the values for all the β slopes are significantly lower than the number of stages of the rings. As the stages in the rings increase, the variance grows. The rings with 101 stages have a β slope of only about 6. Such a result is expected because the cut-off β slope is low enough that the variance of the interference interacts strongly with the variance of the failure signal in all the modes.

Table 6.1 summarizes the results of all the tests. Since the accuracy limitation is directly related to the technology, the most probable cause of the interference is due to PVs in the production of the devices in the FPGAs. The increase in variance in the smaller devices would be due to the increase in ratio of the PVs to the scaling of the technology.

Figure 6.15 shows the increase in variance of the *TTF* data with technology scaling. The points of the plot are taken from the β slopes of the 101 stage rings. We see

Table 6.1 The summary of results for all technologies.

		Technology		
		45 nm	**28 nm**	**16 nm**
	3	3.521	3.089	2.8
Ring size	**5**	5.307	5.035	3.55
	11	10.38	10.763	4.95
	101	20.228	15.181	5.97

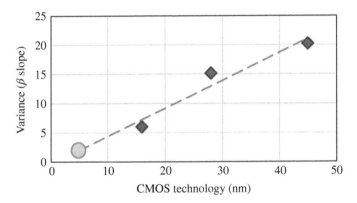

Figure 6.15 Trend of process variations with scaling of the technology. The 5 nm point is extrapolated from the 45, 28, and 16 nm data.

a linear increase in the variance with the scaling of the devices. This further supports the assumption that the cause of the variance is the increase in the PV to channel length ratio. Based on the 45, 28, and 16 nm results, the variance expected in 5 nm technology is projected to have a β slope of about 2. The outcome revealed poses a serious concern of premature BTI failures as the devices become more scaled. In Section 6.3, we established that the β slope in Weibull distributions for an ensemble of RO TTFs is dictated by the number of stages in the rings. This characteristic is well preserved in the small rings in the 45 and 28 nm data. The β slope for 3-stage rings in the 16 nm data is 2.8, which is below 3. That suggests that the β slope of a single element is below 1. One can assume that the β slope for a single element in 7 and 5 nm devices will be 0.7 and below. That will imply a decreasing failure rate (DFR) may be expected.

A practical scenario is examined to demonstrate the severity of the PV effects on BTI. 16 nm devices are operated at conditions of 0.85 V core voltage, 80 °C core temperature, and a frequency of 30 MHz. The TTF for this scenario, based on

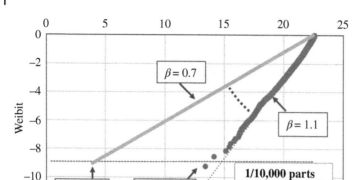

Figure 6.16 A graphic example of a Weibull plot with 10,000 parts shows the first failure times of a plot with a β slope of 1.1 and a plot with a β slope of 0.7.

the reliability profile in Ref. [13], would be about 6×10^9 h. A Weibull distribution of 10,000 parts for this scenario is depicted in Figure 6.16.

With a β slope of 1.1, the first failure out of 10,000 will occur after 73 years. If the β slope were decreased to 0.7, the first failure would occur after 1 week. This example illustrates how a relatively small decrease in the β slope of the distribution will impose very early failures in a significant number of parts. The definition of parts in this study can be used generally to include single elements in a chip. Such a rapid failure of 1/10,000 elements in a chip would be dire and may indicate a need to return to burn-in screening.

Therefore, the issue of increase in variance due to the PV impact on BTI should be considered with caution. A vital expansion of this study is to follow the same procedure on packaged 7 or 5 nm devices. One of the implications of the study is that aggressively scaled technologies may have an issue with DFR. If so, a screening process eliminating devices with excessive performance degradation problems would be advisable.

7

Device Failure Mechanism

It is a basic assumption that every failure has an explainable physical or chemical cause. This cause is called the failure mechanism. It is the goal of reliability physicists to identify the failure mechanism, determine what materials and stresses are involved, and quantify the results. An equation is derived that can be used to predict when failure will occur in the device under use conditions.

A single modern IC may have more than a billion transistors, miles of narrow metal interconnections, and billions of vias or contacts. As circuit entities, metal structures dominate the semiconductor transistors, and their description is as challenging and necessary as that of semiconductors. Transistor oxides have shrunk, and dimensions are now on the order of single Si and SiO_2 molecules thick and approaching 1 nm. Metal and oxide materials failure modes are more significant in deep-submicron technologies.

Besides, reliability tests are designed to reproduce failures of a product which can occur in actual use. The goal of understanding failure mechanisms from the results of reliability testing, which is extremely important to know, if the product is reliable during its actual use. The effect of stresses (temperature, humidity, voltage, current, etc.) on the occurrence of failures can be identified by understanding the failure mechanisms. The product reliability in actual use can be predicted from the results of the reliability tests, which are conducted under accelerated conditions.

Failure mechanisms of electronic devices could be classified as those related to wafer, assembly, mounting, and handling processes.

The relation among failure modes, mechanisms, and factors has been reported in some semiconductor handbooks as shown in Table 7.1.

One of the most important features of electronic systems is the issue of scaling. For more than 50 years, semiconductor circuit device technologies have been improving at a dramatic rate. A large part of the success of the MOS transistor is that it can be scaled to increasingly smaller dimensions, which results in higher performance. The ability to improve performance consistently while decreasing power consumption has made c-MOS architecture the dominant technology for

Reliability Prediction for Microelectronics, First Edition. Joseph B. Bernstein, Alain A. Bensoussan, and Emmanuel Bender.
© 2024 John Wiley & Sons Ltd. Published 2024 by John Wiley & Sons Ltd.

Table 7.1 The relation among failure modes, mechanisms, and factors.

Failure factors		Failure mechanisms	Failure modes
Diffusion	**Substrate**	**Crystal defect**	Decreased breakdown voltage
Junction or active area	Diffused junction 2D gas Isolation	Impurity precipitation Photoresist mask misalignment Surface contamination	Short circuit Increased leakage current R_{DSon} drift (GaN)
Oxide film	Gate oxide film Field oxide film	Mobile ion Pinhole interface state TDDB Hot carrier BTI	Decreased breakdown voltage Short circuit Increased leakage current h_{FE} and/or V_{th} drift IDSS decrease (GaAs) Power slump under high drive stress (HEMT)
Metallization	Interconnection Contact hole Via hole	Scratch or void damage Mechanical damage (cracks) Non-ohmic contact Step coverage Weak adhesion strength Improper thickness Corrosion Electromigration Stress migration	Open circuit Short circuit Increased resistance h_{FE} and/or V_{th} drift Noise deterioration
Passivation	Surface protection film Interlayer dielectric film	Pinhole or crack Thickness variation Contamination Surface inversion	Decreased breakdown voltage Short circuit Increased leakage current
Die bonding	Chip-frame connection Eutectic die attach Epoxy die attach	Die detachment Die crack Voids in die attach Ionic contamination	Open circuit Short circuit Unstable/intermittent operation Increased thermal resistance
Wire bonding	Wire-bonding connection Wire lead	Wire bonding deviation Off-center wire bonding Damage under wire bonding contact Disconnection, loose wire Contact between wires	Open circuit Short circuit Increased resistance On–off high power stress (wire bonding lift-off)
Sealing	Resin Sealing gas	Void No sealing Water penetration Peeling Surface contamination	Open circuit Short circuit Increased resistance IDSS drift, V_{th} drift

Table 7.1 (Continued)

Failure factors		Failure mechanisms	Failure modes
Diffusion	Substrate	Crystal defect	Decreased breakdown voltage
		Insufficient airtightness Impure sealing gas Particles, hydrogen poisoning (outgassing metal in package atmosphere)	
Input/output pin	Static electricity Surge, overvoltage Over current RF overdrive	Diffusion junction breakdown Oxide film damage Metallization defect/destruction	Open circuit Short circuit Increased leakage current Power slump (high RF drive)
Others	Alpha particles, heavy ions, radiations High electric field	Electron–hole pair generation Surface inversion	Soft error, noise Increased leakage current Loss DC and AC characteristics

integrated circuits. The scaling of the c-MOS transistor has been the primary factor driving improvements in microprocessor performance. They include shallow source/drain extension (SDE), profiles for improved short channel effects, the use of retrograde and halo well profiles to improve leakage characteristics, and the effect of scaling the gate oxide thickness are the domains of discussion. The ability to overcome current physical technology limits such as gate oxide thickness and shallow junction formation as well as tradeoffs in circuit design is one of the most challenging areas of engineering.

7.1 Time-Dependent Dielectric Breakdown

Time-dependent dielectric breakdown (TDDB), also known as oxide breakdown, is one of the most important failure mechanisms in semiconductor technology. TDDB refers to the wear-out process in gate oxide of the transistors. TDDB as well as electromigration and hot carrier effects are the main contributors to semiconductor device wear-out. However, the exact physical mechanisms leading to TDDB are still unknown [116, 245, 246].

It is generally believed that TDDB occurs as the result of defect generation by electron tunneling through gate oxide. When the defect density reaches a critical density, a conducting path would form in the gate oxide, which changes the dielectric properties. Defects within the dielectric, usually called traps, are neutral at first but quickly become charged (positively or negatively) depending on their locations in regard to anode and cathode.

Currently, the accepted model for oxide breakdown is the percolation model in which breakdown occurs when a conduction path is formed by randomly distributed defects generated during electrical stress. It is believed that when this conduction path is formed, the soft breakdown happens. When the sudden increase in conductance happens as well as a power dissipation rate more than a certain threshold, hard breakdown occurs. The whole process is very complicated but could be explained by physics of breakdown [247].

7.1.1 Physics of Breakdown

In a MOS transistor (which is a metal oxide semiconductor system), the channel under the gate region is controlled by the voltage on the gate. The gate voltage produces an electric field across the dielectric. The electric field across the oxide has the following form [248]:

$$E = \frac{V_{OX}}{t_{OX}} \tag{7.1}$$

The shrinkage of semiconductor device features, which requires ultra-thin gate oxide, causes a high electric field across the dielectric. Electrons penetrate through the oxide as the result of the above-mentioned electric field. Fowler–Nordheim tunneling, direct tunneling, and trap-assisted tunneling are the three tunneling mechanisms allowing charge to pass through the oxide.

The Fowler–Nordheim tunneling is a quantum mechanical phenomenon in which the electron penetrates through the oxide barrier into the conduction band of the oxide. It can occur in almost any gate oxide and depends on the voltage across the gate oxide. The current density of the electrons at room temperature is given by:

$$J = C \cdot E^2 \cdot exp\left(-\frac{\beta}{E}\right) \tag{7.2}$$

where

$$C = \frac{q^3 \cdot m_0}{16\pi^2 \cdot \hbar \cdot m_{ox} \cdot \Phi_b} \tag{7.3}$$

and

$$\beta = \frac{4(2 \cdot m_{ox})^{\frac{1}{2}}}{3 \cdot q \cdot \hbar} \cdot \Phi_b^{\frac{3}{2}} \tag{7.4}$$

E is the uniform electric field, q is the electron charge, m_0 and m_{ox} are the electron mass-free space and in the oxide, respectively, $\hbar$ is Plank's constant, and Φ_b is the barrier height [249]. Direct tunneling is another type of quantum mechanical tunneling that occurs in thin or ultra-thin oxide films and depends on the thickness of the dielectric. The related current density is:

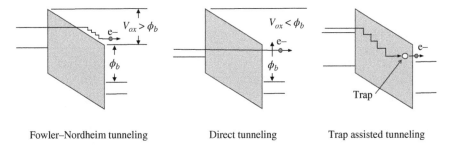

Figure 7.1 Tunneling mechanisms allowing charge to pass the oxide.

$$j = \left(C \cdot E^2\right) \cdot \left[1 - \left(\frac{\Phi_b - q \cdot V_{ox}}{\Phi_b}\right)^{\frac{1}{2}}\right]^{-2} \cdot exp\left(\frac{-\beta}{E} \cdot \frac{\Phi_b^{\frac{3}{2}} - (\Phi_b - q \cdot V_{ox})^{\frac{3}{2}}}{\Phi_b^{\frac{3}{2}}}\right)$$

(7.5)

Trap-assisted tunneling occurs when electrons tunnel into the traps in dielectric layer and then into the silicon layer. The whole process is very complicated but the current density depends on the trap density and electric field. Figure 7.1 shows the concept behind the three tunneling mechanisms [248].

One of the most popular models used in engineering is the statistical model developed by Degraeve [248, 250]. The model called the percolation takes defects as spheres that randomly affect electrical behavior of the oxide as shown in Figure 7.2; as sphere number increases, they make a bridge between the oxide edges, which causes the micro breakdown.

"In the percolation model, the last trap that finally completed the conduction path triggered the soft breakdown. Before this last event, conduction through the oxide does happen, just not at as high a level. Traps that facilitate the pre-breakdown conduction are well known to be reversible, causing the current level to switch up and down. On the other hand, once soft breakdown occurs, it does not switch back to the nonbroken state" [247].

In general, electric field in transistors causes dielectric degradation and conductive path formation in dielectric material, which somehow connects the anode and cathode. Then, continuous stress of electric field on the gate oxide leads to the thermal runaway through the breakdown path (soft breakdown) or energy dissipation (hard breakdown). The oxide breakdown leads to the gate current increase. The whole process could be modeled physically and mathematically; the models give the lifetime function of the device.

7.1.2 Early Models for Dielectric Breakdown

During electric stress, the time to dielectric breakdown depends on electric stress parameters such as the electric field, temperature, and the total area of the dielectric film. From the 1970s until 2000, several models were suggested for the time to break down in dielectric layers.

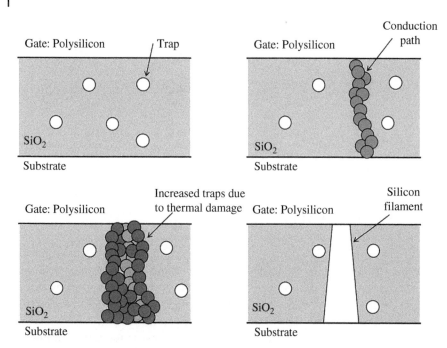

Figure 7.2 Formation of the traps in the dielectric, creation of conduction path, increased traps, and finally the cross section after hard breakdown.

The intrinsic failure, which occurs in defect-free oxide, is modeled in four forms:

- Bandgap ionization model happens in thick oxides; it occurs when the electron energy reaches the oxide bandgap and causes electron–hole pairs.
- Anode hole injection ($1/E$) model happens when the electron injected from cathode gate gets enough energy to ionize the atoms and create hot holes; some of the holes tunnel back to the cathode and either create traps in the oxide or increase the cathode field and eventually the sudden oxide breakdown. The time to breakdown, t_{BD}, has a reciprocal electric field dependence ($1/E$) from the Fowler–Nordheim electron tunneling current:

$$t_D = t_0 \cdot exp\left(\frac{G}{E_{OX}}\right) \cdot exp\left(\frac{E_A}{k_B \cdot T}\right) \tag{7.6}$$

where E_{OX} is the electric field across the oxide, G the field acceleration parameter (~350 MV/cm with a weak temperature dependence) and t_0 is a constant.

- Thermochemical (E) model which relates the defect generation to the electrical field. The applied field interacts with the dipoles and causes the oxygen vacancies and hence the oxide breakdown. The lifetime function has the following form:

$$t_{BD} = t_0 \cdot exp(-\gamma \cdot E_{OX}) \cdot exp\left(\frac{E_A}{k_B \cdot T}\right) \tag{7.7}$$

where t_0 and γ are constant.

- Anode–Hydrogen release model happens when electrons at anode release hydrogen which diffuses through the oxide and generates electron traps.

For thick dielectrics, the dielectric field is an important parameter controlling the breakdown process, while temperature dependence of dielectric breakdown is another key point. The E and $1/E$ models can only fit a part of the electric field, as it is shown in Chapter 4, figure 4.6. There are some articles trying to unify both models as well [120–122]. However, both of those models are not applicable to the ultra-thin oxide layers. For ultra-thin oxide layers (between 2 and 5 nm), other models are used.

7.1.3 Acceleration Factors

Acceleration factors could be given when a kind of stress similar to use stress applied to the device in a very short period of time results in the device failure. Failure may occur because of mechanical or electrical mechanisms. The device lifetime would change with the change of the operating parameters. "This change is quantified as the acceleration factor and is defined as the ratio of the measured failure rate of the device at one stress condition to the measured failure rate of identical devices stressed at another condition" [116]:

$$AF = \frac{MTTF_{use}}{MTTF_{stress}} = \frac{t_{BD_{use}}}{t_{BD_{stress}}} \tag{7.8}$$

where MTTF is mean time to failure of the device.

For thick dielectrics, the TDDB lifetime models are based on exponential law for field or voltage acceleration and Arrhenius law for temperature.

The acceleration factor of E model TDDB is calculated by [116]:

$$AF_{TDDB} = exp[\gamma \cdot (E_{OX} - E_{OXstress})] \cdot exp\left[\frac{E_{A_{TDDB}}}{k_B} \cdot \left(\frac{1}{T} - \frac{1}{T_{stress}}\right)\right] \tag{7.9}$$

Reliability engineering uses several oxide breakdown test configurations such as voltage ramp, current ramp, constant voltage stress test, and constant current stress test [43, 251].

7.1.4 Models for Ultra-Thin Dielectric Breakdown

The dielectric thickness of modern semiconductor devices has been steadily thinning. The early models of thick dielectric breakdown should be revised to give more accurate results for dielectric breakdown.

There are several models to describe the ultra-thin dielectric breakdown both in devices and in circuits. "Voltage-driven model for defect generation and breakdown" is a model that applies the percolation theory to find the breakdown in SiO$_2$ layers and suggests the use of a voltage-driven model instead of an electric field-driven one. It says that the tunneling electrons with energy related to the applied gate voltage are the driving force for defect generation and breakdown in ultra-thin dielectrics. According to this model, despite the thick dielectric layer's time to failure breakdown, the ultra-thin dielectric time to breakdown does not have the Arrhenius behavior anymore [123, 246].

On the other hand, it is said that the exponential law with a constant voltage acceleration factor in ultra-thin dielectrics would lead to two unphysical results:

1) projected lifetime for thinner oxides would be much longer than that of thick oxides at lower voltages and
2) the violation of the fundamental breakdown properties related to Poisson random statistics.

Therefore, the voltage dependence of time to breakdown follows a power law behavior instead of an exponential law. The interrelationship between voltage and temperature dependencies of oxide breakdown shows that the temperature activation with non-Arrhenius behavior is consistent with power law voltage dependence. This dependency for t_{BD} has the following form:

$$t_{BD} \propto V_{Gate}^{-n} \tag{7.10}$$

where n constant is driven from the experimental data. The power law voltage dependence can eliminate the two unphysical results of the exponential law as well [124].

Other experimental results support the power law model but give a wide range of numbers for n for different gate voltages [252].

Further studies show that, while there is neither microscopically information about the breakdown defect and the interaction with stress voltage and temperature, nor a complete explanation of temperature dependence of dielectric breakdown of ultra-thin dielectric layers, there are empirical relations between stress voltage and temperature:

$$\frac{V}{t_{BD}} \frac{\partial t_{BD}}{\partial V} = const = n(T) \tag{7.11}$$

which shows a power law dependence of t_{BD} on voltage and covers experimental data in this regard and

$$\frac{d}{dT} \left(\frac{1}{t_{BD}} \cdot \frac{\partial t_{BD}}{\partial V} \right) \Bigg|_{t_{BD}} = 0 \tag{7.12}$$

which shows the voltage acceleration at a fixed value of t_{BD}.

The third relation gives the t_{BD} temperature dependence for a given voltage:

$$t_{BD} = t_{BD_0}(V) \cdot exp\left(\frac{a(V)}{T} + \frac{b(V)}{T^2}\right) \tag{7.13}$$

$\dfrac{b(V)}{T^2}$ in the above relation shows the non-Arrhenius behavior of temperature [253, 254].

Further articles associate a linear relation to $n(T)$ for simplicity (which $n(T)$ less than 0):

$$n(T) = a' + b'T \tag{7.14}$$

Then the power law relationship between voltage and time to break down is:

$$t_{BD} \propto \left(V_{Gate\ to\ Source}\right)^{a' + b' \cdot T} \tag{7.15}$$

Since according to Weibull distribution, the time to failure has the following relation with cumulative failure probability (F):

$$t_{BD} = \alpha\left[ln\ \frac{1}{1-F}\right]^{\frac{1}{\beta}} \tag{7.16}$$

One can get the following approximation:

$$t_{BD} \propto F^{\frac{1}{\beta}} \tag{7.17}$$

Considering the lifetime dielectric breakdown as a function of gate oxide area as:

$$t_{BD} \propto \left(\frac{1}{W}\right)^{\frac{1}{\beta}} \tag{7.18}$$

where W is the channel width and L is the channel length and taking the temperature dependence as:

$$t_{BD} \propto exp\left[\frac{a(V)}{T} + \frac{b(V)}{T^2}\right] \tag{7.19}$$

the time-dependent breakdown becomes a function of the following terms:

$$t_{BD} = A \cdot \left(\frac{1}{W \cdot L}\right)^{\frac{1}{\beta}} \cdot F^{\frac{1}{\beta}} \cdot \left(V_{Gate\ to\ Source}\right)^{a' + b'T} \cdot exp\left[\frac{a(V)}{T} + \frac{b(V)}{T^2}\right] \tag{7.20}$$

where β, a, b, a', and b' can be driven according to the experimental data [254].

One can use generalized Weibull distribution to get a better adjustment of experimental and simulated data in comparison with the common Weibull one, which leads to improved t_{BD} extrapolation and area scaling [255].

7.1.5 Statistical Model

The statistics of gate oxide breakdown are usually described using the Weibull distribution:

$$F(t) = 1 - exp\left[-(t/\alpha)^{\beta}\right] \tag{7.21}$$

where α is the scale parameter and β is the shape parameter. Weibull distribution is an extreme-value distribution in $ln(x)$ and is a "weakest link"-type of problem. Here F is the cumulative failure probability, α is the scale parameter, and β is the shape parameter. The "weakest link" model was formulated by Sune et al. and described oxide breakdown and defect generation via a Poisson process [149]. In this model, a capacitor is divided into a large number of small cells. It is assumed that during oxide stressing, neutral electron traps are generated at random positions on the capacitor area. The number of traps in each cell is counted, and at the moment that the number of traps in one cell reaches a critical value, breakdown will occur. Dumin and Maddux [156] incorporated this model to describe failure distributions in thin oxides. Figure 7.3 shows from a collection of different studies how the thickness of the oxide is directly proportional to the β parameter. As the oxide increases in thickness, the Weibull slope increases.

It can be found that β is approaching one as T_{ox} is near 1 nm, which means Weibull distribution will become exponential distribution. Log-normal distribution has also been used to analyze accelerated test data of dielectric breakdown.

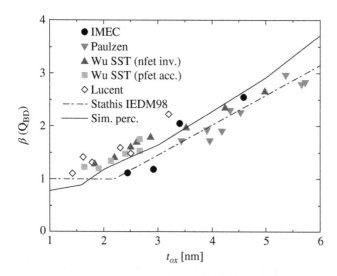

Figure 7.3 A collection of studies showing the comparison of results received on different oxide widths to Weibull slopes. *Source:* Adapted from Ghetti [256].

Although it may fit failure data over a limited sample set, it has been demonstrated that the Weibull distribution more accurately fits large samples of TDDB failures. An important disadvantage of log-normal distribution is that it does not predict the observed area dependence of TDDB for ultra-thin gate oxides [257].

7.2 Hot Carrier Injection

Hot carriers in the semiconductor device are the cause of a distinct wear-out mechanism, the hot carrier injection (HCI). Hot carriers are produced when the source-drain current flowing through the channel attains high energy beyond the lattice temperature. Some of these hot carriers gain enough energy to be injected into the gate oxide, resulting in charge trap and interface state generation. The latter may lead to shifts in the performance characteristics of the device, e.g. the threshold voltage, transconductance, or saturation current, and eventually to its degradation. The rate of HCI is directly related to the channel length, oxide thickness, and the operating voltage of the device. Since the latter are minimized for optimal performance, the scaling has not kept pace with the reduction in channel length. Current densities have been increased with a corresponding increase in device susceptibility to hot carrier effects.

Hot carriers are generated during the operation of semiconductor devices as they switch states. As carriers travel through the channel from source to drain, the lateral electric field near the drain junction causes carriers to become hot. A small part of these hot carriers gains sufficient energy – higher than the Si–SiO$_2$ energy barrier of about 3.7 eV– to be injected into the gate oxide. In n-MOS (negative-channel metal-oxide semiconductor) devices, hot electrons are generated while hot holes are produced in p-MOS (positive-channel metal-oxide semiconductor) devices. Injection of either carrier results in three primary types of damage: trapping of electrons or holes in pre-existing traps, generation of new traps, and generation of interface traps. These traps may be classified by location while their effects vary.

Interface traps are located at or near the Si–SiO$_2$ interface and directly affect transconductance, leakage current, and noise level. Oxide traps are located further away from the interface and affect the long-term MOSFET stability, specifically the threshold voltage. Effects of defect generation include threshold voltage shifts, transconductance degradation, and drain current reduction. NBTI seems to have similar degradation patterns, except for p-MOS.

Hu et al. proposed the "lucky" electron model for hot carrier effects [217]. This is a probabilistic model proposing that a carrier must first gain enough kinetic energy to become "hot," and then the carrier momentum must become redirected perpendicularly so the carrier can enter the oxide. The current across the gate is denoted by I_{gate} and during normal operation its value is negligible. Degradation due to hot carriers is proportional to I_{gate}, making the latter a good monitor of the former. As electrons flow in the channel, some scattering of the electrons in the lattice of the silicon substrate occurs due to interface states and fixed charges (interface defects). As electron scattering increases, the mobility of the hot carriers is reduced, thereby reducing the current flowing through the channel. Over a period of time, hot carriers degrade the silicon bonds with an attendant increase in electron scattering due to an increase in interface and bulk defects. As a result, the transistor slows down over a period of time. The hot carrier lifetime constraints in an n-MOS device limit the current drive that can be used in a given technology. By improving the hot carrier lifetime, the current drive can be increased, thereby increasing the operating speed of a device, such as a microprocessor [43, 151, 217, 218, 258, 259].

7.2.1 Hot Carrier Effects

HCI is the second major oxide failure mechanism that occurs when the transistor electric field at the drain-to-channel depletion region is too high. This leads to the HCI effects that can change circuit timing and high-frequency performance. HCI rarely leads to catastrophic failure but the typical parameters affected are: I_{dsat}, G_m, and V_{th}, weak inversion subthreshold slope, and increased gate-induced drain leakage.

HCI happens if the power supply voltage is higher than needed for the design, the effective channel length is too short, there is a poor oxide interface or poorly designed drain substrate junction, or over-voltage accidentally occurs on the power rail. The horizontal electric field in the channel gives kinetic energy to the free electrons moving from the inverted portion of the channel to the drain; when the kinetic energy is high enough, electrons strike Si atoms around the drain substrate interface, causing impact ionization. Electron–hole pairs are produced in the drain region and scattered. Some carriers go into the substrate, causing an increase in substrate current and the small fraction has enough energy to cross the oxide barrier and cause damage. A possible mechanism is that a hot electron breaks a hydrogen silicon bond at the Si–SiO$_2$ interface. If the Si and hydrogen recombine, then no interface trap is created. If the hydrogen diffuses away, then an interface trap is created.

Once the hot carrier enters the oxide, the vertical oxide field determines how deeply the charge will go. If the drain voltage is positive with respect to the gate voltage, then holes entering the oxide near the drain are accelerated deeper into the oxide and the electrons in the same region will be retarded from living the oxide interface. Electric field in the channel restricts the damage to oxide over drain substrate depletion region, with only a small amount of damage, just outside the depletion layer [260].

In sub-micron range electronic devices, one of the major reliability problems is hot carrier degradation. This problem is related to the continuous increase of the electrical fields in both oxide and silicon. Under the influence of the high lateral fields in short channel MOSFETs, electrons and holes in the channel and pinch-off regions of the transistor can gain sufficient energy to overcome the energy barrier or tunnel into the oxide. This leads to injection of a gate current into the oxide, and subsequently to the generation of traps, both at the interface and in the oxide, and to electron and hole trapping in the oxide, which will cause changes in transconductance, threshold voltage, and drive currents of the MOSFET [258, 261].

7.2.2 Hot Carrier Generation Mechanism and Injection to the Gate Oxide Film

There are different types of hot carrier generation mechanisms:

1) Channel hot electron (CHE) injection occurs when both the gate voltage and the drain voltage are significantly higher than the source voltage, with $V_G \approx V_D$. Channel carriers that travel from the source to the drain are sometimes driven toward the gate oxide even before they reach the drain because of the high gate voltage (Figure 7.4).
2) Drain avalanche hot carrier (DAHC) injection produces the worst device degradation under normal operating temperature range. This occurs when a high voltage is applied at the drain under non-saturated conditions ($V_D > V_G$) and results in very high electric fields near the drain, which accelerate channel carriers into the drain's depletion region. Studies have shown that the worst effects occur when $V_D = 2V_G$. The acceleration of the channel carriers causes them to collide with Si lattice atoms, creating dislodged electron–hole pairs in the process. This phenomenon is known as impact ionization, with some of the displaced e–h pairs also gaining enough energy to overcome the electric potential barrier between the silicon substrate and the gate oxide. Under the

Figure 7.4 CHE injection involves propelling of carriers in the channel toward the oxide even before they reach the drain area.

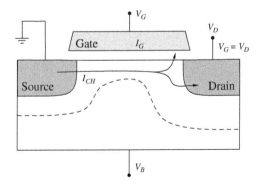

influence of drain-to-gate field, hot carriers that surmount the substrate-gate oxide barrier get injected into the gate oxide layer where they are sometimes trapped. This HCI process occurs mainly in a narrow injection zone at the drain end of the device where the lateral field is at its maximum. Hot carriers can be trapped at the Si–SiO_2 interface (hence referred to as "interface states") or within the oxide itself, forming a space charge (volume charge) that increases over time as more charges are trapped. These trapped charges shift some of the characteristics of the device, such as its threshold voltage (V_{th}) and its conveyed conductance (g_m). Injected carriers which have not been trapped in the gate oxide become gate current. On the other hand, the majority of the holes from the e–h pairs generated by impact ionization flow back to the substrate, and comprise a large portion of the substrate's drift current. Excessive substrate current may therefore be an indication of hot carrier degradation. In gross cases, abnormally high substrate current can upset the balance of carrier flow and facilitate latch-up (Figure 7.5).

3) Substrate hot electron (SHE) injection occurs when the substrate back bias is very positive or very negative, i.e. $|V_B| \gg 0$. Under this condition, carriers of one type in the substrate are driven by the substrate field toward the Si–SiO_2 interface. As they move toward the substrate-oxide interface, they further gain kinetic energy from the high field in surface depletion region. They eventually overcome the surface energy barrier and get injected into the gate oxide, where some of them are trapped (Figure 7.6).

4) Secondary generated hot electron (SGHE) injection involves the generation of hot carriers from impact ionization involving a secondary carrier that was likewise created by an earlier incident of impact ionization. This occurs under conditions similar to DAHC, i.e. the applied voltage at the drain is high or $V_D > V_G$, which is the driving condition for impact ionization. The main difference, however, is the influence of the substrate's back bias in the hot carrier generation. This back bias results in a field that tends to drive the hot carriers generated by the secondary carriers toward the surface region, where they further gain kinetic energy to overcome the surface energy barrier [34] (Figure 7.7).

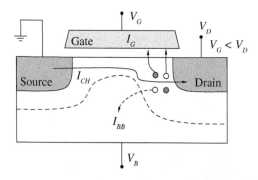

Figure 7.5 DAHC injection involves impact ionization of carriers near the drain area.

Figure 7.6 SHE injection involves the trapping of carriers from the substrate.

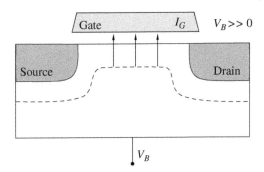

Figure 7.7 SGHE injection involves hot carriers generated by secondary carriers.

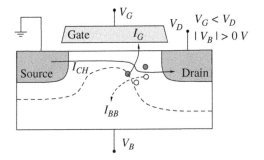

In the case of CHE, electrons flowing from a source are accelerated by the high electric field near a drain, and of these, the "lucky electrons" which do not experience energy-dissipating impacts are injected into the gate oxide film. This mechanism occurs most easily when the gate voltage and the drain voltage are equal ($V_{GS} = V_{DS}$). And in the case of DAHC, electrons flowing from a source undergo impact ionization due to the high electric field near a drain, and generate electron–hole pairs. Of these, the electron or the hole, whichever has the higher energy, is injected into the gate oxide film.

7.2.3 Hot Carrier Models

Lucky electron approach is a widely used model to explain the conduction mechanism of hot carriers to the insulator layer. Channel electrons could reach the gate oxide by gaining sufficient energy from the channel field (getting "hot") and then redirecting their momentum perpendicularly (the Cancannon model is similar to the lucky electron model, but it assumes a non-Maxwellian high energy tail on the distribution function and requires the solution of energy balance equation for carrier temperature).

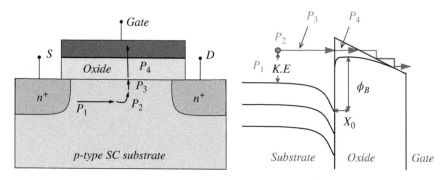

Figure 7.8 Lucky electron model.

As depicted in Figure 7.8, P_1 is probability that the electron gains sufficient energy from the electric field to overcome the potential barrier; P_2 is probability of redirecting collision to occur, to send the electron toward the Si/insulator interface; P_3 is the probability that the electron travels toward the interface without losing energy; P_4 is probability that electron will not scatter in the image potential well. The various probabilities are calculated using [262]:

$$P_1 = \frac{1}{\lambda \cdot E_x} exp\left(-\frac{\varepsilon}{\lambda \cdot E_x}\right) \cdot d\varepsilon \tag{7.22}$$

$$P_2 = \frac{1}{2 \cdot \lambda_r}\left(1 - \sqrt{\frac{\varphi_B}{\varepsilon}}\right) \tag{7.23}$$

$$P_3 = exp\left(-\frac{y}{\lambda}\right) \tag{7.24}$$

$$P_4 = exp\left(\frac{x_0}{\lambda_{OX}}\right) \tag{7.25}$$

where λ is scattering mean-free path, λ_r is redirection mean-free path, and Φ_B is barrier height at the Si–Oxide interface,

$$\Phi_B = \Phi_{B0} - \alpha \cdot E_{ox}^{1/2} - \beta \cdot E_{ox}^{2/3} \tag{7.26}$$

Φ_{B0} is the zero-field barrier height, $\alpha \cdot E_{0x}^{1/2}$ is barrier lowering due to image potential, $\beta \cdot E_{0x}^{2/3}$ accounts for probability of tunneling, λ_{0x} is the mean free path in the oxide (3.2 nm).

The total gate current is then given by:

$$I_g = \int dxdy \int_{\varphi_B}^{\infty} d\varepsilon J_n(x,y)P_1P_2P_3P_4 \tag{7.27}$$

Lucky electron model, in its simplest form, can predict the relationships between the substrate current I_{Sub} and gate current I_G as:

$$I_{Sub} \propto I_D \cdot exp(-\varphi_i/E^*) \qquad (7.28)$$

$$I_G \propto B(E_{OX}) \cdot I_D \cdot exp(-\varphi_b/E^*) \qquad (7.29)$$

$$I_G/I_D \propto (I_{Sub}/I_D)^{\varphi_b/\varphi_i} \qquad (7.30)$$

where $E^* = q \cdot \lambda \cdot E_L$ for lucky electron model and $E^* = k_B \cdot T_e$ for effective temperature model. The simple semi-empirical model for dc degradation of n-MOSFET could be obtained from [121, 217]:

$$\tau \propto \frac{W}{I_D} \cdot \left(\frac{I_{Sub}}{I_D}\right)^{-\frac{\varphi_{it}}{\varphi_i}} \qquad (7.31)$$

In the p-MOSFET, the gate current or the total injected charge is the determining parameter used in modeling instead of the substrate current:

$$\tau \propto \left(\frac{I_G}{W}\right)^{-\frac{\varphi_{it}}{\varphi_i}} \qquad (7.32)$$

7.2.4 Hot Carrier Degradation

Hot carrier degradation is normally limited around the drain junction because of the maximum electric field near the drain. The hot carriers injected into the oxide can get trapped or generate neutral traps in the oxide bulk or produce interface states. Hot holes and hot electrons are the bases of hot carrier degradation which lead to similar degradation mechanisms which are different in magnitudes. A part of the degradation happens under uniform HCI. Under the effect of electron injection, three kinds of damages occur:

1) Electron trapping happens when the electron injected into the gate oxide gets trapped in the oxide bulk-existing traps.
2) Electron trap generation happens at higher oxide field; the electron that got enough energy in the oxide field creates and fills additional electron traps.
3) Electron interface trap generation is a highly temperature-activated process. At the activation temperature, the interface trap-generation rate has an exponential relation with the field over a wide range of field oxide.

Under the same conditions for hole, three other kinds of damages would happen:

4) Hole trapping occurs when the holes injected into the gate oxide get trapped in the oxide bulk-existing traps. The only difference between this process and that of the electrons is the high trapping efficiency of the holes. At lower temperatures (such as 77 K), the hole trapping increases with an efficiency of up to 70% because of a larger effective capture cross section at this temperature.

5) Hole trap generation occurs at large injected hole density when a clear saturation level of the trapping can be seen; it is independent of the oxide field during the injection. The absence of additional hole trap generation as the result of hole injection is just the opposite case of electron trap generation.

6) Hole interface trap generation is a process similar to the electrons; however, hole interface trap generation is more efficient with holes than electrons.

Holes are a few degrees of magnitude more effective at producing interface states than electrons and their trapping rate is very higher than those of electrons. The interface states are mostly produced when a Si—H bond is broken in the interface and electron or hole traps are due to the dangling bonds in the bulk of the oxide. There is a strong dependence of these bulks and interfacial traps on the gate oxidation and subsequent process conditions, such as moisture, H contents of passivation layers and plasma-induced damage [263].

A part of the degradation happens under real operation conditions. There are different degradation mechanisms that get active under stress conditions. Table 7.2 shows the current change and the type of damage generated for n-MOS, p-MOS, and LDD n-MOS [261].

In n-MOS devices, the maximum degradation occurs in the medium voltage range and is caused by mobility degradation because of interface trap generation. At low gate voltages, the current increases due to the efficient trapping of holes, leading to a channel-shortening effect; then interface traps and neutral electron traps are created as well, but the trap holes have more impact on the current. Their influence might become visible, however, after neutralization of these holes, and filling of the neutral traps [261, 263].

The change of the drain current in p-MOSFETs is similar to the hole trapping in n-MOSFET. Electron injection and trapping increase the drain current by channel

Table 7.2 Degradation mechanism for three types of MOSFETs and three stress gate voltage ranges.

V_G range	n-MOS	n-MOS LDD	p-MOS
Low V_G	$I_D \uparrow$	$I_D \uparrow$	$I_D \uparrow$
	h-trapping	h-trapping	e-trapping
	D_{it}-creation	D_{it}-creation	D_{it}-creation
Medium V_G	I_D	$I_D \downarrow\downarrow$	$I_D \uparrow\uparrow$
	D_{it}-creation	e-trapping	e-trapping
		D_{it}-creation	D_{it}-creation
High V_G	$I_D \downarrow$	$I_D \downarrow$	$I_D \downarrow$
	e-trapping	e-trapping	D_{it}-creation

shortening, which has more impact than the generated interface traps. However, this electron trapping occurs for almost all stress gate voltages because of the hotter electrons (than holes) in the same electric field and the smaller electron energy barriers.

The electrons (and holes) that have enough energy to be injected into traps in the gate oxide create interface traps affect the device operation directly. They lower the gate oxide field and result in changes in different parameters in the device such as shift in threshold voltage (V_{th}), which can degrade high frequencies witching and create mismatches in balanced signal path stages. Most semiconductor failure modes exhibit higher degradation rates at higher temperatures. The hot carrier effect, however, accelerates as the temperature decreases [218]. They charge, sub-threshold slope, lowering the transconductance (G_m) and lowering of drain current in the field effect transistor linear region ($I_{D_{lin}}$) and saturation region ($I_{D_{sat}}$) and the result is the lowering of driving current of the device. There are other stress conditions such as high oxide fields, Fowler–Nordheim injection, and radiation exposure that can affect the device severely. But what differentiates the hot carrier-induced damages from the others is the damage caused by HCI is highly localized [218]. Since the maximum electric field in a MOSFET is highly localized close to the drain, the hot carrier degradation of the gate oxide is limited to the ~0.1 μm region around the drain junction and because of that makes the device parameter highly asymmetric when measured at high drain voltage (V_D). Here, we will not discuss the microscopic nature of the oxide degradation but it is enough to say that interface states are probably produced when a Si—H bond is Brocken at the interface, and electrons or hole trap are, among others < due to dangling bonds in the bond s in the bulk of the oxide [263].

The lifetime models for n-MOSFET and p-MOSFET are different but, in general, the time evolution of a selected device parameter shift $Y(t)$ (excluding the saturation effects) follows either a power law:

$$|Y(t))| = C \cdot t^N \tag{7.33}$$

or a log of time dependence:

$$|Y(t)| = C \cdot log(t) \tag{7.34}$$

Typical n-MOSFET HC lifetime models for a given device geometry are given by:

a) Drain-source voltage (V_{DS}) model:

$$t_{target} = t_0 \cdot exp\left(\frac{B}{V_{DS_{use}}}\right) \tag{7.35}$$

b) Substrate current model:

$$t_{target} = C \cdot \left(\frac{I_{B_{use}}}{W}\right)^{-B} \tag{7.36}$$

c) Substrate/drain current ratio model:

$$t_{target} = \frac{H \cdot W}{I_{use}} \cdot \left(\frac{I_{B_{use}}}{I_{C_{use}}}\right)^{-M}$$

(7.37)

For p-MOSFET HC lifetime models are given by:

a) Drain-source voltage (V_{DS}) model:

$$t_{target} = t_0 \cdot exp\left(\frac{B}{V_{DS_{use}}}\right)$$

(7.38)

b) Gate current method:

$$t_{target} = C \cdot \left(\frac{I_{G_{use}}}{W}\right)^{-B}$$

(7.39)

t_{target} is the time needed to reach a given target degradation level at use conditions. In ac operation, some duty cycle assumptions at worst-case operation may be needed for a correct lifetime projection. *HC* model parameters *C*, t_0, and *HW* may depend on L_{EFF} (W_{EFF} is applicable) [264].

7.2.5 Hot Carrier Resistant Structures

The optimum design to minimize hot carrier effects is the best solution for hot carrier problems. Common design techniques for preventing hot carrier effects include:

1) increase in channel lengths;
2) $n+/n-$ double diffusion of sources and drains;
3) use of graded drain junctions;
4) introduction of self-aligned $n-$ regions between the channel and the $n+$ junctions to create an offset gate; and
5) use of buried $p+$ channels.

Figure 7.9 shows the cross section of some of these structures [264]:

7.2.6 Acceleration Factor

Hot carrier phenomena are accelerated by low temperature, mainly because this condition reduces charge detrapping. A simple acceleration model for hot carrier effects is as follows:

$$AF = t_{50(2)}/t_{50(1)}$$

(7.40)

$$AF = exp\left[\left(\frac{E_A}{k_B}\right) \cdot \left(\frac{1}{T_1} - \frac{1}{T_2}\right) + C \cdot (V_2 - V_1)\right]$$

(7.41)

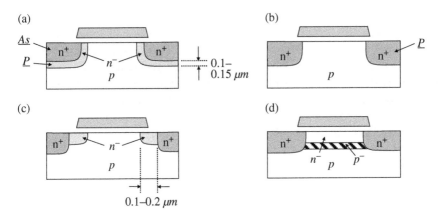

Figure 7.9 Cross section of a few of MOSFET structures: (a) As-P(n^+–n^-) double diffusion; (b) P-drain; (c) offset gate; and (d) buried channel.

where AF = acceleration factor of the mechanism; $t_{50(1)}$ = rate at which the hot carrier effects occur under conditions V_1 and T_1; $t_{50(2)}$ = rate at which the hot carrier effects occur under conditions V_2 and T_2; V_1 and V_2 = applied voltages for R_1 and R_2, respectively; T_1 and T_2 = applied temperatures (K), respectively; E_A = −0.2 to −0.06 eV; and C = a constant [98].

Typically, the assessment of devices with regard to hot carrier effect will take place at low temperatures (−20 to −70°C) due to the negative activating energy associated with this mechanism.

Temperature acceleration is often treated as a minor effect in most HCI models. However, in order to consider possible large temperature excursion, FaRBS includes temperature acceleration effect based on the HCI lifetime model as given below [218]:

$$t_f = B \cdot (I_{sub})^{-n} \cdot exp\left(\frac{E_A}{k_B \cdot T}\right) \text{ for n − type} \qquad (7.42)$$

$$t_f = B \cdot (I_{gate})^{-m} \cdot exp\left(\frac{E_A}{k_B \cdot T}\right) \text{ for p − type} \qquad (7.43)$$

The combination of temperature effect produces a more complicated HCI lifetime model as below [252]:

$$t_f = A_{HCI}\left(\frac{I_{sub}}{W}\right)^{-n} exp\left(\frac{E_{A_{HCI}}}{k_B \cdot T}\right) \qquad (7.44)$$

Then it should be possible to calculate the acceleration factor from the above equation.

7.2.6.1 Statistical Models for HCI Lifetime

There is little discussion in literature about a proper statistical lifetime distribution model for HCI. However, Weibull failure statistics [151, 258] for failure rate and log-normal distribution for HCI lifetime [265] are reported as well. A logical hypothesis for the lifetime distribution would be the exponential one. This is a good assumption because as a device becomes more complex, with millions of gates, it may be considered a system. The failure probability of each individual gate is not most likely an exponential distribution. However, the cumulative effect of early failures and process variability, ensuring each gate has a different failure rate, widens the spread of the device failures. The end result is that intrinsic HCI becomes statistically more random as the failures occur at a constant rate.

The effect of process variation on hot carrier reliability characteristics of MOS-FET is another field of research, as well. A significant variation of hot carrier lifetime across the wafers due to gate length variation is reported. Since the nonuniformity of gate length is always present, it is necessary to consider enough margin of hot carrier lifetime and the accurate evaluation of gate length variation for optimizing device performance should be considered [258].

7.2.6.2 Lifetime Sensitivity

HCI lifetime is sensitive to changes in the input parameters. The acceleration factor for HCI is:

$$AF_{HCD} = exp[B \cdot (1/V_{dd} - 1/V_{dd,max})] \tag{7.45}$$

HCI continues to be a reliability concern as device feature sizes shrink. HCI is a function of internal electric fields in the device and, as such, is affected by channel length, oxide thickness, and device operating voltage. Shorter channel lengths decrease reliability but the oxide thickness and the voltage may also be reduced to help alleviate the reduction in reliability. Another way of improving hot carrier reliability may be by shifting the position of the maximum drain so it is deeper in the channel. This would result in hot carriers being generated further away from the gate and Si–SiO$_2$ interface, reducing the likelihood of injection into the gate. Another method is to reduce the substrate current by using a lightly doped drain (LDD) where part of the voltage drop is across an LDD extension not covered by the gate. Annealing the oxides in NH$_3$, N$_2$O, or NO or growing them directly in N$_2$O or NO improves their resistance to interface state generation by the hot carrier.

7.3 Negative Bias Temperature Instability

Negative bias temperature instability (NBTI) happens to p-MOS devices under negative gate voltages at elevated temperatures. The degradation of device performance, mainly manifested as the absolute threshold voltage V_{th} increase and mobility, transconductance, and drain current t I_{dsat} decrease, is a big reliability concern for today's ultra-thin gate oxide devices [164]. Deal [165] named it "Drift VI" and discussed the origin of the study of oxide surface charges. Goetzberger et al. [166] investigated surface state change under combined bias and temperature stress through experiments that utilized MOS structures formed by a variety of oxidizing, annealing, and metalizing procedures. They found an interface trap density D_{it} peak in the lower half of the band gap and p-type substrates gave higher D_{it} than n-type substrates. The higher the initial D_{it}, the higher the final stress-induced D_{it}. Jeppson and Svensson [167] first proposed a physical model to explain the surface trap growth of MOS devices subjected to negative bias stress. The surface trap growth was described as diffusion controlled at low fields and tunneling limited at height fields. The power law relationship ($t^{1/4}$) was also proposed for the first time. The study of NBTI has been very active in recent years since the interface trap density induced by NBTI increases with decreasing oxide thickness which means NBTI is more severe to ultra-thin oxide devices. New developments in NBTI modeling and surface trap analysis have been reported in recent years. At the same time, effects of various process parameters on NBTI have been studied in order to minimize the NBTI. Schroder and Babcock [102] reviewed pre-2003 experimental results and various proposed physical models together with the effects of manufacturing process parameters. Detailed latest reviews can also be found in Refs. [71, 72, 102, 200, 266]. In this section, the up-to-date research discoveries of NBTI failure mechanism, models, and related parameters will be briefly discussed.

7.3.1 Physics of Failure

Silicon dioxide, the critical component of silicon devices, serves as insulation and passivation layer and is never completely electrically neutral. Mobile ionic charges, oxide trapped electrons or holes, fabrication-process-induced fixed charges, and interface-trapped charges are four main categories of charges inside oxide and at the silicon-oxide interface. The electrical characteristics of a silicon device are very sensitive to the density and properties of those charges. As already known, the threshold voltage of p-MOSFET is given by:

$$V_{th} = V_{FB} - 2\varphi_B - |Q_B|/C_{ox} \tag{7.46}$$

where

$$\varphi_B = (k_B \cdot T/q) \cdot ln(N_D/n_i), \quad |Q_B| = (4 \cdot q \cdot \varepsilon_{Si} \cdot \varphi_B \cdot N_D)^{1/2} \tag{7.47}$$

and C_{ox} is the oxide capacitance per unit area. The flat band voltage V_{FB} is given by:

$$V_{FB} = \varphi_{MS} - \frac{Q_f}{C_{ox}} - \frac{Q_{it}(\varphi_s)}{C_{ox}} \tag{7.48}$$

where Q_f is the fixed charge and Q_{it} the interface trapped charge. From previous equations, it can be found that the only parameters to change the threshold voltage are the Q_f and Q_{it}. Most research works on NBTI failure mechanism have been focused on the generations of Q_{it} and Q_f [212].

7.3.2 Interface Trap Generation: Reaction–Diffusion Model

The reaction–diffusion (R–D) model is the most prevalent among various proposed NBTI models. Jeppson and Svensson [167] were the first to propose the R–D model to explain the generation of interface states at low fields. In this model, it is assumed that the silicon interface contains a large number of defects that are electrically inactive and can be activated through chemical reactions like this:

$$(Surface\ Defect) \leftrightarrow (Surface\ Trap) + (Surface\ Charge)^+ + X_{interface} + e^-$$

$$X_{interface} \leftrightarrow X_{bulk}$$

$X_{inteface}$ is a diffusing species which is formed at the interface in the reaction. Based on the infrared measurements report which showed large numbers of Si—H groups existing in bulk silicon and probably also at the interface, Jeppson et al. proposed this reaction

$$Si_3 \equiv SiH + O_3 \equiv SiOSi \equiv O_3 \leftrightarrow Si_3 \equiv Si \cdot + O_3 \equiv Si^+ + O_3 \equiv SiOH + e^-$$

where $Si_3 \equiv SiH \cdot$ is the surface defect, $Si_3 \equiv Si \cdot$ is the surface trap, $O_3 \equiv Si^+$ is the oxide charge, and $O_3 \equiv SiOH$ is the diffusing X.

When the defect is activated, the H of SiH bond is released by some dissociation mechanisms and reacts with the SiO_2 lattice to form an OH group bonded to an oxide atom, leaving a trivalent Si atom in the oxide to form a fixed charge and one trivalent Si atom at the Si surface to form an interface trap. This chemical reaction is schematically shown in Figure 7.10. The $N_{it} \sim t^{1/4}$ relationship was observed and mathematically proved by assuming that the process is diffusion-limited rather than reaction-rate-limited.

Various mechanisms have been proposed for the dissociation process. Ogawa and Shiono [267] listed three of those:

1) High-electric field dissociation

$$Si_3 \equiv SiH \rightarrow Si_3 \equiv Si \cdot + H_i$$

where H_i, the neutral species X, is an interstitial hydrogen atom.

2) Interstitial atomic hydrogen attack

$$Si_3 \equiv SiH + H_i \rightarrow Si_3 \equiv Si \cdot + H_2$$

molecular hydrogen H_2 is the species X.

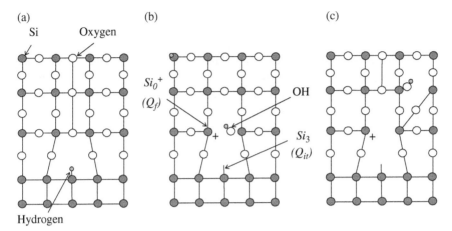

Figure 7.10 Schematic two-dimensional representation of the Si–SiO$_2$ interface, showing: (a) the ≡Si defect, (b) how this defect may be electrically activated during NBTI to form an interface trap, a fixed oxide charge, and a hydroxyl group, and (c) the OH diffuses through the oxide. *Source:* Adapted from Jeppson and Svensson [167].

3) Dissociation involves holes

$$Si_3 \equiv SiH + h^+ \rightarrow Si_3 \equiv Si \cdot + H^+$$

The actual diffusing species X have not yet been identified. Possibilities include interstitial atomic hydrogen (H_i) [267], molecular hydrogen (H_2), hydroxyl (OH) group, and proton (H^+). Rashkeev et al. [268] did first-principle calculations to show that the proton is the only stable charge state of H at the Si–SiO$_2$ interface. The protons can react directly with SiH to form H_2 and leave behind positively charged dangling bonds. Alam and Mahapatra [71] proposed that the H is released as atomic H, then converted to and diffused as molecular H_2, and the measurement delay is the main reason for various diffusion species' observations.

7.3.3 Fixed Charge Generation

The fixed charge Q_f is a positive charge in the oxide and near the Si–SiO$_2$ interface. It cannot be charged or discharged by varying the silicon surface potential. It is primarily due to excess silicon species introduced during oxidation and during post-oxidation heat treatment [269]. Negative bias stress can also increase its value like the generation of interface trap.

Ogawa and Shiono [267] determine fixed oxide charge densities from capacitance-voltage measurements and interface trap densities from conductance measurements of MOS capacitors under low field stress (-1.6 to -5.0 MV/cm). The formulated expressions for N_f and N_{it}:

$$\Delta N_f(E_{ox}, T, t) = A_1 \cdot E_{ox}^{1.5} \cdot t^{0.14} \cdot exp(-0.15/k_B T) \tag{7.49}$$

$$\Delta N_{it}(E_{ox}, T, t, t_{ox}) = A_2 \cdot E_{ox}^{1.5} \cdot t^{0.25} \exp(-0.2/k_B T)/t_{ox} \tag{7.50}$$

where A_1 and A_2 are two constants independent of E_{ox}, T, and t_{ox}. The thickness of oxide in their experiments is ranged between 4.2 and 30 nm, together with an early report [269] which stated no thickness dependence of fixed charges for 40–100 nm oxides, and showed that ΔN_f is independent of oxide thickness in a wide range. ΔN_{it} is inversely proportional to oxide thickness as shown in the above second equation, which means NBTI is worse for thinner oxides.

7.3.4 Recovery and Saturation

Another important phenomenon of NBTI is the recovery of the threshold voltage shift after the negative bias stress is removed [167]. This means NBTI may have different characteristics between DC and AC operations. Abadeer et al. [122] reported a 3X increase in the magnitude of threshold voltage shift under DC operation than that of AC operation. Rangan's experiment [75] showed the recovery is independent of stress voltage, time, and temperature (under 25) and can reach 100% at 25 for gate oxides ranging from 4.5 to 15.0 nm. The mechanisms of recovery are still under investigation. One explanation is that diffusion species X moves back to the Si–SiO$_2$ interface under the influence of positive gate voltage and passivates the Si dangling bond [270]. Another interpretation is the delicate interplay between forward dissociation and reversed annealing rates during the stress and relaxation phases of AC degradation [203]. At each stress interval, the V_{th} degradation at first returns quickly and then continues to degrade more slowly. The ratio of AC to DC degradation is affected by the duty cycle. It was reported that the degradation under AC operation has little or no frequency dependence up to 500 KHz [203, 269, 270], but then decreases further above 2 MHz [122].

There were reports that indicated that NBTI shifts tend to saturate over time [71, 171, 271]. One possible reason is the reaction limitation mechanism. The generation of Si+ decreases as the number of available SiH bonds reduces with time. Another possible reason is if the diffusing species encounters a new interface at which it is not transferred across but reflected [171]. Saturation may have important implications for long-term reliability prediction [271, 272]. However, no agreement on the physical understanding behind the saturation has been reached. Alam and Mahapatra [71] proposed that saturation is an artifact of measurement delay.

7.3.5 NBTI Models

The time dependence of the threshold voltage shift ΔV_{th} is found to follow a power law model:

$$\Delta V_{th}(t) = A \cdot t^n \tag{7.51}$$

where A is a constant which depends on oxide thickness, field, and temperature.

The time exponent n is a sensitive measure of the diffusion species. The theoretical value of the exponent parameter n is 0.25 according to the solution of diffusion equations [167]. Chakravarthi et al. [171] suggested that n varies around 0.165, 0.25, and 0.5 depending on the reaction process and the type of diffusion species. According to Alam et al. [72], $n = 1 = 2$ for proton, $n = 1 = 6$ for molecular H_2, $n = 1 = 4$ for atomic H. n was also reported to change from ~0 : 25 initially (stress time ~100 s) to 0.16 at 10^6 s stress time [266]. NBTI degradation is thermally activated and sensitive to temperature. The temperature dependence of NBTI is modeled by the Arrhenius relationship. The activation energy appears to be highly sensitive to the types of potential reacting species and to the type of oxidation methods used [102]. Reported activation energies range from 0.18 to 0.84 eV [273, 274]. Improved models have been proposed after the simple power law model. Considering the temperature and gate voltage, ΔV_{th} can be expressed as:

$$\Delta V_{th}(t) = B \cdot exp(\beta \cdot V_g) \cdot exp(-E_A/k_B T) \cdot t^{0.25} \tag{7.52}$$

where B and β are constants and V_g is the applied gate voltage. Considering the effects of gate voltage and oxide field, Mahapatra et al. [275] proposed a first-order N_{it} model.

$$\Delta N_{it}(t) = K\left[C_{ox} \cdot \left(V_g - V_{th}\right)\right]^{0.5} \cdot exp(\beta \cdot E_{ox}) \cdot exp(-E_A/k_B T) \cdot t^{0.25} \tag{7.53}$$

where K is a constant.

7.3.6 Lifetime Models

NBTI failure is defined as ΔV_{th} reaches a threshold value. Based on the degradation models such as previous equations, NBTI lifetime can be represented as:

1) Field model

$$\tau = C_1 \cdot E_{ox}^{-n} \cdot exp(E_A/k_B T) \tag{7.54}$$

where C_1 is a constant.

2) Voltage model

$$\tau = C_2 \cdot exp(-\beta \cdot V_g) \cdot exp(E_A/k_B T) \tag{7.55}$$

where C_2 is a constant.

7.4 Electromigration

Electrons passing through a conductor transfer some of their momentum to its atoms. At sufficiently high electron current densities (greater than 10^5 A/cm^2 [276]), atoms may shift toward the anode side. The material depletion at the cathode side causes circuit damage due to decreased electrical conductance and eventual formation of open circuit conditions. This is caused by voids and micro-cracks, which may increase the conductor resistance as the cross-sectional area is reduced. Increased resistance alone may result in device failure, yet, the resulting increase in local current density and temperature may lead to thermal runaway and catastrophic failure [179], such as an open circuit failure. Alternatively, short circuit conditions may develop due to excess material buildup at the anode. Hillocks form where there is excess material, breaking the oxide layer, and allowing the conductor to come in contact with other device features. Other types of damage include whiskers, thinning, localized heating, and cracking of the passivation and interlevel dielectrics [277].

This diffusive process, known as electromigration is still a major reliability concern despite vast scientific research as well as electrical and materials engineering efforts. Electromigration can occur in any metal when high current densities are present. In particular, the areas of greatest concern are the thin-film metallic interconnects between device features, contacts, and vias [277].

7.4.1 Electromigration Physics

At high current densities, the force exerted by electrons on the scattering of the positively charged metal ions becomes stronger than the electrostatic pull force toward the cathode. Thus, the diffusion of the ions is biased in the direction of the electron flow, leading to electromigration.

Its effects are expected to be characteristic of the material, such that the activation energy for electromigration is dependent on the material type, the size and orientation of the grains, stress, temperature, and even the length of the conductor. Even low-concentration doping may have great impact on the EM features. As an example, the EM activation energy of bulk Al is 1.4 eV, while adding small amounts (0.3–5%) of Cu reduces this activation energy by about 0.5–0.8 eV [277] (Figure 7.11).

Grain size and pattern also have substantial impact on the effective EM activation energy of the metal. For instance, the activation energy ranges between 1 and 2 eV for thin films with large grain sizes. For very fine-grained samples, the activation energy may be as low as 0.4–0.6 eV. Thus, mass transport-induced damage is more severe at grain boundaries and is greatest where three or more grains meet.

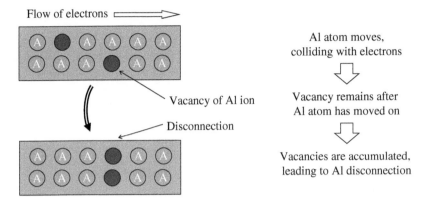

Flow of electrons

Vacancy of Al ion

Disconnection

Al atom moves, colliding with electrons

⇩

Vacancy remains after Al atom has moved on

⇩

Vacancies are accumulated, leading to Al disconnection

Figure 7.11 EM failure mechanism.

Figure 7.12 Lattice diffusion, grain boundary diffusion, and surface diffusion of polycrystalline Al.

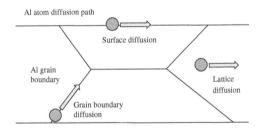

Al atom diffusion path

Surface diffusion

Al grain boundary

Grain boundary diffusion

Lattice diffusion

When small-dimension conductors are used, columnar growth of the metal lowers the grain boundary density and increases the electromigration lifetime (Figure 7.12).

Stress gradients also affect electromigration since they can induce atomic motion within the metal. Atoms migrate from regions of compressive stress to regions of tensile stress. When a conductor is shorter than a critical length, L_c, known as the "Blech Length," the stress-induced flow of atoms counters the EM driving force and EM is eliminated [278–280].

Temperature gradients, caused by high current Joule heating, also affect electromigration. While these gradients may only span a temperature change of some tens of degrees, the temperature change over a few microns results in large gradients [278]. Since EM is a thermally activated process, the temperature gradients produce flux divergences like those found at contacts or other device features.

Increasingly, low-resistivity Cu interconnects have been made use of in ICs since Cu has a lower atomic diffusivity than Al. However, the surface self-diffusion in Cu appears to be faster than grain-boundary self-diffusion. Thus, Cu does not provide the desired solution and the reliability of Cu interconnects may be improved by suppressing the interface and surface diffusion [180].

7.4.2 Lifetime Prediction

Modeling electromigration median time to failure (MTTF) from the first principles of the failure mechanism is difficult. While there are many competing models attempting to predict time to failure from first principles, there is no universally accepted model.

Currently, the favored method to predict time to failure is an approximate statistical one given by Black's equation, which describes the MTTF by:

$$MTTF = A \cdot (j_e)^{-n} \cdot exp(E_A/k_B T) \tag{7.56}$$

where j_e is the current density and E_A is the EM activation energy.

Failure times are described by the log-normal distribution [181]. The symbol A is a constant, which depends on a number of factors, including grain size, line structure and geometry, test conditions, current density, thermal history, etc. Black determined the value of n to equal 2. However, n is highly dependent on residual stress and current density [8 m] and its value is highly controversial.

A range of values for the EM activation energy, E_a, of aluminum (Al) and Al alloys are also reported. The typical value is $E_A = 0.6 \pm 0.1$ eV. The activation energy can vary due to mechanical stresses caused by thermal expansion. Introduction of 0.5% Cu in Al interconnects may result in $n = 2.63$ and activation energy of $E_A = 0.95$ eV. For multi-level Damascene Cu interconnects, an activation energy of $E_A = 0.94 \pm 0.11$ eV at a 95% confidence interval (CI) and a value of the current density exponent of $n = 2.03 \pm 0.21$ (95% CI) were found [184].

7.4.3 Lifetime Distribution Model

Traditionally, the EM lifetime has been modeled by the log-normal distribution. Most test data appear to fit the log-normal distribution, but these data are typically for the failure time of a single conductor [185].

Through the testing of over 75 000 Al(Cu) connectors, Gall et al. [185] showed that the electromigration failure mechanism does follow the log-normal distribution. This is valid for the TTF of the first link with the assumption that the first link failure will result in device failure. The limitation is that a log-normal distribution is not scalable. A device with different numbers of links will fail with a different log-normal distribution. Thus, a measured failure distribution is valid only for the device on which it is measured. Gall et al. also showed that the Weibull (and thus the exponential) distribution is not a valid model for electromigration. Even though the log-normal distribution is the best fit for predicting the failure of an individual device due to EM, the exponential model is still applicable for modeling EM failure in a system of many devices where the reliability is determined by the first failure of the system.

7.4.4 Lifetime Sensitivity

The sensitivity of the electromigration lifetime can be observed by plotting the lifetime as a function of the input parameters. For EM, the most significant input parameters corresponding to lifetime are the temperature (T) and current density (j_e). The lifetime may be normalized using an acceleration factor. Substituting Black's equation and assuming an exponential failure distribution into

$$AF = \lambda_{rated}/\lambda \qquad (7.57)$$

provides the acceleration factor for EM,

$$AF_{EM} = \left(j_e/j_{e,rated}\right)^{-n} \cdot exp[(E_{A_{EM}}/k_B) \cdot (1/T - 1/T_{rated})] \qquad (7.58)$$

Obviously, T has a much greater impact on A_f than j_e.

As device features continue to shrink and interconnect current densities grow, EM will remain a concern. New technologies may reduce the EM impact of increasing densities but new performance requirements emerge that require increased interconnect reliability under conditions of decreased metallization-inherent reliability [278]. Thus, EM will remain a design and wear-out issue in future semiconductor designs.

7.5 Soft Errors due to Memory Alpha Particles

One of the problems which hinder development of larger memory sizes or the miniaturization of memory cells is the occurrence of soft errors due to alpha particles. This phenomenon was first described by May and Woods [281]. U (Uranium) and Th (Thorium) are contained in very low concentrations in package materials and emit alpha particles that enter the memory chip and generate a large concentration of electron/hole pairs in the silicon substrate. This causes a change in the electric potential distribution of the memory device amounting to electrical noise, which, in turn, can cause changes in the stored information. Inversion of memory information is shown in Figure 7.13. The generated holes are pulled toward the substrate with its applied negative potential. Conversely, electrons are pulled to the data storage node with its applied positive potential. A dynamic memory filled with charge has a data value of 0. An empty or discharged cell has a value of 1. Therefore, a data change of 1–0 occurs when electrons collect in the data storage node. Such a malfunction is called "memory cell model" of a soft error.

The "bit line model" occurs due to change in the bit line electric potential. The bit line's electric potential varies with the data of the memory cell during readout and is compared with the reference potential, resulting in a data value of 1 or 0. A sense amplifier is used to amplify the minute amount of change. If α particles

(a)

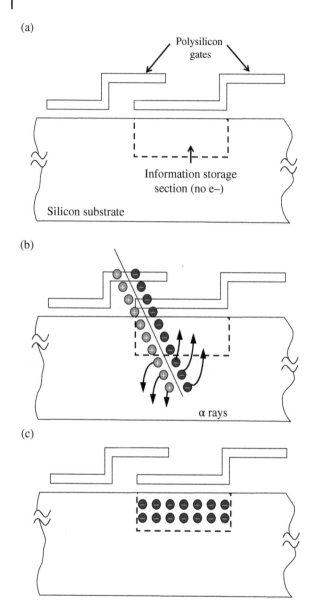

(b)

(c)

Figure 7.13 Memory cell model of soft error. (a) Memory information, generating electron/hole pairs, (b) α rays penetrates, generating electron/hole pairs. Electrons drawn to information storage section holes drawn by substrate supply, and (c) reversal of memory information 1 (electron empty) to 0 (electron full). *Note*: α particles penetrating into the silicon chip cause a high density of electrons and holes, which results in information reversal of the memory cell.

penetrate the area near the bit path during the minimal time between memory readout and sense amplification, the bit path potential changes. An information 1–0 operation error results when the bit path potential falls below the reference potential. Conversely, if the reference potential side drops, an information 0–1 operation error results. The memory cell model applies only to information 1–0 reversal, while the bit path model covers both information 1–0 and 0–1 reversals. The generation rate of the memory cell model is independent of memory cycle time because memory cell data turns over. Since the bit path model describes problems that occur only when the bit line is floating after data read-out, increased frequency of data read-out increases the potential for soft errors, i.e. the bit path model occurrence rate is inversely proportional to the cycle time. In product, the "mixed model" combined model describes the combination of the memory cell and bit path models.

8

Reliability Modeling of Electronic Packages

The scope of this book is primarily focused on the reliability of microelectronic devices. Therefore, packaging is included as a peripheral topic. The literature is filled with books and resources relating to the packaging of electronic devices. We will summarize packaging hazards and discuss methodologies for expanding reliability estimates to include packaging concerns.

Electronic packages contain many electrical circuits and interconnects which are used to connect individual circuits to form functional entities. Mechanical support and environmental protection are required for the integrated circuits (ICs) and their interconnections. The packaging system must also provide an adequate means for removal of heat. Previous chapters detail the contribution of high temperatures to the harmful effects of almost all failure mechanisms in the devices. Progress in packaging has evolved significantly throughout the last 20 years starting from the original wire bonds which were replaced by solder joints for better performance and I/O pin density. An example of a plastic ball grid array (*BGA*) is shown in Figure 8.1. The silicon die is mounted on an organic substrate through adhesives, and then the electrical signals are connected through gold (Au)-bonding wires from the silicon to substrates and then to the printed circuit boards (PCBs). The silicon device as well as the bonding wires are protected by the molding compound. The package is also capable of dissipating heat to ensure stable thermal characteristics. Additional performance and throughput are achieved with flip-chip bonding. Lately, structure solutions for high-density IC packaging called 2.5D/3D IC packaging have been developed. Fanout wafer-level packaging, FOWLP is one solution. FOWLP is bonded directly onto a circuit board without the need for silicon vias.

A step beyond FOWLP is the fan-out panel-level packaging (FOPLP) solution. FOPLP is an extension of wafer level fan out that has a larger substrate size of about 600 mm compared to that of FOWLP which is about 300 mm. This increases the throughput and lowers the cost-per-device margin compared with FOWLP [282].

Reliability Prediction for Microelectronics, First Edition. Joseph B. Bernstein, Alain A. Bensoussan, and Emmanuel Bender.
© 2024 John Wiley & Sons Ltd. Published 2024 by John Wiley & Sons Ltd.

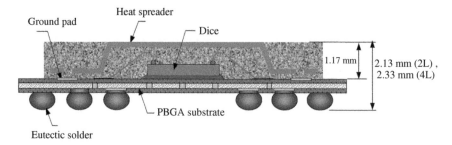

Figure 8.1 Schematic diagram of an electronic package. *Source:* Courtesy of ASE.

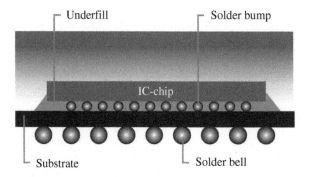

Figure 8.2 Flip-chip BGA. *Source:* Courtesy of GE Global Research.

The purpose of electronic packaging is to achieve design performance within the specified application environment for a certain period of time. Reliability prediction must account for all failure mechanisms acting on all parts of the package. That includes board to substrate interfaces, the substrate itself, and substrate to chip interfaces in the case of a flip-chip BGA as presented below (Figure 8.2).

Reliability of electronic packages is defined as the probability of operating a package for a given time period under specified conditions without failures. The reliability study normally uses the statistical approach for the failure data analysis and the distribution functions to describe reliability models and failure models. The objective of the package reliability study is to understand the physics-of-failure during the application and improve the product reliability performance. The reliability improvement can be made by changing packaging materials and/or modifying the package design. Beyond that, the significant purpose of the reliability study is to predict the package life and failure rates in the field application.

Most package reliability studies adopt established test types defined in Joint-Electron-Device-Engineering-Council (JEDEC) test standards [283–285], including temperature/humidity and temperature cycling tests. However, the qualification or reliability testing data obtained from traditional stress-based approaches is not sufficient to predict the reliability and the failure rate of the products in field applications. This is due to the lack of failure data available at the completion of the tests. The failure data obtained during technology design and engineering stages is also not fully effective because improvements in the development process make the intermediate results inaccurate. Test results with few failure data points have any significant statistical meaning for any models developed. Table 8.1 shows an example of package qualification results, although some failure data should be available through the package or product development cycles.

A physics-based test approach must be established to enable benefit from the failure data making it useful for the reliability models. Moreover, it is also necessary to generate parameters for reliability models for today's electronic packages with new materials, processes, and structures. Indeed, the calculation of accelerated factors used for lifetime prediction is often based on the activation energy (E_A) and parameters from models which might not be suitable for current package structures and materials [286]. The calculation of the acceleration factor (AF) for multiple failure mechanisms should be recalibrated for the field life prediction instead of using a simple product of individual AFs for each stress type. One issue for today's package reliability study is concerning using available failure data which could be obtained from different processes or materials. The knowledge gained from a reliability/qualification test is seldom passed on for a future reliability prediction in the field. The main reason is the lack of predictive models

Table 8.1 Examples of package qualification results presentation.

Test types	Test conditions	Test duration	Sample size	Results
Preconditioning test	30°C/60%RH	192 hours	Sum of samples for TC and HAST	Pass
Temperature cycling	−65°C to 150°C air to air	1000 cycles	45 units/Lot and 3 lots	Pass
Temperature and humidity test (no bias)	85°C/85%RH	1000 hours	45 units/Lot and 3 lots	Pass
Pressure cooker test (PCT)	121°C/2atm/ 100%RH	168 hours	45 units/Lot & 3 lots	Pass
Highly accelerated stress test (HAST)	130°C/85%RH	100 hours	45 units/Lot & 3 lots	Pass
Thermal shock	−55°C to 125°C liquid to liquid	1000 cycles	45 units/Lot and 3 lots	Pass
High temperature storage	150°C	1000 hours	45 units/Lot and 3 lots	Pass

Table 8.2 Example of field return failures (parts qualified but failed in the field).

Items	Failure mechanisms (%)	Description
Broken wires/bonds and lifted wires	32.19	Failures seen at the second stitch bonds on the lead frame or substrates
Die cracking	15.54	Die chipping, passivation cracking or metal traces cracking in the die
Delamination	12.71	Any interface delamination, such as mold/die interface
Die damage/wafer defects	12.15	Die surface damage or scratch
Package/substrate cracking	10.17	Organic substrate cracks and solder mask cracks
Others	17.23	Other failures including solderability, foreign materials

for extrinsic failure mechanisms and random failure mechanisms which would need the failure data from the population to obtain. The end-of-life wear-out failure mechanisms have been studied often and thoroughly, although they are not typically observed in field applications. In truth, many failures observed in the field occur in the first 0–3 years of used, which cannot constitute end-of-life. It has been shown that field failures were often related to defect-driven failure mechanisms instead of end-of-life failures. Table 8.2 listed some key package failure mechanisms seen in the field supporting the viewpoints.

The failure mechanisms shown in the table above are observed in the field although they passed qualification. They are still dominant failure mechanisms today [287].

Reliability studies for packaged devices based on physics of failure have been used for many years to characterize silicon intrinsic failure mechanisms. There are being implemented in the packaging industry for technology certification in recent years [288, 289]. Using that approach, the required test conditions and durations can be estimated based on product use conditions. In addition, the failure data gathered and failure models developed can be used to predict early failures in the field and the number of failures of the population instead of the mean time to failure (MTTF). The failure rate models developed can be used to accurately predict the failure rate of the packages in the field. The ultimate goal is to predict the failure rate in the field and reduce the field return as much as possible by understanding the failure mechanisms and implementing the corrective actions. The efforts will focus on validating and developing the AFs for multiple failure mechanisms seen in the packages. In summary, this chapter will:

1) Describe the failure mechanisms seen in electronic packages.
2) Describe the reliability prediction models associated with failure mechanisms.
3) Introduce the predictive reliability study and failure rate modeling in the packaging industry.

8.1 Failure Mechanisms of Electronic Packages

The reliability study for electronic packages has two objectives, one is to understand the weakest link and then make changes to improve their reliability, and the other is to predict the time to failure in the field, often based on the models using an assumed E_A (activation energy) and the AF. The reliability testing methods were usually referred to JEDEC or MIL-STD-883 standards as shown in Table 8.3.

Most of the tests can be used to detect the dominant failure mechanisms, understand the root causes of the failures and investigate key driving factors affecting the failure mechanisms. The popular physics-of-failure approach proposed [290] is often used to assess the reliability and improve the design and/or changing materials. On the other hand, the predictive reliability approach is applied to predict the failure rate or the time to failure in the field while understanding root causes for failure, i.e. the failure mechanisms. Each stress test will likely accelerate more than one type of failure mechanism. For instance, the temperature cycling test will accelerate cracking failures, interfacial delamination failures, as well as fatigue failures. One will not succeed in finding the dominant failure mechanism at used conditions by performing a thermal cycling test on a given device without understanding the un-dermining physical processes involved. This is because the conditions for failure are different for different failure mechanisms. It boils down to understanding the root cause of the failures and connecting them to the failure data.

Care must be taken in the selection of the test conditions. For instance, the cracking failure is a dominant failure mechanism in the conditions caused by temperature cycling to certain extremes and hold times. However, a delamination/ bond lift failure is seen as dominant for the temperature cycling of different stress conditions [291]. Additionally, bond lift failures (observed in low-stress conditions) are likely to be seen in the field but could be missed in the higher-stress testing conditions. Higher test acceleration and shorter test times are accessible using a combination of the TC and HAST stress conditions to accelerate interface

Table 8.3 Standard stress test types used in electronic package study [JEDEC, 2004/2005/2006].

Test	Conditions	Target failure mechanisms
Preconditioning	JESD 22A 113	Cracking, delamination, interconnect damage failures
Unbiased and biased highly accelerated stress testing (HAST)	JESD22A118	Corrosion, delamination, contamination, and migration; polymer aging failures
High temperature storage	JESD22A103	Diffusion, oxidation, degradation of material properties, IMC, creep failures
Temperature humidity bias (or no bias) (THB)	JESD22A101	Corrosion, contamination, and migration failures
Temperature cycling (TC)	JESD22A104	Cracking, deamination, fatigue failures
Power thermal cycling	JESDA105	Cracking and delamination, fatigue, material degradation failures
Mechanical shock (drop test)	JESD22B104	Cracking and delamination and fatigue, brittle fracture failures
Vibration	JESD22-B103B	Solder joint failures. Cracking and impact failures
Bending (monotonic and cyclic)	JESD22B113; JS9702	Package, solder joint failures, cracking, and delamination
Thermal shock (TS)	JESD22A106	Cracking, delamination, and fatigue; brittle fracture failures
Autoclave (PCT)	JESD22A102	Corrosion, delamination, and migration; interface contamination failures

delamination between mold and die surface [292]. However, combining stress testing types can introduce new failure mechanisms which might not be seen in the field.

Test plans should be designed with the understanding that different test types can reveal the same failure mechanisms. Some tests are more effective in isolating the failure mechanisms than the others. For instance, HAST (130C/85%RH) is demonstrated to be more effective than traditional temperature and humidity test (85C/85%RH) with no extreme damage to the samples [293]. It means HAST can help dramatically reduce the testing time and the weakest failure modes can be quickly found. It is also reported that PCT tests could significantly activate bond pad corrosion failures than the 85C/85%RH test (Table 8.4).

Both extrinsic and intrinsic failures can be explored with appropriate reliability testing. Primarily, the reduction of the failure rate in the early stage or "extrinsic"

Table 8.4 Comparison of failure mechanisms detected by HAST and TC.

	Failure mechanisms HAST (130C/85%)	Failure mechanisms TC (−65°C to 150°C)
Moisture penetration	59%	NA
Corrosion	21%	6%
Passivation cracks	8%	9%
Bond failures	8%	37%
Die cracks	4%	18%
Package cracks	NA	15%
Other		15%

Source: Adapted from Scalise [294].

region of the life distribution is key to achieving high reliability in the field (Figure 8.3) [295]. Table 8.5 summarizes the observed failure mechanisms seen in the electronic packages and systems.

In summary, the failure mechanisms can be classified into defect-driven, early "wear-out" and end-of-time wear-out failures.

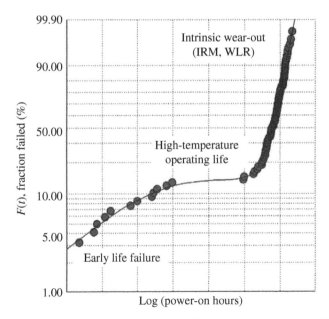

Figure 8.3 Example of cumulative failures, *F(t)*. *Source:* Adapted from Carulli and Anderson [296].

Table 8.5 Failure mechanisms seen in electronic packages.

Failure mechanisms of electronic packages	Failure mechanism descriptions	Driving forces	References
Die cracking; thin film cracking; passivation cracking	Failures often were shown to be open failure mode	Temperature cycling; power cycling; thermal shock and preconditioning test. Example conditions are (−55°C +125°C and 65°C +150°C)	[297]; [298]; [299]; [300]; [301]; [302]; [303]; [304]; [291]; [305]; [306]; [307]; [308]
Interface delamination and induced micro-cracks	Delamination and cracking inside the die or any other interfaces in the package	Temperature cycling and thermal shock; HAST; temperature and humidity test; pressure cooker test and mechanical bending test; in stacked die chip-scale packages (CSPs)	[309]; [292]; [310]; [311]; [312]; [313]; [314]; [315]; [316]; [317]; [318]; [319]; [320]; [321]; [322]; [323]; [324]
Bond pad crack	Gap between the epoxy and die top, detecting wire bond inter-layer dielectric crack by dark-field imaging	Low-k device bond pad, crack post temperature cycle; root cause for Al bond pad crack post TC; local compressive/tensile loading during wire bonding impact/ vibration step	[325]; [326]; [327]
Package cracking; substrate cracking; underfill cracking	Package body or internal "element" cracking	Temperature cycling, such as (−65°C +150°C); impact of package geometry on delamination	[300]; [328]; [329]; [330]; [331]; [93]; [332]
Solder joint fatigue/cracking; BGA and PoP balls failure	Solder joint cracking and solder creep fatigue damages	Temperature cycling; power cycling; vibration fatigue testing	[300]; [333]; [334]; [335]; [336]; [337]; [338]; [339]; [323]; [340]; [341]; [342]
Wire lifting/ broken bond/ heel broken of stitch bonds	IMC cracks or wire heel cracking and bond degradation	High temperature storage (150°C, 170°C); power cycling and thermal cycling	[343]; [344]; [304]; [291]; [345]; [346]; [347]

Table 8.5 (Continued)

Failure mechanisms of electronic packages	Failure mechanism descriptions	Driving forces	References
Corrosion	Due to the impacts from moisture and contaminants; due to the residues present on the electronic device (PCBs)	Temperature and humidity test; pressure cooker test; HAST; PCT	[348]; [349]; [310]; [290]; [350]; [351]
Electromigration	Damages seen at interconnects or solder bumps with high current applications	Current density; temperature; directionality of EM failure at high current density and temperature conditions	[352]; [353]; [354]; [355]; [356]; [357]

8.2 Failure Mechanisms' Description and Models

Multiple failure mechanisms of device packages are observed frequently during the qualification stage and field applications at different sites and interfaces. Some failure mechanisms are specific to certain package technologies. The primary risk observed in organic flip chip packages is die cracking, and delamination failures with increased mechanical loads. In the following section, failure mechanisms and their associated models will be discussed.

The failure mechanisms discussed in the following section include wire bond failures, package interface delamination failures, die cracking and package cracking failures, and solder joint fatigue failures, and corrosion and electromigration failures. The failure conditions are charaterized using failure models. Ghaffarian [358] discussed some aspects of Weibull and reliability models of *BGA* and *CSP* packages under temperature cycling testing. The prediction of time to failure by calculation is not realistic since many parameters in thermal cycling tests are not comperable to use conditions. In addition, the characteristic life in Weibull distribution is sometimes misleading. Considerable details and discussion will be provided in the following sections to elaborate on more and less successful testing methods for the discussed failure mechanisms.

8.2.1 Wire Bond Failures (Wire Lifting, Broken Wires, Bond Fracture, etc.)

Wire fatigue failures have been widely studied in less modern devices. The failure mechanisms observed include bond breakage and delamination. At times, wires break at the heel of wedge bonds due to their reduced cross-sectional area. In experiments conducted by Boettge [326], modules failed after approximately 20,000 cycles by bonding wire-lift-off, which is a typical failure mode under power cycling. Residues of bond wire material can still be detected on the surface of the chip metallization, indicating a crack formation and propagation within the aluminum (Al) material of the bonding wire. Figure 8.4 supports this assumption by a cross-sectional analysis of a remaining bond wire contact. The crack formation does not occur directly at the interface between bond wire and chip metallization, but it emerges partially in the volume of the bond wire as shown. Likewise, wires also fracture at the neck of a wire bond. This is different from BGA failures where the fractures resulting are either from tensile or shear forces induced by thermal stress or the flow of encapsulant during molding. At times, wire fatigue failures result from the interface delamination between the molding compound and the die [343]. One influencing factor of failures seen in Au/Al bonds is the degradation

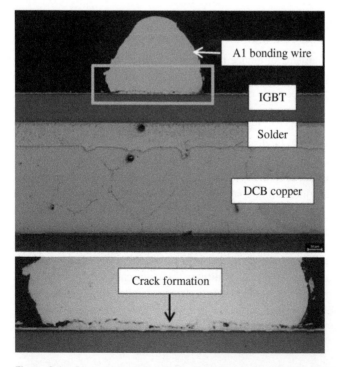

Figure 8.4 Cross-section of a wire bond wedge with crack formation after power cycling (light microscopic images) [323]. Bianca Boettge et.al., 2018 / from IEEE.

of bonding strength and electrical resistance increase. The increase of resistivity is the intermetallic compound (IMC) formation at the Au/Al bonds as well as diffusion at the interfaces of Au wire-Ag plating on lead frames. The factors reducing the IMC growth rate are not always found to have a dominant impact on the bond reliability; instead, the ball bond reliability is dependent on bond parameters and the bond pad bendability [344]. However, the presence of the compounds will cause the interface to degrade. Khan et al. [359] reported the presence of halogenated organic residues causing increased Au—Al wire bond failures through the degradation of the intermetallic, in which case the activation energy is calculated to be between 0.7 and 1.0 eV for various resins. Park et al. [345] studied the degradation of the Au—Al bonding under an HTS condition using different molding compounds. The lifetime of Au—Al bonding encapsulated with bi-phenyl (BP) epoxy resin is much longer than that of the o-cresol novolac (OCN) epoxy resin, the failure is attributed to the appearances of Sb at the interfaces and the Br originated from the EMC.

Wire fatigue failures had been widely studied and failure mechanisms observed included the bond breakage and the bond lifting-off. Wires break at the heel of wedge bonds due to reduced cross-section of wires. The cross-section in the heel of a ball bond arise as a result of excessive flexing of the wire during loop formation. Fatigue failures by crack propagation also occur at the heel. A wire can also break at the neck of a wire bond, leading to an electrical open. Ball bond fractures cause bond lift-off failures. The fracture is a result of either tensile or shear forces induced by thermal stress or the flow of encapsulant during molding. Sometimes the wire fatigue failures are the results of the interface delamination between the molding compound and the die [343]. One influencing factor of failures seen in Au/Al bonds is the degradation of bonding strength and an electrical resistance increasing. The increase of resistivity is considered to be the IMC formation at the Au/Al bonds as well as diffusion at the interfaces of Au wire-Ag plating on lead frames. The factors reducing the *IMC* growth rate do not always have a dominant impact on the bond reliability, instead, the ball bond reliability is dependent on bond parameters and the bond pad bondability [344]. However, the presence of the compounds cause the interface strength to degrade. Khan [359] reported that the presence of halogenated organic residues cause increased Au–Al wire bond failures through the degradation of the intermetallic, in which case the activation energy is calculated to be between 0.7 and 1.0 eV for various resins. Park [345] studied the degradation of the Au—Al bonding under an HTS condition using different molding compounds. The lifetime of Au—Al bonding encapsulated with BP is much longer than that of the OCN resin. The failure is attributed to the appearance of Sb at the interfaces and the Br originated from the EMC.

As mentioned above, key failure mechanisms were wire brakage or bond fracture due to the stress loads. Nevertheless, the formation of IMC has a significant impact on the bonding interface failures. The wire fracture oftenly occurs at low cycles as well.

8.2.2 BGA and Package-on-Package Failures

The 3D packaging consists of stacking of packaged devices called package-on-package (*PoP*) and stacking of die within a package. In 2019, Ghaffarian [342] studied various 3D PoP packaging assembly configurations and reliability characterizations under accelerated thermal cycling (ATC) and accelerated extreme harsh thermal shock cycling (ASTC) conditions (−55°C to +125°C, and −100°C to +125°C respectively). Four methods of packaging stack assemblies and their effect on the stack packaging assembly reliability were covered. The key failure occurred at the bottom package with solder joint at package/PCB interfaces.

Figure 8.5 displays optical a scanning electron microscope (SEM) view of the *PoP* showing the existence of voids and microcracks at solder joints.

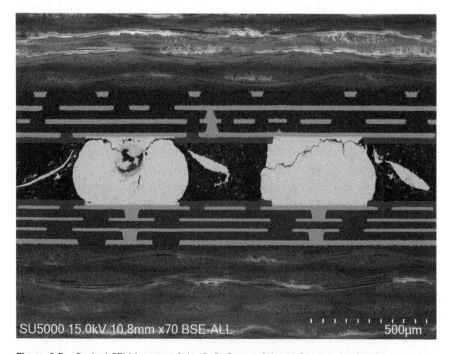

Figure 8.5 Optical SEM images of the PoP. *Source:* Private Communication from ELEMCA (2023).

8.2.3 Die Cracking Failures

A high percentage of die cracking failures is due to defects in the die coupled with the high stress applied during the testing process or application, such as surface cracking or scratching formed at the processing stages. The pre-existing cracks have a high probability of causing catastrophic fractures. Micro-cracks at the die top surface propagate vertically to cause active circuit damage and even failures. Edge cracks induced from the wafer sawing process are likely to propagate at the corners of the die or thin film layers inside the silicon devices. Voids in the die attach materials as well as the fillet height of die attachment during the die attachment process might induce die cracking failures.

Figure 8.6 depicts a solder ball bond and a die cracking failures seen at die and shows the cross-section. With the fast adoption of the low-k die and the copper (Cu) metallization, die cracking failures were increasing seen in the qualification and in the field.

Localized passivation cracks on the die were associated with passivation defects from the manufacturing processes [297]. Nguyen [306] described the thin film cracking due to the thermal expansion mismatch between silicon substrates and thin films under loads of fast temperature cycling – power cycling with a 10% duty cycle. The passivation processing defects (e.g. holes, micro-cracks, and physical damage) were the main reason for early failures as well [293]. Chip backside contamination is found to be a main driving force for package cracking/interface delamination failures after 85C/85%RH testing [328]. For defect-driven failure mechanisms, screening using physics-of-failure approach can help improve the reliability in the field as recommended by [295] in a study by Parker et al. They discussed several fatigue failures seen in temperature cycling tests. The failures that resulted, lift bonds and passivation cracks in the corner of the die, were examined using Coffin–Manson equations.

Kwon et al. [320] studied the interface degradation (bumping and interface dielectrics and adjacent materials) in flip chip packages under temperature cycling loads. The weakening and delamination of the Benzocyclobutene polymer (BCB) passivation layer at the die corner are the main reasons for the reduction of die adhesion strength after thermal cycling. The sliding traces on fracture surface after three-point bending fracture revealed that cyclic shear displacement in a polymeric BCB layer fatigued the interface bonding or interconnect bump failure upon thermal cycling. Syndergaard and Young [360] studied the reliability performance of PQFP packages under temperature cycling testing (−65°C +150°C). The key failure mechanisms are metallization cracks at or crushed at or near the die corner. The damage started to happen when the plastic started to delaminate from the surface of the die.

Raghunathan et al. [361] focused on the effects of thermal stresses on underfill materials of solder bumps to the chip, substrates, and other components in the

(a)

(b)

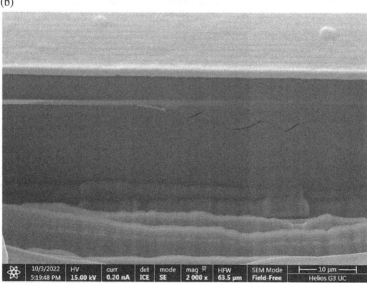

Figure 8.6 (a) Cross-section view of a series of micro-cracks in a solder ball seen in the field. *Source:* Private communication from ELEMCA (2023), (b) Focused Ion Beam (FIB) cross-section view of die cracking failures. *Source:* Private communication from ELEMCA (2023).

packages. Die cracking is a prime failure mechanism observed due to thermal stress on the underfill materials. It was found that the die–substrate thickness ratio had a high influence on the die backside stress. When a micro-crack is located at the die center, the die has the highest chance of reaching a catastrophic crack failure. As the crack length is increased, the chance for die to crack increases. The critical die cracking length in FCBGA packages is about 17 μm. The optimization of die thinning and die polishing processes are recommended process optimizations to reduce the chances of die cracking.

The unoptimized die attach process introduce die cracking failures. Increasing Bromine content also accelerate the degradation in both molding compounds and wire bonds. Likewise, the molding compound can affect the mechanism of degradation [362].

8.2.3.1 Die Cracking Failure Mechanisms

Coffin–Manson models were typically applied to model the die-cracking failures. Cory [291] applied a typical Coffin–Manson equation in bond failure and passivation cracking failure mechanism studies, the number of cycles to failure is expressed as where the m value can be different depending on the failure mechanisms applied. A related study by Zelenka et al. [300] developed a stress model for interlayer dielectric cracking failure mechanism. The number of cycles to failure used in this model is $N_{5\%}$. The m is found to be 7 for the failure mechanism. Blish [363] compiled and studied the exponent m in the Coffin–Manson equation for various failure mechanisms. It is found that m for ductile metal fatigue is 1–3, for commonly used IC metal alloys and intermetallic is 3–5 and m for brittle fracture is 6–8.

Additionally, the study above [300] shows the interlayer dielectric cracking failure mechanism seen during temperature cycling test (−65°C to +150°C and −40°C to +125°C) with TO-220 packages. The accumulated failure rate is represented by a Weibull distribution and a power law model is fit to the failures and is a function of the applied stresses. The stress applied to the die surface is controlled by the elastic modulus and thermal expansion coefficient of the plastic and the coldest temperature of the cycle. Even for different molding materials, the Weibull plot had a shape parameter of 2.7.

The study mentioned above [360] developed the temperature cycling failure rate models by considering the die size and is shown as:

$$F = 1 - exp[-exp(Z)] \tag{8.1}$$

where d is the diagonal measure of the die in mm, and

$$Z = A + B \cdot \ln\left[c\left(\frac{d}{163.81}\right)\right]$$

where A, B, c, and d empiric parameters can be driven according to the experimental data.

This model is different from traditional Coffin–Manson models instead of focusing on the process parameters and structures of the packages. It is more a rate- or structure-related model.

Paris–Erdogan equation [364] established a relationship between the fatigue crack growth rate and the variation in the cyclic stress intensity factor, expressed as

$$\frac{da}{dN} = A \cdot (\Delta K)^{m_p} \tag{8.2}$$

where da/dN is the crack growth rate, A and m_p are material constants, $K = G \cdot \sigma \cdot \sqrt{2\pi \cdot a}$ is the new intensity factor, a is the crack length, σ is the nominal stress, and the factor G is a function of geometry, $\Delta K = G\sigma_r\sqrt{2\pi a}$. σ_r is the nominal stress range. The fatigue life is dominated by crack initiation.

Shirley et al. [365] developed a model for moisture penetration through passivation micro-cracks. The models applied in the study are a combination of Arrhenius model and power models, shown as

$$t_f = C \cdot exp\left(\frac{-E_A}{k_B T}\right) \cdot (RH)^b \tag{8.3}$$

where $E_A = 0.79$ eV and $b = 4.64$ for passivation cracking failures. The model is similar to Peck's but with different values for the parameters.

8.2.4 Interface Delamination

Interface delamination failures were a key trigger for other failure mechanisms seen in electronic packages. Interface delaminations are the separation of any interfaces in the package structures due to loss of the interface adhesion strength or damage to the interface structure. Interface delamination failures introduce package cracking, die cracking failures as well as wire bonding failures.

At each interface of electronic packages, the existence of moisture contents can hydrolyze the epoxy, and then degrade the interfacial chemical bonds. Surface cleanliness is a crucial requirement for good adhesion, oxidized surfaces, or contaminated surfaces often lead to interface delamination. With the adoption of low-k dielectrics and Cu interconnect technology, interface delamination on the silicon is becoming a major reliability concern. Figure 8.7 shows an example of interface delamination between the molding compound and the substrate.

Merrett et al. also studied the delamination failure between the molding compound and the die seen after temperature and humidity tests (HAST and PCT) [297]. Passivation defects were the main reason for delamination failures.

(a)

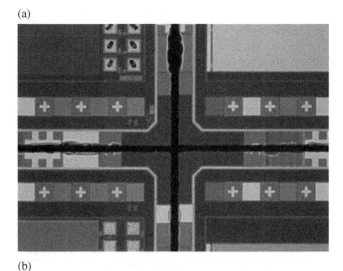

(b)

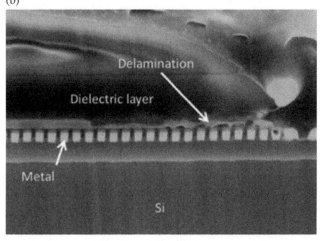

Figure 8.7 Chipping (a) and delamination (b) in device layers of a die induced by dicing. *Source:* Wei-Sheng Lei and Ajay Kumar [366] / Reproduced from John Wiley & Sons, Inc.

The interface delamination is found to be dependent on relative humidity (RH) and temperature and the delamination is not always a precondition for corrosion failures.

With the development of flip chip packages, Braun et al. presented a study of their reliability under high temperature conditions [323]. Temperature cycles with a higher upper temperature and higher delta temperature led to an increased area of interface delamination. The high temperatures led to an increased delamination rate by higher thermal mismatch and shear forces at the interface combined with a loss of interface adhesion strength. The delamination was often seen at the chip/ underfill interface and resulted in the cracking of solder bumps and electrical failures.

Kwon et al. [322] showed that the properties of anisotropic conductive film (ACF) had a significant role in causing interface delamination and solder joint reliability (SJR) problems by temperature cycling testing (−40°C +125°C). The lifetime of the thermal cycling performance depended on the failure rate and failure criteria defined. The MTTF was strongly dependent on the distance of the solder joint from the chip center. The degradation was initiated from the edge of the silicon die and then evolved to the inner area of the assemblies. A high coefficient of thermal expansion (CTE) and low modulus of the ACF exhibited significant interface delamination failures. In addition, it was found that the cumulative failure rate did not correspond to the Weibull function value.

Aihara et al. [314] studied critical material properties to enhance the soldering cracking resistance for electronic packages. Lowering moisture absorption and improving the interface adhesion property will help to mitigate cracking failures. Ferguson et al. [367] demonstrated that moisture preconditioning strongly influenced the interfacial adhesion, decreasing the adhesion by approximately one-half for both underfill/solder mask interfaces after 725 hours of exposure at 85 °C/85%RH.

A study from Chung et al. [316] reported delamination failures on the interfaces between the die attach and the backside of the die using the Bismaleimide-Triazine (BT) substrate-based CSP (chip scale package) under the pressure cooker test (PCT) condition. The molding compound formation influenced the amount of moisture uptake, which obviously would affect the PCT reliability. Tanaka et al. [313] highlighted the factors affecting interface delamination by considering the swelling of the molding compound due to moisture absorption. The authors found the delamination-occurrence temperature drops sharply when the moisture absorption temperature is high. The change in delamination-occurrence temperature in the package corresponded to the change in the true adhesion strength due to the moisture absorption, and so did the residual stress and the stress intensity factor. The change is due to moisture absorption which then causes the swelling of the molding compound. Interface delamination in a moisture package occurred when the stress intensity factor exceeded the time adhesion strength after exposure to moisture.

Lin et al. [317] studied the reliability performance of stacked-die packages using a lead frame platform. Full delamination at the lead-frame paddle/mold compound interface is seen after 100 temperature cycles (−65°C +150°C) although there are no electrical failures. The die attach voids were major factors affecting the package integrity and could introduce the delamination initiation at the edge of the die attach paste due to a high stress concentration and a low adhesion strength. The delamination failures could be controlled by reducing the die-attach voids and enhancing the interface adhesion strength.

Chen et al. [368] studied the effect of underfill materials on the reliability of low-k flip chip packages with Cu silicon with seven layers of metallization and substrates with six build-up layers. The failures observed were due to delamination caused by poor interface adhesion and a high low-k shear stress. Low-k cracking is also found for certain underfill materials at the metallization layers. Wang et al. [57] investigated various factors and their influences on the interface delamination failures widely observed in the Cu/low-k flip chip packages. The delamination typically started at the die/underfill interface from the die corner in a flip chip package. The thermal stress induced by the packaging in interconnects is primarily tensile and normal to the interconnect interfaces. The scaling did not have a large influence on the interface delamination failures. However, larger thermal stresses at the die surface for the lead-free solder package compared with the high lead and lead eutectic solder package resulted in the highest driving force for interconnect delamination in lead-free packages. Underfill materials can dramatically affect the interface delamination for the better or for worse. Increasing the CTE of the underfill materials enhances the thermal mismatch between the underfill and solder bumps, underfill with higher CTE had a lower Young's modulus that will put a weaker protection for solder bumps. Increasing the CTE of the underfill materials increasing the [337] crack driving force for low-k interfacial delamination under the critical solder bumps. Tsao et al. [321] found low-k inter-level dielectric (ILD) cracking failures at the chip corners seen under temperature cycling and preconditioning test using cavity down packages. Molding compound with the right CTE and modulus combination could help eliminate the delamination failures. Metal peeling is found at the die edge indicating an initial point of delamination. Zhai et al. [369] also investigated delamination failures in high performance organic flip chip packages with a Cu/low-k backend technology. The lower modulus of low-k dielectrics result in a higher crack driving force. The corner delamination is less sensitive to the modulus of the ILD materials than the near-bump delamination. The presence of the underfill fillet formed at the die periphery play an important role in the delamination at the die corner or the die edge. When cracks grow, the total energy release rate increases. Once the crack growth occurs, it becomes unstable and leads to catastrophic failures. Wang et al. [337] studied the interfacial delamination failures of low-k structures (SiLK) and methylsilsesquioxane (MSQ). They found that the increasing die size in a package increased the cracking driving force for low-k materials. Packaging effects were smaller for Cu/MSQ structures than Cu/SILK structures due to the high modulus of MSQ materials.

8.2.5 Package Cracking Failure

Many vendors design their products for application conditions which have a thermal mismatch of packaging materials and die large enough to make the interfaces susceptible to delamination and cracking. The cracks can start at various interfaces, including die top surface and molding interfaces, die attach/lead frame or substrate interfaces, as well as interfacial layers in the substrates. The cracked package is susceptible to corrosion and contamination failures by the migration of external ions along the surface of cracks.

Dias et al. [329] reported temperature cycling resulting from metal line breaks due to cracks in the flip chip organic substrate layers. Saitoh et al. [318] observed edge delamination at the interface of die pad on the lead frame and encapsulant materials under the temperature cyclic loading between −65°C and 150°C. The package resin cracking was then observed from the lower wedge of the die pad perpendicular to the bottom surface of the package as the delamination extended. Ray et al. [370] reported the resin layer cracking failures in flip chip packages. The cracks propagated to damage the traces underneath the BC4 layers. Optimization of the design ground rules is needed to addess the cracking problem.

Lin et al. [331] studied the critical stress characteristic of low-cost over-mold flip chip packages under a −65°C to 150°C temperature cycling load. Delamination occurred and cracks were initiated at the interfaces and then propagated into the package when the stresses surpassed the bonding strength and strength limits of the materials. On the substrates, a high stress was located around the metal pads on the top of the substrate. Moreover, the curing of the mold compound was the most stringent process for the reliability of the substrates. The high stress initiated from the bottom corner of the pad with a max distance from the neutral point (DNP) drove the cracks to propagate toward the middle of the substrate. The high stress is induced by high CTE mismatch and cure shrinkage of the molding compound. The authors also described that the higher the underfill fillet, the lower the stress in the substrates. The stress in the substrate was minimal when die size is about 80% of the package size.

8.2.6 Solder Joint Fatigue Failure

During any temperature cycling conditions, solder joints experienced a complex stress and strain history as a result of the CTE mismatch of the packaging materials. The cracks initiated at the solder joints lead to the failures of the joints. The thermal fatigue of solder joints deepen due to a number of parameters related to materials, configuration and manufacturing. The stress and strain in the solder joints result from the global mismatch of the coefficient of thermal expansion between packages and substrates and the total CTE mismatch between solder and Cu pads/leads.

Lin et al. [371] proposed to improve the SJR by using a Cu post and increasing the solder tub depth. The diameter of Cu post is critical to enhance the solder fatigue life under cycling load. Liu et al. [372] studied the solder fatigue life of BGA packages under a high cycle vibration load. The primary solder crack always started at the inner corner of the component side, and the secondary cracking always started at the outer corner of the joint and showed up at the solder/Ni interface at component sides. The delamination between the solder mask and solder joint gives impetus to the development of the primary crack, which emanates from the termination point of the delamination. However, the crack growth rate is not sensitive to the frequency variations in the range tested.

Noctor et al. [373] presented a reliability study for thin small outline package (TSOP) solder joints under a temperature cycling load. Even with optimized soldering processes, the key failure mechanisms were the total separation on the TSOP sides and individual solder joint cracks. The cracks in the solder joint that propagated rapidly to separate the whole side of the package from boards were unusual. This is contrast to more conventional failures where the cracking of an individual solder joints is observed. The thermal expansion mismatch between the package and the PCBs is a driving force causing solder attachment failures. For many components, the predicted failure probability depends on the assembly parameters (solder joint dimensions, solder alloy, board CTE, size, and thickness), the intended field environment and the intended product design life. ATC tests and predictive modeling showed that TSOP solder joint life during test is about five times longer with Cu than with alloy 42 lead frames.

Tu et al. [333] revealed the majority of cracks formed during cycling testing initiated and propagated along the IMC η-phase/solder joint interface. The thicker the intermetallic layer in the joint, the shorter the fatigue lifetime of surface-mounted assemblies. The intermetallic layer and highly brittle region formed along the solder/IMC interfaces caused cracks to propagate along the IMC η-phase/bulk solder interface during the cycling load. The lifetime was predicted by monitoring the thickness growth inside solder joints during the operation of electronic assemblies. The study included the impact of IMC on the SJR. It pointed out that the cracks propagate along the IMC/bulk solder interface due to the CTE mismatch inducing a greater shear stress in the solder joints during the thermal cycling. The presence of an IMC help cause cracks and affect the fatigue lifetime of solder joints.

Jeon et al. [374] studied the relationship between SJR and the properties of pad surface finish including electroless nickel immersion gold (ENIG) and organic solderability preservative (OSP). The IMC thickness was controlled by the thickness of an electroless Cu layer. Amagai [375] demonstrated that the higher CTE and elastic modulus die attach film increase the life of solder joints.

In summary, numerous components are classified for portable applications by exposure to shock and vibration environments. Dynamic loads (often high cycle fatigue loads) also have significant effects on the solder joint fatigue life. The failure modes are observed as cracks on the solder joints as well. Wang et al. [337] studied the vibration fatigue failures of solder joints. The fatigue failure were related to the devices' location in the PCB as well as the bump location on the package. Often the cracks induced by vibration fatigue were created in metal compound layers or solder materials nearby. Cracks were first observed in the bottom round angle area of the joint and then appeared at the top of the joint along the progress of the vibration. The solder joints' fatigue is also connected with the mass of the chips, the stiffness of the chips, and the shape and number of solder joints.

8.3 Failure Models

We mentioned in the previous section that the primary stress factors for interface delamination failures include temperature, moisture contents (humidity), and shear or tensile loads. The interface delamination could further propagate and form cracking at critical interfaces, such as thin film layers, the molding compound and the die surface, and the solder joints.

The models for the interface delamination failure considering temperature, T and moisture, RH, the time to failure could be described as:

$$t_f = A \cdot (\mathrm{RH})^n \cdot exp\left(\frac{E_A}{k_B T}\right) \tag{8.4}$$

where A is a constant and k_B is the Boltzmann constant. The equation can also be formatted as follows [297]:

$$t_f \propto exp\left(4.4 \cdot 10^{-4} \cdot \mathrm{RH}^2 + \frac{E_A}{k_B T}\right) \tag{8.5}$$

where the temperature range of 85°C–110°C.

8.3.1 IMC Diffusion Models

The growth rate of the IMCs usually follows a parabolic relationship.

$x = kt^{1/2}$ where x is the IMC thickness and t is the time. k is the rate constant, and $k = Ce^{-\frac{E_A}{kT}}$ where E_A is the activation energy, ranging from 0.2 to 1 eV for various materials and stresses. Kim [376] reported an IMC growth rate of Cu wires on Al pads described as

$$x = 0.004658 \cdot t \cdot exp\left(\frac{-13046.179}{T}\right) \tag{8.6}$$

Uno and Tatsumi [377] further listed the activation energy for bonding under various molding compound compositions including BP and OCN. It is detailed in the following formulas:

$$BP\ (< 450\,\text{k}), k = 2.0 \cdot 10^{15}\ exp\left(\frac{-2.0\,\text{eV}}{k_B T}\right) \tag{8.7}$$

$$BP\ (> 450\,\text{k}), k = 5.1 \cdot 10^{19}\ exp\left(\frac{-1.5\,\text{eV}}{k_B T}\right) \tag{8.8}$$

$$OCN\ (> 430\,\text{k}), k = 5.5 \cdot 10^{19}\ exp\left(\frac{-2.3\,\text{eV}}{k_B T}\right) \tag{8.9}$$

A paper review from Xu et al. [346] relates growth kinetics of *IMC* in solder joints during thermal cycling. As an example of IMC thickness after aging at different temperatures, the apparent activation energy E_A is 0.76 eV, the time exponent n is 0.51, the initial IMC thickness x_0 is 1.35 µm, and the pre-exponential factor A_0 is 4.92.

It is clear from literature the compressive stress induced by CTE mismatch during temperature variation promotes the IMC growth at the interface and deteriorates the residual strength of solder joints. This means that the mechanism for accelerated IMC growth is closely related to both stress and temperature.

The accumulated dwelling time in the thermal cycles seems to be reasonably equivalent to the isothermal time despite the proposed effective aging time in literature. In ATC tests, the temperature variation in the thermal cycling will induce stress variation on the solder joints. The kinetics model based on the equivalent isothermal time does not reflect the contribution of stress state or the temperature range below the upper soak temperature. Therefore, the thermal cycles cannot be replaced by an equivalent time with a high isothermal temperature. Further studies regarding the IMC growth kinetics account for the coupled temperature and stress effects based on finite element analysis for the board-level packaging structures.

Regarding the isolated IMC evolution in the solid state, after comparing the solid and liquid states, Tang et al. [347] found that the area distribution of the IMC particles can be well described by the three-parameter Weibull distribution. Also, the average sizes of the isolated IMC particles of Ag_3Sn and Cu_3Sn in the solder joints annealed at a temperature higher than the melting temperature are greater. This means that the input thermal energy effectively drives the atom diffusion around the isolated IMC particles. Therefore, a higher temperature with a sufficient duration will justify the size growth of isolated IMCs in solder joints.

8.3.2 Fracture Models Due to Cyclic Loads

Bond fracture failures due to cyclic load during the application can be described by power law models, which could be expressed as some format of Coffin–Manson equation, e.g.:

$N_f \propto C \cdot (\Delta T)^n$ where ΔT is the temperature range and can be expressed as $T_{max} - T_{min}$. C and n are constants and could be determined by experiments.

Heleine [378] presented a wire bond reliability model using a defect modeling approach. It was observed that the weak link was not the interface of the wire to the Au pad, but rather, the heel of the wire bond. The variable was the wiring bond width, depending upon the width of the bonds, the failure mode was either a wire bond lift or a heel crack. The failure mode observed was crack initiation and propagation at the heel of the wire bond. The probability of failure for given bond widths is calculated as

$$p(f/BW) = a \cdot exp(b \cdot BW) \tag{8.10}$$

where a and b are constants derived experimentally from the experimental stress data and bandwidth (BW) is the bond width.

8.3.3 Die Cracking Failure Models

Similar failure models are compatable for both die-cracking and package-cracking failure mechanisms. However, the model parameters would be different. Each failure mechanism will need to calculate the parameters based on failure data collected from experiments.

Cui et al. [93] studied the substrate via cracking failures found after temperature cycling testing. Coffin–Manson failures models were used to calculate the AF between various TC conditions. The delta T ranges from 125°C to 195°C. The shape parameters were about 1.2–1.8 and activation energy is calculated to be 0.42 eV.

Paris's law described the relationship between the fatigue crack growth rate, da/dN, and the range of the energy release rate, delta G per cycle, shown below. The model is another way to calculate the cracking growth rate as well as the influencing factors.

$$\frac{da}{dN} = B \cdot (\Delta G)^C \tag{8.11}$$

ΔG is the range of energy release, da/dN is the crack growth rate, G is the energy release rate (cracking driving force) B is an empirically determined material constant, and C is the function of the phase angle. Reducing G or increasing C would reduce the risk of crack initiation.

8.3.4 Solder Joint Fatigue Failure Models

In general, the solder joint fatigue models depended on the temperature range, the dwell time, and solder/package materials. The Coffin–Manson model is the commonly used model to estimate the time or cycles,

$$N_f = C \cdot \left(\frac{\Delta\gamma}{2 \cdot \varepsilon_f}\right)^{1/B} \tag{8.12}$$

N_f is the mean number of cycles to failure, $\Delta\gamma$ is the inelastic strain range depending on the geometry of solder joint and the change in temperature, B is the Coffin-Manson exponent coefficient, ε_f is the fatigue ductile coefficient in shear,

$$C = -0.442 - 6 \cdot 10^{-4}T_s + 1.74 \cdot 10^{-2}Ln\left(1 + \frac{1}{t_0}\right) \tag{8.13}$$

where T_s is the mean cycle temperature of the solder in C and t_0 is the dwell time in minutes at max temperature.

The first reported comprehensive model for thermal fatigue of tin-lead solder interconnects was presented by Norris and Landzberg in 1969 [379]. The model uses the cyclic frequency and maximum temperature as follows:

$$\frac{N_{lab}}{N_{machine}} = \left(\frac{f_L}{f_M}\right)^{1/3} \cdot \left(\frac{\Delta T_M}{\Delta T_L}\right)^2 \cdot \Phi(T_{max}) \tag{8.14}$$

where:

$\dfrac{f_L}{f_M}$ and $\dfrac{\Delta T_M}{\Delta T_L}$ are the maximum fatigue life and temperature change ratios under isothermal conditions,

$$\Phi(T_{max}) = exp\left(\frac{E_a}{k_B} \cdot \left(\frac{1}{T_{L_{max}}} - \frac{1}{T_{M_{max}}}\right)\right) \tag{8.15}$$

Using lognormal distribution, the failure rate (FR) at any time t is given as:

$$FR = \frac{1}{R(t) \cdot \sigma \cdot t \cdot (2\pi)^2} exp\left\{-\frac{1}{2}\left[\frac{ln(t) - ln(t_{50})}{\sigma}\right]^2\right\} \tag{8.16}$$

where $R(t)$ is the reliability function of the distribution.

John Lau [380] emphasized the Coffin–Manson equation using the following form:

$$N_f = C \cdot \left(\Delta \varepsilon_{eq}\right)^n \tag{8.17}$$

and C is noted to be 2.37 and $n = -2.564$ for 63Sn37Pb eutectic solders. Other studies show different values for the parameters such as: $C = 8.138$ and $n = -0.7935$ [379]. Chen et al. [381] used a little bit different format to describe the relationship between strain and number of cycles,

$$\left[N_f \cdot f^{(k-1)}\right]^m \cdot \Delta \varepsilon = C \tag{8.18}$$

where f is the frequency factor and k is the frequency exponent, where

$$k = 0.919 - 1.765 \cdot 10^{-4} \cdot T - 8.634 \cdot 10^{-7} \cdot T^2$$

$$m = 0.731 - 1.63 \cdot 10^{-4} \cdot T + 1.392 \cdot 10^{-6} \cdot T^2 - 1.151 \cdot 10^{-8} \cdot T^3$$

$$C = 2.122 - 3.57 \cdot 10^{-3} \cdot T + 1.329 \cdot 10^{-5} \cdot T^2 - 2.502 \cdot 10^{-7} \cdot T^3$$

Strifas et al. [382] investigated the crack growth in solder joints under cycling conditions and the characteristic joint fatigue life is computed from:

$$t_f = N_0 + \frac{a}{da/dN} \tag{8.19}$$

where a is the crack length and t_f is the number of cycles to 63.2% population failure. N_0 is cycles to crack initiation. da/dN is the crack growth rate per cycle. The expression for N_0 is defined as:

$$N_0 = K_1 \cdot \Delta W_{avg}^{K_2} \tag{8.20}$$

where ΔW_{avg} is the volume average viscoplastic strain energy density accumulated through the third thermal cycle. K_1 and K_2 are constants. The expression for da/dN is:

$$\frac{da}{dN} = K_3 \cdot \Delta W_{avg}^{K_4} \tag{8.21}$$

K_3 and K_4 are constants.

Based on Paris's law, a model under high-frequency vibration load is presented:

$$N_f = \int_{a_0}^{a_f} \frac{da}{C \cdot (\Delta K)^m} \tag{8.22}$$

$$\Delta K = \Delta K_{II} = \sqrt{E \cdot \frac{\Delta J}{1 - \nu^2}} \tag{8.23}$$

where:

a – the length of the crack

ν – Poisson ratio

ΔJ – J integral range

N – Number of cycles

E – Modulus

8.4 Electromigration

Electromigration is caused by metal transport in a conductor or the interconnection metallurgy. The electro-transport rate increases as the potential gradient or the current density increases for semiconductor devices. Mass transport resulting from the passage of DC current is the cause of failures in small interconnects with high current density. Electromigration is one of several transport phenomena which occur in solids. Traditionally, electromigration is a very low-risk failure mechanism for packages due to the large dimension involved compared to silicon technology. However, with the requirements for high performance and small form factor packages, the dimensions in packages are becoming smaller and the power is increasing. There is a high risk of causing electromigration in connects, such as solder bumps in flip chip technology. Another immerging failure mechanism associated with electromigration is thermomigration, which is thought to be a more benign mechanism and much rarer than electromigration. However, with the scaling down of the interconnect dimensions, especially the solder bumps in flip chip applications, thermomigration coupled with electromigration is causing serious reliability issues (Figure 8.8).

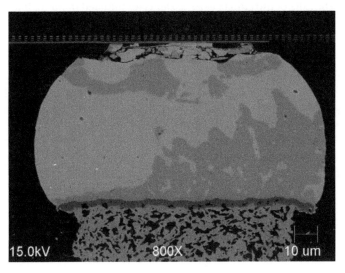

Figure 8.8 Solder joint cracking due to electromigration. *Source:* Min Ding et al. [356]/Reproduced from American Institute of Physics.

8.4.1 Electromigration Failure Description

Basaran et al. [355] studied the failure modes of flip chip solder joints under high electrical current density. Three failure modes were observed and all were related to the high temperature. Among them, one of the modes is due to void nucleation and growth in solder joint interfaces during the current stressing. The combined effects of electromigration and thermomigration were causing the ultimate failures of the solder joints. The pre-existing voids in the solder joints catalyize the failures. When the direction of the thermomigration is the same as the electromigration, the damage is very severe. When the directions of the thermomigration and electromigration were opposite, the thermomigration forces dominated, yet the total damage was smaller. The Ni UBM–solder joint interface is a site with a high tendency for void nucleation and growth. The contaminants in the interface also accelerated the void nucleation process.

Wu et al. [383] presented a study that shows that bump temperature has a more significant influence than current density on bump failures. Sn/95Pb solder bumps were observed to have 13-times higher MTTF than that of eutectic Sn/37Pb. The increase of the measured bump resistance is from bumps with an electrical current flowing upward into UBM/bump interfaces while bumps having opposite current polarity caused only a minor resistance change. The structural damage at the region of UBM and UBM/bump interfaces is observed as solder cracking or delamination. Effects of current polarity and crowding were key factors in observed electromigration behaviors of flip chip packages. Damages initiated in bump fracture or the solder voiding at solder/UBM interface were a result of current crowding, which is also the primary failure mechanism of flip chip interconnects. In a different study, Wu et al. [352] studied the electromigration failures of Sn/37Pb, Sn/36Pb/2Cu, and Sn/36Pb/2Ni solder joints, Pb is found to be the dominant diffusion species and migrated along the electron flow. Cu-doped SnPb bumps show an improved electromigration resistance than eutectic SnPb interconnects. A significant degradation of the electromigration failure is observed when employing an Ni-doped SnPb solder.

Ding et al. [356] studied the electromigration failures in Pb-free solder joints, and the activation energy was determined to be 0.64–0.72 eV for Cu UBM and 1.03–1.11 eV for Ni UBM structures. In addition, electromigration failures were only observed in solder bumps with an electron current flow from the UBM to the substrate. Failure analysis showed the dissolution of the UBM as a result of the Ni migration and a subsequent solder cracking or de-wetting for solder bumps using the Ni UBM. The Cu UBM's failure mechanism is temperature dependent. At high temperature, the Cu UBM dissolved continuously while at low temperature, open failures are caused by a crack formation at the Cu_3Sn/Cu_6Sn_5 interface with little damage to the UBM. The electromigration life of Pb-free solder is much better than that of the eutectic solder but worse than that of the high lead solder at the same temperature.

Wu et al. [384] proposed that a thicker UBM layer would delay an EM failure and prolong the MTTF. The unique EM behavior is the polarity effect on IMC growth at the anode and IMC dissolution at the cathode. At the end of EM stage, voids increase the contact resistance to a crowd joule heat at the interconnect interface. Shao et al. [354] also showed that electromigration failures were on the anode/chip side and large $(Cu/Ni)_6Sn_5$ IMC were observed on the interface of the UBM and the solder bump. Nickel atoms were migrated by an electron flow from the substrate side to the chip side to form the $(Cu, Ni)_6Sn_5$ IMCs.

8.4.2 Electromigration Failure Models

The critical factors influencing the electromigration performance of packages are solder bump dimension, line, and trace space as well as the materials and geometry. There are a small number of publications available for electromigration models of package interconnects. However, some extensive work has been done to characterize the failure mechanisms based on the electromigration models available for silicon metallization.

The application of the failure models for the electromigration failure is mainly focused on the SJR, including various UBM structures and solder materials. The typical time to failure due to the electromigration could be expressed as:

$$t_f = A \cdot J^{-n} \cdot exp\left(\frac{E_A}{k_B T}\right) \tag{8.24}$$

where J is the current density, T is the temperature at the site, and E_A is the activation energy. A, E_A, and n are determined by experimental results.

8.5 Corrosion Failure

The corrosion could be defined as the reaction of a metallic material with its environment. It is a process of chemical or electrochemical degradation of metallic interconnects. The rate of corrosion depends on the component materials, the availability of an electrolyte, the ionic concentration, the geometry, and the local electric field.

In electronic packages, corrosion failures were often seen in the presence of moisture and contaminants. It could show up as a failure or a potential to cause failures. The bond corrosion might not result directly in failures, and it increases the electrical resistance of interconnects until the device becomes nonfunctional. For the lead frame products, any cracks or open voids in the plating could initiate corrosion in the presence of moisture and contaminants.

McGarvey et al. [385] discussed the results under PCT and 85 °C/85%RH testing for plastic packages. The failure rate is affected by the die size, the package size, and the package geometry. Process defects were the key reasons for failures. Corrosion is the key reason for failures observed in the 85 °C/85%RH testing. PCT detected many more failures in the early stage of the testing, which is strongly related to the passivation glass on the surface of the die. Different failure modes were observed with these two different tests. For PCT testing, a large failure rate is observed early. So, we had a small beta but a long characterization life. For THB, the beta is larger than 1 and the characterization life is much shorter. It pointed out that it is meaningless to look at the characterization of life blindly.

Emerson et al. [310] studied the die corrosion resistance of a ceramic dual inline package (CerDIP) package under a 140 °C/85%RH test condition with a 40 V bias. Silicone gel-coating materials demonstrated an excellent HAST performance and only a few failures were shown at the 1100 hours readouts. Besides pad corrosion failures, the bond lift is observed too. Tran et al. [350] mentioned the detection of pad corrosion failures by PCT instead of a temperature cycling test. In addition, by improving the fabrication process, the pad corrosion failures could be resolved. The study by Gestel mentioned above [292] presented the corrosion failures seen from HAST testing with PLCC 68 packages and test chips. It was found the humidity existed causing the adhesion between die surface and mold to diminish. At delaminated area, a thin moist film may arise on the surface of the die. The film then served as an electrolyte and caused bond pads used for applying the bias to corrode. The effect of the moisture penetrating the interconnection layers will cause corrosion on the pads.

8.5.1 Corrosion Failure Models

The basic corrosion reliability models still remained tied down to a few key variables, such as the RH, the temperature, and the voltage across conductors, contaminants, and catalysts. In most cases, an Arrhenius exponential form is sufficient to represent the corrosion rate dependence on the temperature, except that the strength of such dependence is dictated by the activation energy characteristic of the rate-controlling mechanism responsible for the failure.

The growth of corrosion layers on the bonding interface, the width of corrosion layers W is given as kt, [377] where

$$k = k_0 \cdot exp\left(-\frac{E_A}{k_B T}\right) \tag{8.25}$$

The growth rate of the corrosion k is found to be related to the molding compound materials. Considering the Au_4Al alloy, the activation energy is 1.6 eV for BP resin and 2.3 eV for OCN resin.

The RH is one of the strongest factors in the corrosion process. Giacomo [386] depicted the time-to-failure expression of common metallization used in thin-film circuitries as well as thick films and interconnects,

$$t_f = K \cdot \frac{(1 - RH)[1 + (b - 1)RH]}{RH} \tag{8.26}$$

For $b \ll 1$ (most of the cases), the time to failure due to corrosion could be written as

$$t_f = K \cdot \frac{(1 - RH)^2}{RH} \tag{8.27}$$

where K and b are the constants and the RH is the relative humidity. K and b could be determined by experimental results.

Other curve-fitting models were those of a vapor-pressure type with the RH plus the temperature. The time to median failures could be expressed as the equation below, respectively.

$$t_f \propto exp\left(\frac{B}{V_p}\right) \tag{8.28}$$

where

$$V_p = RH \cdot V_{s0} \cdot exp\left[-\frac{10500}{k_B}\left(\frac{1}{T} - \frac{1}{T_0}\right)\right] \tag{8.29}$$

V_{s0} is the saturated water vapors at reference temperature T_0.
or

$$t_f \propto exp[-B \cdot (t + RH)] \tag{8.30}$$

where the growth of filaments due to temperature, humidity, and voltage loads are defined as follows:

$$t_f = \frac{a \cdot f \cdot \left(1000 \cdot L_{eff}\right)^n}{V^m \cdot (M - M_t)} \text{ when } M > Mt \tag{8.31}$$

$t_f = \infty$ when $M < M_t$, a is the filament formation AF, f is the multilayer correction factor, from 1 to 3, n is the geometry AF, V is the applied voltage bias, m is the voltage AF, M is the percentage moisture contents, M_t is the percentage of the threshold moisture content, L_{eff} is the effective length between conductors. $L_{eff} = kL$, k is the shape factor, from 0.5 to 2.

Chlorine-induced corrosion of the Al alloy is found to be the dominating failure mechanism in HAST testing. The temperature- and humidity-biased accelerated life test data fits into the form of [387]:

$$t_f = A \cdot exp\left(-\frac{E_A}{k_B T}\right) \cdot exp\left[-B \cdot (RH)^2\right] \tag{8.32}$$

Striny et al. [348] described a similar model to depict the failures of Al corrosion under temperature and humidity conditions.

$$t_f = A \cdot exp\left(\frac{E_A}{k_B T} + \frac{B}{RH}\right) \tag{8.33}$$

Peck et al. [94] summarized many failure data available and presented the relationship between failure time and stress factors as

$$t_f = A \cdot (RH)^n \cdot exp\left(\frac{E_A}{k_B T}\right) \tag{8.34}$$

and E_A is about 0.79 eV and n is -2.66.

8.6 Failure Rate and Acceleration Factors

The package or product reliability is measured by a failure unit (FIT) in production or in the field. The FIT is a rate, defined as the number of expected device failures per billion-part hours. A system reliability model is a prediction of the expected mean time between failures (MTBF) for an entire system as the sum of the FIT rates for every component. A point estimate of a failure rate could be defined as

$$FR = \frac{Number\ of\ Fails}{Total\ Unit\ Hours} = \frac{r}{\sum t_i} = \frac{r}{n \cdot h} \tag{8.35}$$

To calculate the failure rate for conditions, there would be no failure. This equation could be modified and expressed as the following equation:

$$FR = \frac{Number\ of\ Fails}{Total\ Unit\ Hours} = \frac{r+1}{\sum t_i} = \frac{r+1}{n \cdot h} \tag{8.36}$$

The failure rate of electronics items stayed constant with time and followed an exponential distribution [388]. The initial failures and wear-out failures during use time could be reduced by broadening and deepening the knowledge of materials. Further, the failure rate of the total packages or a system under various conditions is influenced significantly by the AFs and can be expressed as:

$$FR_{total} = FR_{temperature} + FR_{temperature\ \&\ humidity} + FR_{Voltage} + FR_{cyclic} + ... \tag{8.37}$$

When more than one mechanism exists in a system, the relative acceleration of each one must be identified and averaged under the applied conditions. Every potential failure mechanism should include its unique AF that should then be calculated for each mechanism at given stress factors so that the FIT can be approximated for each mechanism separately.

Each AF needs to be accurately modeled and validated so that the FR is properly added to the summation for the total failure rate.

One important task in reliability modeling is the choosing of an appropriate value for the AF based on the physics of the dominant failure mechanisms that would occur in the field. A physical acceleration model is chosen by identifying the failure mode and what stresses were relevant, and lining up fitting stress modes based on a particular model for this mechanism. Some of the AF models of the individual stress factors are listed below:

1) Temperature acceleration

A commonly accepted high-temperature acceleration model is the Arrhenius model. The AF due to temperature is determined from the Arrhenius equations as:

$$\mathrm{AF}_{temperature} = exp\left[\left(\frac{E_A}{k_B}\right) \cdot \left(\frac{1}{T_{field}} - \frac{1}{T_{test}}\right)\right] \tag{8.38}$$

The key parameter in this model is the activation energy. The value of activation energy can either be determined from reliability studies at multiple temperatures or by choosing activation energy that is commonly equated with a particular type of failure mechanism.

2) Voltage acceleration

An AF due to voltage stress can be expressed as

$$\mathrm{AF}_{voltage} = exp\left[B \cdot \left(V_{test} - V_{field}\right)\right] \text{ or } \mathrm{AF} = \left(\frac{V_{test}}{V_{field}}\right)^n \tag{8.39}$$

3) Humidity acceleration

The AF due to humidity for electronics packages can be determined as follows:

$$\mathrm{AF}_{Humidity} = exp\left[0.08 \cdot \left(RH_{test} - RH_{field}\right)\right] \text{ or } \mathrm{AF} = \left(\frac{RH_{test}}{RH_{field}}\right)^n \tag{8.40}$$

4) ΔT acceleration

Low cycle fatigue data is described by the Coffin–Manson equation. It could be used for both ductile and brittle materials. The equation could be expressed as below:

$$N_f = A_0 \cdot \left[\frac{1}{\Delta \varepsilon_p}\right]^B \tag{8.41}$$

or

$$N_f = \left(\frac{A}{\Delta \varepsilon_p}\right)^B \cdot f^m \cdot exp\left(-\frac{E_A}{k_B T_{max}}\right) \tag{8.42}$$

For temperature cycling with plastic deformation, the Coffin–Manson equation became:

$$N_f = C_0 \cdot [\Delta T - \Delta T_0]^{-n} \tag{8.43}$$

ΔT_0 is the portion of the temperature cycle range in the elastic region and ΔT is the entire temperature cycle range for the package.

Modified Coffin–Manson equation [379] could be shown as:

$$\mathrm{AF} = \left(\frac{\Delta T_{test}}{\Delta T_{field}}\right)^n \cdot \left(\frac{F_{field}}{F_{test}}\right)^m \cdot exp\left[1200\left(\frac{1}{T_{max_{field}}} - \frac{1}{T_{max_{test}}}\right)\right] \tag{8.44}$$

8.6.1 Creep

Choice of aging tests and test conditions in device packages is specific to the failure mechanims targeted. In TOC tests, the temperature cycle range selected and test duration will vary significantly depending on the expected faliures. Selection of harsh conditions might not be acceptable for the material set found in a specific application. The determination of an AF of accelerated stress tests is critical for the failure rate prediction and the time-to-failure estimation in the field application. The AF for a single failure mechanism (often referring to the dominant wear-out failure mechanism) under various stress factors has been studied and modeled. The assumption is that only the dominant failure mechanism would be accelerated by single or multiple stress factors. The AF is calculated as the product of the individual AF under each stress factor. For instance, AF of solder joint fatigue failure can be expressed as the modified Coffin–Manson equation (8.44)

Another example is the AF calculation for the corrosion failure mechanism, where the AF is expressed as [389]

$$\mathrm{AF} = \left\{\left(\frac{RH_{field}}{RH_{lab}}\right)^{3.16} \cdot exp\left[1150 \cdot \left(\frac{1}{T_{max_{field}}} - \frac{1}{T_{max_{lab}}}\right)\right]\right\}^{-1.13} \tag{8.45}$$

or [390]

$$\mathrm{AF} = \left(\frac{RH_{field}}{RH_{test}}\right)^3 \cdot exp\left(\frac{0.9}{k_B} \cdot \left(\frac{1}{T_{field}} - \frac{1}{T_{test}}\right)\right) \tag{8.46}$$

The determination of AFs should be based on the number of failure mechanisms observed, stress conditions and appropriate acceleration models. It clearly showed that the total AF for the test is the product of individual AFs of single failure mechanism interested. For instance, the AF due to temperature and humidity for a package could use the Eq. (8.47) [391]. Table 8.6 showed a summary of applicability of models to complex products.

Table 8.6 Models to complex products.

	Acceleration factor model	Note
Temperature (Arrhenius model)	$\mathrm{AF}(T) = exp\left[\dfrac{E_a}{k_B}\left(\dfrac{1}{T_{use}} - \dfrac{1}{T_{stress}}\right)\right]$	E_A: activation energy k_B: Boltzmann constant $= 8.617 \times 10^{-5}\,\mathrm{eV/k}$ T: Temperature (K)
Temperature and voltage (Eyring model)	$\mathrm{AF}(T, V) = \mathrm{AF}(T) \cdot exp$ $(B(V_{stress} - V_{use}))$	B depends on mechanism, default $B = 1$ V: voltage (V)
Temperature and relative humidity (Model for corrosion failures in plastic packages__Peck-model)	$\mathrm{AF}(T, RH) = \mathrm{AF}(T) \cdot \left(\dfrac{\mathrm{RH}_{stress}}{\mathrm{RH}_{use}}\right)^n$	$n = 3$, $E_A = 0.9\,\mathrm{eV}$ RH: Relative humidity (%)
Temperature cycling (Model for mechanical fatigue failures of solder/ other contacts__Coffin–Manson)	$\mathrm{AF}(\Delta T) = \left(\dfrac{\Delta T_{stress}}{\Delta T_{use}}\right)^C$	C depends on material mechanical properties ΔT: temperature interval, K or C

Source: Adapted from JEDEC [285].

$$\mathrm{AF}_{total} = \mathrm{AF}_{temperature} \cdot \mathrm{AF}_{Humidity} \cdot \mathrm{AF}_{Voltage} \tag{8.47}$$

Bernstein et al. [33] showed that if multiple failure mechanisms were accelerated, the AF of the package or system could not be treated as the product of the individual stress AFs. Each failure mechanism should be modeled as an individual "element" in the system and the AF would be based on a combination of competing failure mechanisms with weights on each AF from each failure mechanism.

In addition, an effective AF could be developed based on each AF of individual failure mechanism [392]. For example, effective Arrhenius activation energy could be calculated assuming cumulative failure is expressed as $1 - R(t,T)$. Then

$$\frac{1 - R(t_L, T)}{1 - R(t_L, T_0)} = exp\left[\left(-\frac{E_{A_{eff}}}{k_B}\right) \cdot \left(\frac{1}{T} - \frac{1}{T_0}\right)\right]$$

$$\text{and } ln\left(\frac{1 - R(t_L, T)}{1 - R(t_L, T_0)}\right) = -\frac{E_{A_{eff}}}{k_B}\left(\frac{1}{T} - \frac{1}{T_0}\right) \tag{8.48}$$

t_L test duration and T_0 is ordinary operational temperature.

Cooper [393] presented a model constructed using Weibull distributions for failure rates where multiple failure mechanisms are seen. The assumption is many systems did not have a predominant failure mechanism and several failure mechanisms exist with differing Arrhenius activation energies. Reliability could be expressed as

$$R(t) = \sum_{i=1}^{N} n_i \cdot exp\left[-\left(\frac{1}{t_i}\right)^{\beta_i}\right] \text{ where } t_i = \eta_i \tag{8.49}$$

and then

$$R(t) = \sum_{i=1}^{N} n_i \cdot exp\left\{\frac{t_i}{t_i(T_0)} \cdot exp\left[\left(-\frac{E_{A_i}}{k_B T_0}\right) \cdot \left(\frac{T_0}{T} - 1\right)\right]\right\}^{\beta_i}, \quad t_i(T_0) = \eta_i(T_0) \tag{8.50}$$

n_i – fractional subpopulation for each failure mechanism; N – number of failure mechanisms.

Each failure mechanism or subpopulation had its own Weibull shape parameter and it is own time scale parameters where $\sum_{i=1}^{N} n_i = 1$. In field applications, the failure rate is often treated as constant and β is approximately 1 [286].

Besides the constant failure rate approach, the failure intensity concept has been discussed in the reference [394]. The author used the failure at certain time intervals and the failure rate could be considered constant in that range. It meant at certain time interval t, the failure rate could be treated as constant, where $FI(t) = \dfrac{n}{N \cdot \Delta t}$ is the failure intensity factor, n is observed failures during observation period Δt and N is the population and Δt is the time interval, $Ii\ (t)$ is the failure intensity of device I that is part of the system in ΔT.

System failure intensity can be expressed as

$$FI(t) = \prod_{i=1}^{n} exp[-I_i(t)] = exp\left[\sum_{i=1}^{n} I_i(t)\right] \tag{8.51}$$

In 1992, Moura [395] proposed a similar approach which did not require the assumption of a constant system failure rate. An average failure rate is needed in certain time interval of the interest. The average failure rate in the exact time interval of the interest is estimated for each component at the use conditions.

$$\lambda_T = \sum_{i=1}^{m} n_i \cdot \lambda_i \tag{8.52}$$

n_i is the quantity of components of type I and λ_i is the average failure rate of component type i.

$$AF = \sum_{i=1}^{m} n_i \cdot \lambda_i \cdot AF_i / \sum_{i=1}^{m} n_i \cdot \lambda_{ui} \tag{8.53}$$

The equivalent AF is a weighted average of the AF of each technology type in the subassembly.

The failure rate calculation is expected to use data from field return data about similar products/packages. Data from the accelerated test and the "monitoring" test [396]. The MTTF, expressed as a reverse of failure rate, is not appropriate to describe the reliability life [388]. Although the failure rates of many items follow an exponential distribution, there is a point of inflection after some time. The resulting distribution resembled a hockey stick line which lead to the assumption that the initial failures of all components are eliminated. The authors proposed a BX life concept which is changing according to the shape parameter in the Weibull distribution in spite of the same MTTF.

8.7 Reliability Prediction of Electronic Packages

In this section, we will review the reliability functions, failure models, and failure rate calculations which were related to the reliability prediction of electric packages.

8.7.1 Reliability and Failure Description

The shape parameter of Weibull distributions in failure data is an important trend indicator for failure mechanisms. Weibull distributions with a β value close to one indicates a failure trend relatively constant with time. A failure rate that decreases with test duration corresponds to the early life portion of the failure curve, and the shape parameter must be between 0 and 1. β is approximately 1 for defects with random failures. For failures that are end-of-life or ware-out failures, β is considerably larger than 1. Table 8.7 summarizes some of the values of shape parameters for package failure mechanisms available.

Products that sport failure rate trends that strictly follow the bathtub curve, will be difficult to sell in the market due to excess early failures in the field. In practice, the initial failures must be minimized, or even reduced to zero. In addition, random failures must be minimized as well as failures due to wear and tear during the product's operating life. It should extend product life by delaying the onset of wear-related failures.

8.8 Reliability Failure Models

In defining the probability of failure, the focus is on the statistical distribution of failures rather than the physical or chemical phenomena that cause the failures. In the bathtub curve, the flat middle section has a constant failure rate due to random failures, whose expected time period of occurrence is unpredictable

Table 8.7 Beta (β) value of Weibull distribution for some failure mechanisms.

Sources	Failure mechanism/test types	Description
[93]	Substrate via cracking (Temperature cycling)	$\beta = 1$–2
[309]	Plastic-die interface delamination (Temperature cycling)	$\beta = 0.8798$
[336]	Solder joint cracking (Temperature cycling)	$\beta = 2 - 7$ for various solder alloys, surface finish and test conditions
[382]; [381]; [397]	Solder joint crack failures (Temperature cycling)	$\beta = 2.6 - 5.0$
[340]	Solder joint fatigue failures (Aging + temperature cycling)	$\beta = 5.8 - 7.1$
[398]	Component failures	$\beta = 0.3 - 0.7$
[399]	Die cracking and thin film delamination failures	$\beta = 0.4 - 1.5$

and independent of prior use. Wear-out failures are usually a life-limiting physical or chemical process, which is inherently related to the design of the part and its manner of application. Wear-out failure generally arises from interaction of design-related factors and one or more environmental parameters, such as the temperature and humidity, and the thermal and mechanical cycling loads.

A failure mode is the recognizable electrical symposium by which failure is observed. A failure mechanism is the specific physical, chemical, metallurgical, or environmental phenomenon or process that causes the device degradation or malfunction. Failure modes and mechanisms are the end results of the degenerative processes initiated by interactions of the designed and manufactured configuration with the operational and environmental stresses imposed during its period of operation.

Two different stress models were used in calculating the reliability and the time to failure of electronic package. One is the power law model and the other is the Arrhenius model. Sometimes these two models were combined to predict the life of components in field application reliability study.

8.8.1 Inverse Power Law Models

A flexible model (that has been very successful in many applications) for the expected number of failures in the first t hours, $t(t)$, is given by the polynomial:

$$t(t) = a \cdot t^b \quad \text{for } a, b > 0 \tag{8.54}$$

Coffin–Manson models or their modifications were power models. One of the critical tasks during the reliability stress study is to calculate the exponent used in Coffin–Manson or Norris–Landzberg models. A summary of the constants from literature is listed in Table 8.8. It is obvious that the exponent parameter varies with the materials used and the dominant failure mechanisms detected.

8.8.2 Arrhenius Models

Temperature had significant influences on the package reliability. The time to failure could be expressed as:

$$t_f = A \cdot exp\left(-\frac{E_A}{k_B T}\right) \tag{8.55}$$

Table 8.8 Exponent parameters used in Coffin–Manson models.

Authors/sources	Mechanism/materials	Exponent for Coffin–Manson models
[379]	Solder (97Sn/3Sn)	1.9
[400]	Solder (37Pb/63Sn)	2.27
[401]; [402]; [403]	Solder (37Pb/63Sn)	1.2–2.7
[404]	Cu and lead frame alloys	2.7
[405]	Al wire bonds	3.5
[392]	Au4Al fracture in WB	4.0
[406]	PQFP delamination/bond failure	4.2
[291]		5.3
[363]	Au wire down bond heel crack	5.1
[300]	Interlayer dielectric cracking	5.5 ± 0.7
[407]	Silicon fracture	5.5
[408]	Barrel cracking	2.0
[93]	Substrate via cracking	4.2
[408]	IPN cracking in substrates	1.25
[392]	Si fracture	7.1
[299]; [408]	Thin film cracking	8.4 6.0
[288]	NA	3.0–10: brittle materials 1.5–2.5: ductile 1–2: hard metal

The Arrhenius model is often used to model the impacts of temperature on failure mechanisms. The activation energy E_A has significant value on the final AF and the failure rate. However, E_A should be used with care to model a device failure rate since it is often insufficient to depend on the activation energy along to characterize devices in interest [409]. Through the years, a large amount of data has been accumulated for the activation energy for various failure mechanisms. Some of the activation energy values are shown in Table 8.9.

In practice, the entire concept of the application of a single Arrhenius energy to analyze and extrapolate experimental failure rate data derived from the life testing of ICs had been called into question. There is a concern that many systems did not have a predominant failure mechanism, but several failure mechanisms with differing Arrhenius activation energies. Cooper et al. [411] proposed a model to calculate an effective Arrhenius AF which could be determined by the ratio of the

Table 8.9 Failure mechanisms and activation energy (E_A) obtained.

Failing components and failure mechanisms		Activation energy, E_A (eV)	Driving stress	Sources
Bonds and their interfaces	Neck broken	0.70	Temperature; Δt	
	Lifted bonds	1.26		
	Intermetallic degradation	0.8		
Corrosion of bond pads and metal traces		0.53–0.7 0.6–1.0	RH, temperature	[348]; [410]; [288]
Die cracking		NA	Delta T	
Passivation defect failure		0.79; 0.56	Temperature, RH	[94]; [297]
Thin film cracking		NA	Delta T	
Solder electromigration		0.64–0.72 for pb-free solders with Cu UBM; 1.03–1.11 with Ni-UBM	Current and temperature	
Inter-layer dielectric		0.68	Delta T	
Micro cracking		0.4–0.95	Delta T	
Thermal interface degradation		0.45	Temperature and humidity	[408] [347]
IMC		0.76	Stress and temperature	[346]
Oxidation		1.3–2.0	Temperature	[408]

cumulative failures at an elevated temperature and an optimal temperature at the end of the stress testing. The effective Arrhenius AF could be expressed as

$$\frac{1 - R(t, T)}{1 - R(t, T_0)} = exp\left[-\frac{E_{A_{eff}}}{k_B}\left(\frac{1}{T} - \frac{1}{T_0}\right)\right] \tag{8.56}$$

$$Ln\left(\frac{1 - R(t, T)}{1 - R(t, T_0)}\right) = -\xi_{eff}\Gamma \tag{8.57}$$

ξ_{eff} is the effective normalized activation energy.

The reliability models must be obtained for their specific operating conditions. Reliability modeling is the ultimate task of accelerated testing and the failure physics of evaluation. Reliability models consisted of a combination of a life distribution and the life-stress relationship.

8.8.3 Arrhenius–Weibull Models

At an absolute temperature T, the product life had a Weibull distribution. The Weibull shape parameter β is a constant (independent of temperature). The natural log of the Weibull characteristic life is a linear function of the inverse of T:

$$Ln[\alpha(T)] = \gamma_0 + \frac{\gamma_1'}{T} \tag{8.58}$$

The parameters γ_0, γ_1, and β were the characteristics of the product and test methods, and were estimated from the data. At an absolute temperature T, the cumulative distribution function could be expressed as:

$$F(t, T) = 1 - exp\left[-\left(\frac{t}{\eta(T)}\right)^{\beta}\right]$$

$$F(t, T) = 1 - exp\left\{-\left[t\,exp\left[-\gamma_0 - \left(\frac{\gamma_1'}{T}\right)\right]\right]^{\beta}\right\} \tag{8.59}$$

A high β value corresponded to a narrow distribution of Ln life, and a low β value corresponded to a wide distribution of Ln life. The $100P^{th}$ percentile could be expressed as:

$$\tau_p(T) = \eta(T)[-Ln(1 - P)]^{\frac{1}{\beta}}$$

$$\tau_p(T) = exp[\gamma_0 + \gamma_1(1000/T)] \cdot [-Ln(1 - P)]^{1/\beta} \tag{8.60}$$

The inverse power relationship is widely used to model the product life as a function of accelerating stress and can be expressed as:

$$t(V) = \frac{A}{V^n} \tag{8.61}$$

For the Power-Weibull model, life had a Weibull life distribution whose characteristic life is a power function of stress. The assumptions were:

a) At stress level V, the product life had a Weibull distribution
b) The Weibull shape parameter β is a constant.
c) The Weibull characteristic life is an inverse power function of V,

$$\eta(V) = exp\left(\frac{\gamma_0}{\gamma_1}\right) \tag{8.62}$$

In summary, we have encapsulated a broad range of packaging failure modes, reliability testing methods, and factors of acceleration. As mentioned earlier, a comprehensive review of all packaging studies is beyond the scope of this work. The microelectronics community is undergoing substantial transition with the introduction of a myriad of new advanced packaging assemblies. This reality will undoubtedly change the face of packaging reliability. Advanced packaging reliability studies are planned for future work, so please stay tuned, stay safe and be reliable.

References

1 DoD, *MIL-HDBK-217, Military Handbook for Reliability prediction of Electronic Equipement*, Washington, DC: DoD, 1991.

2 G. H. Ebel, "Reliability Physics in Electronics: A Historical View," *IEEE Transactions on Reliability*, vol. 47, no. 3, pp. 379–389, 1998.

3 M. Pecht and F. Nash, "Predicting the reliability of electronic equipment," *Predicting the IEEE*, vol. 82, no. 7, pp. 992–1004, 1994.

4 A. Goel and R. Graves, "Electronic System Reliability: Collating Prediction Models," *IEEE Transactions on Device and Materials Reliability*, vol. 6, no. 2, pp. 258–265, 2006.

5 Relyence Corp., *"Reliability Prediction Analysis: More Than MTBF,"* 2021. [Online]. Available: https://relyence.com/wp-content/uploads/2021/10/ Reliability-Prediction-More-Than-MTBF.pdf. [Accessed 10 October 2023].

6 Reliability Prediction Standard Development WG, *IEEE Guide for Selecting and Using Reliability Predictions Based on IEEE 1413*," IEEE Std 1413.1-2002, pp. 1–106, 2003.

7 W. Denson, "The History of Reliability Prediction," *IEEE Transaction on Reliability*, vol. 47, no. 3-SP, pp. 321–328, 1998.

8 DoD MIL Standard, *MIL-HDBK-338 B: Military Handbook Electronic Reliability Design Handbook*, Philadelphia, PA: Standardization Documents DoD, 1998.

9 M. Kwiatkowska, G. Norman and D. Parker, "PRISM 4.0: Verification of Probabilistic Real-time Systems," in *Proc. 23rd International Conference on Computer Aided Verification (CAV'11), volume 6806 of LNCS* - Springer, pp. 585–591, 2011.

10 B. Foucher, J. Boullié, B. Meslet and D. Das, "A Review of Reliability Prediction Methods for Electronic Devices," *Microelectronics Reliability*, vol. 42, pp. 1155–1162, 2002.

11 J. Marin, "Experience Report on the FIDES Reliability Prediction Method," in Alexandria, VA, *Reliability and Maintainability Symposium* - Proceedings Annual, 2005.

Reliability Prediction for Microelectronics, First Edition. Joseph B. Bernstein, Alain A. Bensoussan, and Emmanuel Bender.
© 2024 John Wiley & Sons Ltd. Published 2024 by John Wiley & Sons Ltd.

12 DGA, "FIDES Guide - Reliability Methodology for Electronic Systems, version A," French Standard UTE C80811, 2009. [Online]. Available: http://www.fides-reliability.org. [Accessed 10 October 2023].

13 J. Jones and J. Hayes, "A Comparison of Electronic-Reliability Prediction Models," *IEEE Transactions on Reliability*, vol. 48, no. 2, pp. 127–134, 1999.

14 J. Bowles, "A Survey of Reliability-Prediction Procedures for Microelectronic Devices," *IEEE Transactions on Reliability*, vol. 41, no. 1, pp. 2–12, 1992.

15 M. Cushing et al., "Comparison of Electronic-Reliability Assessment Approaches," *IEEE Transactions on Reliability*, vol. 42, no. 4, pp. 542–546, 1993.

16 S. Salemi, L. Yang, J. Dai, J. Qin and J. Bernstein in Utica, NY, Nicholls, D. (ed.), *Physics-of-Failure Based Handbook of Microelectronic Systems* - Reliability Information Analysis Center, 2008.

17 U.S. Nuclear Regulatory Commission Office of Nuclear Regulatory Research Washington, DC 20555-0001, Report NUREG-1774: A Survey of Crane Operating Experience at U.S. Nuclear Power Plants from 1968 through 2002, July 2003. [Online]. Available: https://www.nrc.gov/reading-rm/doc-collections/nuregs/staff/sr1774/sr1774.pdf. [Accessed 3 October 2023].

18 H. Jones, "Common Cause Failures and Ultra Reliability." American Institute of Aeronautics and Astronautics, NASA Ames Research Center, Technical report N0 20160005837, in International Conference on Environmental Systems, 11 p., Moffett Field, CA, 2016.

19 S. Morris and J. Reilly, "MIL-HDBK-217 - A Favorite Target," in Proceeding Annual Reliability and Maintainability Symposium, Atlanta, CA, 1993, pp. 503–509.

20 S. Morris, "Use and application of MIL-HDBK-217," *Solid State Technology*, vol. 33, no. 8, p. 65+, 1990.

21 J. Qin and J. B. Bernstein, "Non-Arrhenius Temperature Acceleration and Stress-Dependent Voltage Acceleration for Semiconductor Device Involving Multiple Failure Mechanisms," in IEEE International Integrated Reliability Workshop Final Report, pp. 93–97, South Lake Tahoe, CA, 2006.

22 L. Baum and T. Petrie, "Statistical Inference for Probabilistic Functions of Finite State Markov Chains," *The Annals of Mathematical Statistics*, vol. 37, no. 6, pp. 1554–1563, 1966.

23 M. Davis, "Piecewise-Deterministic Markov Processes: A General Class of Non-Diffusion Stochastic Models," *Journal of the Royal Statistical Society: Series B (Methodological)*, vol. 46, no. 3, pp. 353–376, 1984.

24 A. Cabarbaye, A. Cabarbaye, A. Bensoussan, O. Gilard, L. S. How and F. Coccetti, "Characterization of the Wear of Electronic Components and Estimation of Their Remaining Useful Life (RUL)," in 22e Congrès de Maîtrise des Risques et Sûreté de Fonctionnement λμ22, Le Havre (France), 2020.

25 A. Kleyner and R. Knoell, "Calculating Probability Metric for Random Hardware Failures (PMHF) in the New Version of ISO 26262 Functional Safety -

Methodology and Case Studies," in SAE Technical Paper 2018-01-0793, Detroit, MI, 2018.

26 P. O'Connor and A. Kleyner, *Practical Reliability Engineering;* 5 Ch.2 Edition, New York, NY: Wiley, 2012.

27 E. J. Gumbel, *Statistics of Extremes*, New York: Columbia University Press, 1958.

28 A. Kolmogorov, "On the Logarithmically Normal Law of Distribution of the Size of Particles Under Pulverization," *Doklady Akademii Nauk SSSR*, vol. 31, no. 99, pp. 281–284, 1941.

29 J. Bisschop, "Reliability methods and standards Microelectronics Reliability," *Microelectronics Reliability*, vol. 47, no. 9-11, pp. 1330–1335, 2007.

30 Toshiba, *Reliability Handbook*, Tokyo, JPN: Toshiba Electronic Devices & Storage Corporation, 2017.

31 ESA-ESCC-ESCIES, "https://ecss.nl/standard," ESA ECSS, 2009. [Online]. Available: https://ecss.nl/standard/ecss-e-st-32-10c-rev-1-structural-factors-of-safety-for-spaceflight-hardware/. [Accessed 2022].

32 Renesas, *Semiconductor Reliability Handbook*, R51ZZ0001EJ0250 Rev. 2.50, http://www.renesas.com, Renesas Electronics, 2017.

33 J. Bernstein, M. Gurfinkel, X. Li, J. Walters, Y. Shapira and M. Talmor, "Electronic Circuit Reliability Modeling," *Microelectronics Reliability*, vol. 46, no. 12, pp. 1957–1979, 2006.

34 M. Williams and M. Throne, "The Estimation of Failure Rates for Low Probability Events," *Progress in Nuclear Energy*, vol. 31, no. 4, pp. 373–476, 1997.

35 F. Coolen and P. Coolen-Schrijner, "On Zero-Failure Testing for Bayesian High-Reliability Demonstration," *Proceedings of the Institution of Mechanical Engineers, Part O: Journal of Risk and Reliability*, vol. 220, no. 1, pp. 35–44, 2006.

36 F. Coolen, "On Probabilistic Safety Assessment in the Case of Zero Failures," *Proceedings of the Institution of Mechanical Engineers, Part O: Journal of Risk and Reliability*, vol. 220, no. 1, pp. 105–114, 2006.

37 P. Tobias and D. Trindade, *Applied Reliability*, Boca Raton, FL: CRC Press, Chapman & Hall, 2011.

38 D. Meade, "Failure Rate Estimation in the Case of Zero Failures," in Austin, TX, *SPIE 3216, Microelectronic Manufacturing Yield, Reliability, and Failure Analysis III*, 1997.

39 M. Krasich, "How to Estimate and Use MTTF/MTBF Would the Real MTBF Please Stand Up?" in Fort Worth, TX, *IEEE Annual Reliability and Maintainability Symposium*, pp. 353–359, 2009.

40 Microsemi, "Microsemi, RT0001 Reliability Report Microsemi FPGA and SoC Products, Revision 17.0," Microsemi, 2019. [Online]. Available: https://www.microsemi.com/document-portal/doc_download/131371-rt0001-microsemi-fpga-and-soc-products-reliability-report. [Accessed 23 March 2023].

41 H. Caruso and A. Dasgupta, "A Fundamental Overview of Accelerated-Testing Analytic Model," in Anaheim, CA, *Annual Reliability and Maintainability*

Symposium - Proceedings International Symposium on Product Quality and Integrity, 1998.

42 W. Vigrass, "Calculation of Semiconductor Failure Rates," Renesas, 2001. [Online]. Available: https://www.renesas.com/eu/en/document/qsg/calculation-semiconductor-failure-rates. [Accessed June 2023].

43 JEDEC, *JEP122H, Failure Mechanisms and Models for Semiconductor Devices*, Arlington, VA: JEDEC Solid State Technology Association, 2016.

44 JEDEC, *JESD85, (July 2001, reaffirmed January 2014), Methods for Calculating Failure Rates in Units of FITs*, Arlington, VA: JEDEC Solid State Technology Association, 2014.

45 M. White, *Product Reliability and Qualification Challenges with CMOS Scaling*, AVSI Consortium, 2005.

46 F. Proschan, "Theoretical Explanation of Observed Decreasing Failure Rate," *American Statistical Association and The America Society for Quality; Technometrics*, vol. 5, no. 3, pp. 375–383, 1963.

47 R. Barlow, A. Marshall and F. Proschan, "Properties of Probability Distributions with Monotone Hazard Rate," *Annals of Mathematical Statistics*, vol. 34, no. 2, pp. 375–389, 1963.

48 J. Gurland and J. Sethuraman, "Reversal of Increasing Failure Rates When Pooling Failure Data," *American Statistical Association and the American Society for Quality Control, Technometrics*, vol. 36, no. 4, pp. 416–418, 1994.

49 L. Gleser, "The Gamma Distribution as a Mixture of Exponential Distributions," *American Statistician*, vol. 43, no. 2, pp. 115–117, 1989.

50 D. Cox, "Regression Models and Life-Tables," *Journal of the Royal Statistical Society. Series B (Methodological)*, vol. 34, no. 2, pp. 187–220, 1972.

51 J. Mi, "Limiting Behavior of Mixtures of Discrete Lifetime Distributions," *Naval Research Logistics*, vol. 43, no. 3, pp. 365–380, 1996.

52 H. Block and H. Savits, "Burn in," *Statistical Science*, vol. 12, no. 1, pp. 1–19, 1997.

53 H. Block and H. Joe, "Tail Behavior of the Failure Rate, Functions of Mixtures, Lifetime Data Analysis," *Lifetime Data Analysis*, vol. 3, pp. 269–288, 1997.

54 K. Balakrishnan, *The Exponential Distribution. Theory, Methods and Applications*; 1st Edition, London, UK: Routledge. Gordon and Breach Publishers, 1995.

55 R. Drenick, "The Failure Law of Complex Equipment," *Journal of the Society for Industrial and Applied Mathematics*, vol. 8, no. 4, pp. 680–690, 1960.

56 D. Kececioglu, *Reliability Engineering Handbook*, Ch. 13, Englewood Cliffs: Prentice-Hall, NJ, 1991.

57 K. Murphy, C. Carter and S. Brown, "The Exponential Distribution: the Good, the Bad and the Ugly. A Practical Guide to Its Implementation," in Seattle, WA, *Proceeding Annual Reliability and Maintainability Symposium*, 2002.

58 W. Denson, "The History of Reliability Prediction," *IEEE Transactions on Reliability*, vol. 47, no. 3, pp. 321–328, 1998.

59 J. Stathis, "Reliability Limits for the Gate Insulator in CMOS Technology," *IBM Journal of Research and Development*, vol. 46, no. 2/3, pp. 265–286, 2002.

60 A. Haggag et al., "Reliability Projections of Product Fail Shift and Statistics Due to HCI and NBTI," in Phoenix, AZ, *International reliability Physics Symposium (IRPS)*, pp. 93–106, 2007.

61 W. Bornstein et al., "Field Degradation of Memory Components Due to Hot Carriers," in Arlington, VA, *International reliability Physics Symposium (IRPS)*, pp. 294–298, 2006.

62 J. Black, "Electromigration - A Brief Survey and Some Recent Results," *IEEE Transactions on Electron Devices*, vol. ED-16, p. 388, 1969.

63 G. La Rosa et al., "NBTI-Channel Hot Carriers Effects in PMOSFETs in Advanced CMOS Technologies," in Denver, CO, *International reliability Physics Symposium (IRPS)*, pp. 282–286, 1997.

64 Q. Zhu, *Power Distribution Network Design for VLSI*, Tempe, AZ: Wiley Interscience, 2004.

65 I. Blech, "Electromigration in Thin Aluminum Films on Titanium Nitride," *Journal of Applied Physics*, vol. 47, no. 4, pp. 1203–1208, 1976.

66 M. White, D. Vu, D. Nguyen, R. Ruiz, Y. Chen and J. Bernstein, "Product Reliability Trends, Derating Considerations and Failure Mechanisms with Scaled CMOS. Final Report," NASA-JPL IIRW, Pasadena, AZ, 2006.

67 L. Condra and G. Horan, *"Impact of Semiconductor Technology on Aerospace Electronic System Design, Production, and Support," in National Software and Complex Electronic Hardware Standardization Conference*, VA: Norfolk, 2005.

68 L. Escobar and W. Meeker, "A Review of Accelerated Test Models," *Institute of Mathematical Statistics*, vol. 21, no. 4, pp. 552–577, 2006.

69 P. Levy, *Theorie de l'Addition des Variables Aleatoires*; 2nd Edition, Paris: Gauthier-Villars, 1954.

70 Microsemi, *RT0001 Reliability Report Microsemi FPGA and SoC* Products rev. 17.0, Microsemi, 2018.

71 M. Alam and S. Mahapatra, "A comprehensive model of PMOS NBTI degradation," *Microelectronics Reliability*, vol. 45, no. 1, pp. 71–81, 2005.

72 M. Alam, H. Kufloglu, D. Varghese and S. Mahapatra, "A comprehensive model for PMOS NBTI degradation: Recent progress," *Microelectronics Reliability*, vol. 47, no. 6, pp. 853–862, 2007.

73 H. Kufluoglu and M. Alam, "A Geometrical Unification of the Theories of NBTI and HCI Time Exponents and Its Implications for Ultra-Scaled Planar and Surround-Gate MOSFETs," in San Francisco, CA, *Proceeding International Electron Devices Meeting (IEDM)*, 2004.

74 S. Mahapatra, D. Saha, D. Varghese and P. Kumar, "On the generation and recovery of interface traps in MOSFETs subjected to NBTI, FN and HCI stress," *IEEE Transactions on Electron Devices*, vol. 53, no. 7, pp. 1583–1592, 2006.

75 S. Rangan et al., "Universal Recovery Behavior of Negative Bias Temperature in Stability," in Washington, DC, *Proceeding International Electron Devices Meeting (IEDM)*, pp. 1.1.1–1.1.8, 2003.

76 M. Ershov et al., "Degradation Dynamics, Recovery, and Characterization of Negative Bias Temperature Instability," *Microelectronics Reliability*, vol. 45, no. 1, pp. 99–105, 2005.

77 V. Huard, F. Cacho and X. Federspiel, "Technology Scaling and Reliability Challenges in the Multicore Era," in Monterey, CA, *International Reliability Physics Symposium (IRPS)*, pp. 282–286, 2012.

78 H. Yunm, A. Kahirdeh, M. Christine and M. Modarres, "Entropic Approach to Measure Damage with Applications to Fatigue," in Reno, NV, *Annual Reliability and Maintainability Symposium (RAMS)*, 2018.

79 E. Wigner, "The Transition State Method," *Transactions of the Faraday Society (London)*, vol. 34, no. 0, pp. 29–41, 1938.

80 M. Evans and M. Polanyi, "Inertia and Driving Force of Chemical Reaction," *Transactions of the Faraday Society (London)*, vol. 34, no. 0, pp. 11–29, 1938.

81 H. Eyring, S. Lin and S. Lin, *Basic Chemical Kinetics*, New York-Chichester-Brisbane-Toronto: John Willey & Sons, 1980.

82 S. Glasstone, K. J. Laider and H. Eyring, *The Theory of Rate Processes: The Kinetics of Chemical Reactions, Viscosity, Diffusion and Electrochemical Phenomena*, New York: Mc Graw-Hill, 1941.

83 G. Hammond, "A Correlation of Reaction Rates," *Journal of the American Chemical Society*, vol. 77, no. 2, pp. 334–338, 1955.

84 E. Snow, A. Grove, B. Deal and C. Sah, "Ion Transport Phenomena in Insulating Films," *Journal of Applied Physics*, vol. 36, no. 5, pp. 1664–1673, 1965.

85 J. McPherson, R. Khamankar and A. Shanware, "Complementary Model for Intrinsic TDDB in SiO_2 Dielectrics," *Journal of the Semiconductor Science and Technology*, vol. 88, no. 9, pp. 5351–5359, 2000.

86 C. Kittel. Introduction to Solid State Physics. 8th ed. Wiley 2005.

87 J. McPherson and D. Baglee, "Acceleration Factors for Thin Gate Oxide Stressing," in New York, NY, *International Reliability Physics Symposium (IRPS)*, pp. 1–5, 1985.

88 J. McPherson, "Stress Dependent Activation Energy," in Anaheim, CA, *International reliability Physics Symposium (IRPS)*, pp. 12–18, 1986.

89 J. W. McPherson, *Reliability Physics and Engineering - Time-to-Failure Modeling*, TX: Springer Nature Switzerland AG, 2019.

90 P. Lall, M. Pecht and E. B. Hakim, *Influence of Temperature on Microelectronics and System Reliability*, Boca Raton, NY: CRC Press, 1997.

91 National Academy of Sciences, *Panel on Reliability Growth Methods for Defense; Reliability Growth: Enhancing Defense System Reliability*, Washington, DC. ISBN 978-0-309-31474-9: The National Academies Press, 2008.

92 M. Yoder, "Ohmic Contacts in GaAs," *Solid State Electronics*, vol. 23, no. 2, pp. 117–119, 1980.

93 C. Lee, B. Welch and W. Fleming, "Reliability of AuGe/Pt and AuGe/Ni Ohmic Contacts on GaAs," *Electronics Letters*, vol. 17, no. 12, pp. 407–408, 1981.

94 H. Cui, "Accelerated Temperature Cycle Test and Coffin-Manson Model for Electric Packaging," in Alexandria, VA, *Proceeding of Annual Reliability and Maintainability Symposium*, 2005.

95 D. Peck, "Comprehensive Model for Humidity Testing Correlation," in Anaheim, CA, *IEEE 23rd International Reliability Physics Symposium (IRPS)*, 1986.

96 I. Chen, S. Holland and C. Hu, "A Quantitative Physical Model for Time-Dependent Breakdown in SiO$_2$," in Orlando, CA, *Proceeding 23rd International Reliability Physics Symp. (IRPS)*, 1985.

97 M. Dai, C. Gao, K. Yap, Y. Shan, Z. Cao, K. Liao, L. Wang, B. Cheng and S. Liu, "A Model with Temperature-Dependent Exponent for Hot-Carrier Injection in High-Voltage nMOSFETs Involving Hot-Hole Injection and Dispersion," *IEEE Transactions on Electron Devices*, vol. 55, no. 12, pp. 1255–1258, 2008.

98 E. Takeda, Y. Nakagome, H. Kume and S. Asai, "New Hot-Carrier Injection and Device Degradation in Submicron MOSFET's," *IEEE Proceedings*, vol. 130, no. 3, pp. 144–149, 1983.

99 E. Takeda, H. Kume, T. Toyabe and S. Asai, "Submicron MOSFET Structure for Minimizing Channel Hot-Electron Injection," in Maui, HI, *Symposium on VLSI Technology*, 1981.

100 K. Decker, "GaAs MMIC Hydrogen Degradation Study," in Philadelphia, PA, *GaAs Reliability Workshop*, 1994.

101 M. Delaney, T. Wiltsey, M. Chiang and K. Yu, "Reliability of 0.25 µm GaAs MESFET MMIC Process: Results of Accelerated Lifetests and Hydrogen Exposure," in Philadelphia, PA, *GaAs Reliability Workshop*, 1994.

102 M. Ciappa, F. Carbognani and W. Fichtner, "Lifetime Modeling of Thermomechanics-Related Failure Mechanisms in High Power IGBT Modules for Traction Applications," in Cambridge, UK, *ISPSD '03, IEEE 15th International Symposium on Power Semiconductor Devices and ICs*, 2003.

103 D. Schroder and J. Babcock, "Negative Bias Temperature Instability: Road to Cross in Deep Submicron Silicon Semiconductor Manufacturing," *Journal of Applied Physics*, vol. 94, no. 1, pp. 1–18, 2003.

104 B. Agarwala et al., "Dependence of Electromigration-Induced Failure Time on Length and Width of Aluminium Thin-Film Conductors," *Journal of Applied Physics*, vol. 41, no. 10, p. 3954, 1970.

105 M. White and J. B. Bernstein, *Microelectronics Reliability: Physics-of-Failure Based Modeling and Lifetime Evaluation; JPL Publication 08-5*, Pasadena, CA: Jet Propulsion Laboratory/California Institute of Technology, 2008.

106 J. Bernstein, A. Bensoussan and E. Bender, "The Correct Hot Carrier Degradation Model," in Monterey, CA, *IEEE International Reliability Physics Symposium (IRPS)*, 2023.

107 H. Eyring, "The Activated Complex in Chemical Reactions," *Journal of Chemical Physics*, vol. 3, no. 2, p. 107, 1935.

108 S. Tyaginov, I. Starkov, H. Enichlmair, J. Park, C. Jungemann and T. Grasser, "Physics-Based Hot-Carrier Degradation Models," *ElectroChemical Society Transaction*, vol. 35, no. 4, p. 321, 2011.

109 S. Rauch, G. Guarin and G. La Rosa, "High-VGS PFET DC Hot-Carrier Mechanism and its Relation to AC Degradation," *IEEE Transactions on Device and Material Reliability*, vol. 10, no. 1, pp. 40–46, 2010.

110 H. Hess, L. Register, W. McMahon, B. Tuttle, O. Aktas, U. Ravaioli, J. Lyding and I. Kizilyall, "Theory of Channel Hot-Carrier Degradation in MOSFETs," *Physica B: Condensed Matter*, vol. 1, no. 4, pp. 527–531, 1999.

111 O. Penzin, A. Haggag, W. McMahon, E. Lyumkis and K. Hess, "MOSFET Degradation Kinetics and Its Simulation," *IEEE Transactions On Electron Devices*, vol. 50, no. 6, pp. 1445–1450, 2003.

112 A. Bravaix, C. Guerin, V. Huard, D. Roy, J. Roux and E. Vincent, "Hot-Carrier Acceleration Factors for Low Power Management in DC-AC Stressed 40nm NMOS Node at High Temperature," in Montreal, QC, *International Reliability Physics Symposium (IRPS)*, 2009.

113 E. Takeda, N. Suzuki and T. Hagiwara, "Device Performance Degradation to Hot-Carrier Injection at Energies Below the Si-SiO$_2$ Energy Barrier," in Washington, DC, *Proceeding International Electron Devices Meeting (IEDM)*, 1983.

114 D.-Y. Jeon, Y. Koh, C.-Y. Cho and K.-H. Park, "Impact of Temperature-Dependent Series Resistance on the Operation of AlGaN/GaN High Electron Mobility Transistors," *AIP Advances*, vol. 11, 115203; https://doi.org/10.1063/5.0064823, pp. 1-5, 2021.

115 K. Kim, "Reliable CMOS VLSI Design Considering Gate Oxide Breakdown," in Rome, Italy, *5th International Conference on Advances in Circuits, Electronics and Microelectronics, CENICS*, 2012.

116 J. Suñé, "Ultra Thin Gate Oxide Reliability: Physical Models, Statistics, and Characterization," *IEEE Transactions on Electron Devices*, vol. 49, no. 6, p. 958, 2002.

117 J. Walter and J. Bernstein, *Semiconductor Device Lifetime Enhancement by Performance Reduction*," Tech. report, Maryland, CA: University of Maryland, ENRE, 2003.

118 T. Garba-Seybou, X. Federspiel, A. Bravaix and F. Cacho, "New Modelling Off-state TDDB for 130nm to 28nm CMOS Nodes," in Dallas TX, *IEEE International Reliability Physics Symposium - IRPS*, 2022.

119 S. Kupke, S. Knebel, S. Rahman, S. Slesazeck, T. Mikolajick, R. Agaiby and M. Trentzsch, "Dynamic Off-State TDDB of Ultra Short Channel HKMG nFETS and Its Implications on CMOS Logic Reliability," in Waikoloa, HI, *International Reliability Physics Symposium (IRPS)*, 2014.

120 J. McPherson, "Time Dependent Dielectric Breakdown Physics - Models Revisited," *Microelectronics Reliability Journal*, vol. 52, no. 9-10, pp. 1753–1760, 2012.

121 Y. Yeo, Q. Lu and C. Hu, "MOSFET Gate Oxide Reliability: Anode Hole Injection Model and Its Applications," *International Journal of High Speed Electronics and Systems*, vol. 11, no. 3, pp. 849–886, 2001.

122 C. Hu, "A Unified Gate Oxide Reliability Model," in Dallas, TX, *International Reliability Physics Symposium (IRPS)*, 1999.

123 W. W. Abadeer, A. Bagramian, D. W. Conkle, C. W. Griffin, E. Langlois, B. F. Lloyd, R. P. Mallette, J. E. Massucco, J. M. McKenna, S. W. Mittl and P. H. Noel, "Key Measurements of Ultra Thin Gate Dielectric Reliability and In-Line Monitoring," *IBM Journal of Research and Development*, vol. 43, no. 3, pp. 407–416, 1999.

124 P. Nicollian et al., "Experimental Evidence for Voltage Driven Breakdown Models in Ultrathin Gate Oxide," in San Jose, CA, *38th Annual International Reliability Physics Symposium (IRPS)*, 2000.

125 E. Wu and J. Suñé, "Power-Law Voltage Acceleration: A Key Element for Ultra-Thin Gate Oxide Reliability," *Microelectronics Reliability Journal*, vol. 45, no. 12, pp. 1809–1834, 2005.

126 A. Strong, E. Wu, R.-P. Vollertsen, J. Suñé, G. La Rosa, S. Rauch III and T. Sullivan, *Reliability Wearout Mechanisms in Advanced CMOS Technologies*, Hoboken, NJ: Institute of Electrical and Electronics Engineers, Inc.; John Wiley & Sons, Inc., 2009.

127 D. DiMaria and J. Stasiak, "Trap Creation in Silicon Dioxide Produced by Hot Electrons," *Journal of Applied Physics*, vol. 65, no. 6, pp. 2342–2356, 1989.

128 J. Stathis and D. DiMaria, "Reliability Projections for Ultra-Thin Oxides at Low Voltage," in San Francisco, CA, *Proceeding International Electron Devices Meeting (IEDM)*, pp. 167–170, 1998.

129 J. Stathis, "Physical and Predictive Models of Ultra Thin Oxide Reliability in CMOS Devices and Circuits," in Orlando, FL, *International Reliability Physics Symposium (IRPS)*, pp. 132–149, 2001.

130 J. Suñé and E. Wu, "Hydrogen Release Mechanisms in the Breakdown of Thin SiO_2 Films," *Physical Review Letters*, vol. 92, no. 8, 087601, 2004.

131 W. McMahon, A. Haggag and K. Hess, "Reliability Scaling Issues for Nanoscale Devices," *IEEE Transactions on Nanotechnology*, vol. 2, no. 1, pp. 33–38, 2003.

132 G. Ribes, S. Bruyere, M. Denais, D. Roy and G. Ghibaudo, "Modeling Charge-to-Breakdown Using Hydrogen Multivibrational Excitation (Thin SiO_2 and High-K

Dielectrics)," in Lake Tahoe, CA, *IEEE International Integrated Reliability Workshop*, Final Report, 2004.

133 A. Haggag, N. Liu, D. Menke and M. Moosa, "Physical Model for the Power-Law Voltage and Current Acceleration of TDDB," *Microelectronics Reliability Journal*, vol. 45, no. 12, pp. 1855–1860, 2005.

134 P. Nicollian, A. Krishnan, C. Chancellor and R. Khamankar, "The Traps that Causes Breakdown in Deeply Scaled SiON Dielectrics," in San Francisco, CA, *Proceeding International Electron Devices Meeting (IEDM)*, pp. 743–746, 2006.

135 E. Wu, "Processes—Part I: Statistics, Experimental, and Physical Acceleration Models - Facts and Myths of Dielectric Breakdown," *IEEE Transactions on Electron Devices*, vol. 6, no. 11, pp. 4523–4534, 2019.

136 P. Nicollian, A. Krishnan, C. Chancellor, R. Khamankar, S. Chakravarthi, C. Bowen and V. Reddy, "The Current Understanding of the Trap Generation Mechanisms that Lead to the Power Law Model for Gate Dielectric Breakdown," in Phoenix (AZ), *IEEE 45th Annual International Reliability Physics Symposium*, 2007.

137 I. Chen, S. Holland and C. Hu, "Hole Trapping and Breakdown in Thin SiO_2," *IEEE Electron Device Letters*, vol. 7, no. 3, pp. 164–167, 1986.

138 I. Chen, S. Holland, K. Young, C. Chang and C. Hu, "Substrate Hole Current and Oxide Breakdown," *Applied Physics Letters*, vol. 49, no. 11, pp. 669–671, 1986.

139 K. Schuegraf and C. Hu, "Hole Injection SiO2 Breakdown Model for Very Low Voltage Lifetime Extrapolation," *IEEE Transactions on Electron Devices*, vol. 41, no. 5, pp. 761–767, 1994.

140 J. Bude, B. Weir and P. Silvermann, "Explanation of Stress-Induced Damage in Thin Oxides," in San Francisco, CA, *Proceeding International Electron Devices Meeting (IEDM)*, p. 179, 182, 1998.

141 K. Schuegraf and C. Hu, "Effects of Temperature and Defects on Breakdown Lifetime of Thin SiO_2 at Very Low Voltages," in San Jose, CA, *IEEE Annual International Reliability Physics Symposium*, pp. 126–135, 1994.

142 M. Alam, J. Bude and A. Ghetti, "Field Acceleration for Oxide Breakdown: Can an Accurate Anode Hole Injection Model Resolve the E versus 1/E Controversy?" in San Jose, CA, *IEEE Annual International Reliability Physics Symposium*, pp. 21–26, 2000.

143 A. Padovani, D. Gao, A. Shluger and L. Larcher, "A Microscopic Mechanism of Dielectric Breakdown in SiO2 Films: An Insight from Multi-Scale Modeling," *Journal of Applied Physics*, vol. 121, no. 15, pp. 155101-1 to 155101-10, 2017.

144 J. Noguchi, N. Ohashi, J. Yasuda, T. Jimbo, H. Yamaguchi and N. Owada, "TDDB Improvement in Cu Metallization under Bias Stress," in Dallas, TX, *IEEE Annual International Reliability Physics Symposium*, pp. 339–343, 2000.

145 N. Suzumura, S. Yamamoto, D. Kodama, K. Makabe, E. Murakami, S. Maegawa and K. Kubota, "A New TDDB Degradation Model Based on Cu Ion Drift in Cu

Interconnect Dielectrics," in San Jose, CA, *IEEE International Reliability Physics Symposium Proceedings*, 2006.

146 F. Chen, O. Bravo, K. Chanda, P. McLaughlin, T. Sullivan, J. Gill, J. Lloyd, R. Kontra and J. Aitken, "Comprehensive Study of Low-k SiCOH TDDB Phenomena and Its Reliability Lifetime Model Development," in San Jose, CA, *IEEE International Reliability Physics Symposium (IRPS)*, 2006.

147 J. Lloyd, E. Liniger and T. Shaw, "Simple Model for Time-Dependent Dielectric Breakdown in Inter- and Intralevel Low-k Dielectrics," *Journal of Applied Physics*, vol. 98, no. 8, pp. 084109:1–084109:10, 2005.

148 J. Lloyd, M. Lane, X.-H. Liu, E. Liniger, T. Shaw, C. Hu and R. Rosenberg, "Reliability Challenges with Ultra Low-k Interlevel Dielectrics," *Microelectronics Reliability Journal*, vol. 44, no. 9-11, pp. 1835–1841, 2004.

149 R. Degraeve, G. Groeseneken, R. Bellens, M. Depas and H. Maes, "A Consistent Model for the Thickness Dependence of Intrinsic Breakdown in Ultra-Thin Oxides," in Washington, DC, *Proceedings of International Electron Devices Meeting*, pp. 863–866, 1995.

150 J. Suñé, I. Placencia, N. Barniol, E. Farrés, F. Martín and X. Aymerich, "On the Breakdown Statistics of Very Thin SiO₂ Films," *Thin Solid Films*, vol. 185, no. 2, pp. 347–362, 1990.

151 D. Dumin, S. Mopuri, S. Vanchinathan, R. Scott, R. Subramoniam and T. Lewis, "High Field Related Thin Oxide Wearout and Breakdown," *IEEE Transactions on Electron Devices*, vol. 42, no. 4, pp. 760–772, April 1995.

152 S. Mahapatra, C. Parikh, V. Rao, C. Viswanathan and J. Vasi, "Device Scaling Effects on Hot-Carrier Induced Interface and Oxide-Trapping Charge Distributions in MOSFETs," *IEEE Transactions on Electron Devices*, vol. 47, no. 4, pp. 789–796, 2000.

153 J. Maserjian and N. Zamani, "Observation of Positively Charged State Generation Near the Si/SiO₂ Interface During Fowler-Nordheim Tunneling," *Journal of Vacuum Science and Technology*, vol. 20, no. 3, pp. 743–746, 1982.

154 R. Rofan and C. Hu, "Stress-Induced Oxide Leakage," *IEEE Electron Device Letters*, vol. 12, no. 11, pp. 632–634, 1991.

155 P. Olivio, T. Nguyen and B. Ricco, "High Field Induced Degradation in Ultra-Thin SiO₂ Films," *IEEE Transactions on Electron Devices*, vol. 35, no. 12, pp. 2259–2265, 1988.

156 N. Matsukawa, S. Yamada, K. Amemiya and H. Hazama, "A Hot Hole Induced Low-Level Leakage Current in Thin Silicon Dioxide Films," *IEEE Transactions on Electron Devices*, vol. 43, no. 11, pp. 1924–1929, 1996.

157 D. Dumin and J. Maddux, "Correlation of Stress-Induced Leakage Current in Thin Oxides with Trap Generation Inside the Oxides," *IEEE Transactions on Electron Devices*, vol. 40, no. 5, pp. 986–993, 1993.

158 M. Kato, N. Miyamoto, H. Kume, A. Satoh, M. Ushiyama and K. Kimura, "Read-Disturb Degradation Mechanism due to Electron Trapping in the Tunnel Oxide for Low-Voltage Flash Memories," in San Francisco, CA, *Proceeding International Electron Devices Meeting (IEDM)*, pp. 45–48, 1994.

159 A. Chou, K. Lai, K. Kumar, P. Chowdhury and J. Lee, "Modeling of Stress-Induced Leakage Current in Ultrathin Oxides with the Trap Assisted Tunneling Mechanism," *Applied Physics Letters*, vol. 70, no. 25, p. 3407, 1997.

160 T. Wang, T. Chang, L. Chiang, C. Wang, N. Zous and C. Huang, "Investigation of Oxide Charge Trapping and Detrapping in a MOSFET by Using a GIDL Current Technique," *IEEE Transaction on Electron Devices*, vol. 45, no. 7, pp. 1511–1517, 1998.

161 H. Belgal, N. Righos, I. Kalastirsky, J. Peterson, R. Shiner and N. Mielke, "A New Reliability Model for Post-Cycling Charge Retention of Flash Memories," in Dalla, TX, *International Reliability Physics Symposium (IRPS)*, 2002.

162 J.-D. Lee, J.-H. Choi, D. Park and K. Kim, "Degradation of Tunnel Oxide by FN Current Stress and its Effects on Data Retention Characteristics of 90 nm NAND Flash Memory Cells," in Dallas, TX, *International reliability Physics Symposium (IRPS)*, 2003.

163 J.-D. Lee, J.-H. Choi, D. Park and K. Kim, "Effects of Interface Trap Generation and Annihilation on the Data Retention Characteristics of Flash Memory Cells," *IEEE Transactions on Devices and Materials Reliability*, vol. 4, no. 1, pp. 110–117, 2004.

164 R. Yamada, Y. Mori, Y. Okuyama, J. Yugami, T. Nishimoto and H. Kume, "Analysis of Detrap Current Due to Oxide Traps to Improve Flash Memory Retention," in San Jose, CA, *International Reliability Physics Symposium (IRPS)*, 2000.

165 Y. Lee, N. Mielke, M. Agostinelli, S. Gupta, R. Lu and W. McMahon, "Prediction of Logic Product Failure Due To Thin-Gate Oxide Breakdown," in San Jose, CA, *International Reliability Physics Symposium (IRPS)*, 2006.

166 B. Deal, "The Current Understanding of Charges in the Thermally Oxidized Silicon Structure," *Journal of The Electrochemical Society*, vol. 121, no. 6, pp. 198–205, 1974.

167 A. Goetzberger, A. Lopez and R. Strain, "On the Formation of Surface States During Aging of Thermal Si-SiO$_2$ Interfaces," *Journal of Electrochemical Society*, vol. 120, no. 90, pp. 90–96, 1973.

168 K. Jeppson and C. Svensson, "Negative Bias Stress of MOS Devices at High Electric Fields and Degradation of MNOS Devices," *Journal of Applied Physics*, vol. 48, no. 5, pp. 2004–2014, 1977.

169 J. Stathis, S. Mahapatra and T. Grasser, "Controversial Issues in Negative Bias Temperature Instability," *Microelectronics Reliability*, vol. 81, pp. 244–251, 2018.

170 T. Grasser, B. Kaczer, W. Goes, H. Reisinger, T. Aichinger, P. Hehenberger, P.-J. Wagner, F. Schanovsky, J. Franco, M. Toledano Luque and M. Nelhiebel,

"The Paradigm Shift in Understanding the Bias Temperature Instability: from Reaction-Diffusion to Switching," *IEEE Transaction on Electron Devices*, vol. 58, no. 11, pp. 3652–3665, 2011.

171 T. Grasser, "Stochastic Charge Trapping in Oxides: From Random Telegraph Noise to Bias Temperature Instabilities," *Microelectronics Reliability*, vol. 52, pp. 39–70, 2012.

172 S. Chakravarthi, A. Krishman, V. Reddy, C. Machala and S. Krishman, "A Comprehensive Framework for Predictive Modeling of Negative Bias Temperature Instability," in Phoenix, AZ, *IEEE International Reliability Physics Symposium Proceedings*, 2004.

173 D. Patra, A. Reza, M. Katoozi, E. Cannon, K. Roy and Y. Cao, "Accelerated BTI Degradation Under Stochastic TDDB Effect," in Burlingame, CA, *IEEE International Reliability Physics Symposium (IRPS)*, 2018.

174 T. Asuke, R. Kishiday, J. Furuta and K. Kobayashi, "Temperature Dependence of Bias Temperature Instability (BTI) in Long-term Measurement by BTI-Sensitive and -Insensitive Ring Oscillators Removing Environmental Fluctuation," in Chongqing, China, *IEEE 13th International Conference on ASIC (ASICON)*, 2019.

175 N. Choudhury, U. Sharma, H. Zhou, R. Southwick, M. Wang and S. Mahapatra, "Analysis of BTI, SHE Induced BTI and HCD Under Full VG/VD Space in GAA Nano-Sheet N and P FETs," in Dallas, TX, *IEEE International Reliability Physics Symposium (IRPS)*, 2020.

176 A. Islam, H. Kufluoglu, D. Varghese, S. Mahapatra and M. Alam, "Recent Issues in Negative-Bias Temperature Instability: Initial Degradation, Field Dependence of Interface Trap Generation, Hole Trapping Effects, and Relaxation," *IEEE Transactions on Electron Devices*, vol. 54, no. 9, pp. 2143–2154, 2007.

177 E. Bender and J. Bernstein, "Microchip Health Monitoring System Using the FLL Circuit," *Sensors (Basel)*, vol. 21, no. 7, p. 2285, 2021.

178 E. Bender, J. Bernstein and D. Boning, "Mitigation of Thermal Stability Concerns in FinFET Devices," *Electronics*, vol. 11, no. 20, p. 3305, 2022.

179 E. Bender, J. Bernstein and D. Boning, "The Effects of Process Variations and BTI in Packaged FinFET Devices," in Monterey (CA), *IEEE International Reliability Physics Symposium (IRPS)*, 2023.

180 D. Young and A. Christou, "Failure Mechanism Models for Electromigration," *IEEE Transactions on Reliability*, vol. 43, no. 2, pp. 186–192, 1994.

181 C. Hau-Riege and C. Thompson, "Electromigration in Cu Interconnects with Very Different Grain Structures," *Applied Physics Letters*, vol. 78, no. 22, pp. 3451–3453, 2001.

182 A. Fischer, A. Abel, M. Lepper, A. Zitzelsbergr and A. von Glasgow, "Modeling Bimodal Electromigration Failure Distributions," *Microelectronics Reliability*, vol. 41, no. 3, pp. 445–453, 2001.

183 A. Budiman, C. Hau-Riege, W. Baek, C. Lor, A. Huang, H. Kim, G. Neubauer, J. Pak, P. Besser and W. Nix, "Electromigration-Induced Plastic Deformation in

Cu Interconnects: Effects on Current Density Exponent, and Implications for EM Reliability Assessment," *Journal of Electronic Materials*, vol. 39, no. 11, pp. 2483–2488, 2010.

184 DoD MIL Standard, *MIL-PRF-19500 Standard*, Columbus (HO): DoD, DSCC/VAC, 2010.

185 S. Yokogawa, N. Okada, Y. Kakuhara and H. Takizawa, "Electromigration Performance of Multilevel Damascence Copper Interconnects," *Microelectronics Reliability*, vol. 41, no. 9-10, pp. 1409–1416, 2001.

186 M. Gall, C. Capasso, D. Jawarani, R. Hernandez and H. Kawasaki, "Statistical Analysis of Early Failures in Electromigration," *Journal of Applied Physics*, vol. 90, no. 2, pp. 732–740, 2001.

187 A. Bensoussan, P. Coval, W. Roesch and T. Rubalcava, "Reliability of a GaAs MMIC Process Based on 0.5μm Au/Pd/Ti Gate MESFETs," in San Jose, CA, *32nd Annual Proceeding International Reliability Physics Symposium (IRPS)*, 1994.

188 P. Ladbrooke and M. M. I. C. Design, *GaAs FETs and HEMTs*, Boston and London: Artech House, 1989.

189 E. Suhir, "Statistics-Related and Reliability-Physics-Related Failure Processes in Electronics Devices and Products," *Modern Physics Letters B*, vol. 28, no. 13, p. 9 pages, 2014.

190 E. Suhir, *Applied Probability for Engineers and Scientists*, New York: McGraw-Hill, 1997.

191 M. Alam, "Reliability and Process-Variation Aware Design of Integrated Circuits," *Microelectronics Reliability*, vol. 48, no. 8-9, pp. 1114–1122, 2008.

192 B. Tudor, J. Wang, Z. Chen, R. Tan, W. Liu and F. Lee, "An Accurate MOSFET Aging Model for 28 nm Integrated Circuit Simulation," *Microelectronics Reliability*, vol. 52, no. 8, pp. 1565–1570, 2012.

193 C. Nunes, P. Butzen, A. Reis and R. Ribas, "BTI, HCI and TDDB Aging Impact in Flipflops," *Microelectronics Reliability*, vol. 53, no. 9-11, pp. 1355–1359, 2013.

194 V. Huard, F. Cacho, Y. Mamy Randriamihaja and A. Bravaix, "From Defects Creation to Circuit Reliability – A Bottom-up Approach," *Microelectronic Engineering*, vol. 88, no. 7, pp. 1396–1407, 2011.

195 J. B. Bernstein, M. Gabbay and O. Delly, "Reliability Matrix Solution to Multiple Mechanism Prediction," *Microelectronics Reliability Journal*, vol. 54, no. 12, pp. 2951–2955, 2014.

196 J. Bernstein, A. Bensoussan and E. Bender, "Reliability Prediction with MTOL," *Microelectronics Reliability*, vol. 68, pp. 91–97, 2017.

197 F. Cacho, W. Arfaoui, P. Mora, X. Federspiel, V. Huard and E. Dornel, "Modeling of Hot Carrier Injection Across Technology Scaling," in South Lake Tahoe, CA, *IEEE International Integrated Reliability Workshop Final Report (IIRW)*, 2014.

198 C. Moonju Cho, R. Ritzenthaler, R. Krom, Y. Higuchi, B. Kaczer, T. Chiarella, G. Boccardi, M. Togo, N. Horiguchi, T. Kauerauf, et al., "Negative Bias

Temperature Instability in p-FinFETs With 45° Substrate Rotation," *IEEE Electron Device Letters*, vol. 34, no. 10, pp. 1211–1213, 2013.

199 G. Delarozee, "Introduction to Reliability," *Microelectronic Engineering*, vol. 49, no. 1-2, pp. 3–10, 1999.

200 M. H. Hsieh, "The Impact and Implication of BTI/HCI Decoupling on Ring Oscillator," in Monterey, CA, *IEEE International Reliability Physics Symposium (IRPS)*, 2015.

201 V. Huard, M. Denais and C. Parthasarathy, "NBTI Degradation: From Physical Mechanisms to Modeling," *Microelectronics Reliability*, vol. 46, no. 1, pp. 1–23, 2006.

202 J. Bernstein, *Reliability Prediction from Burn-in Data Fit to Reliability Models*, Elsevier, Ed., London: Academic Press, 2014.

203 J. Bernstein, "Reliability Prediction for Aerospace Electronics," in Big Sky (Montana), *IEEE Aerospace Conference; paper number 2042*, 2013.

204 D. Varghese et al., "Hole Energy Dependent Interface Trap Generation in MOSFET Si/SiO$_2$ Interface," *IEEE Electron Device Letters*, vol. 26, no. 8, pp. 572–574, 2005.

205 L. Fang et al., "A Unified Aging Model of NBTI and HCI Degradation Towards Lifetime Reliability Management for Nanoscale MOSFET Circuits," in San Diego, CA, *IEEE/ACM International Symposium on Nanoscale Architectures*, pp. 175–180, 2011.

206 A. Bensoussan et al., "M-STORM: Multi-Physics Multi-Stressors Predictive Reliability Model Applied to DSM Technologies," in Paris, France, *Automotive Power electronics - SIA editor*, 2017.

207 C. Liu et al., "Experimental Study on BTI Variation Impacts in SRAM Based on High-k/Metal Gate FinFET: From Transistor Level Vth Mismatch, Cell Level SNM to Product Level Vmin," in Washington, DC, *IEEE International Electron Devices Meeting (IEDM), 2015*, 2015.

208 K. Lee, W. Wang, E.-A. Chung, G. Kim, H. Shim, H. Lee, H. Kim, M. Choe, N.-I. Lee, A. Patel, J. Park and J. Park, "Technology Scaling on High-K & Metal-Gate FinFET BTI Reliability," in Monterey, CA, *IEEE International Reliability Physics Symposium (IRPS)*, 2013.

209 C.-L. Lin, P.-H. Hsiao, W.-K. Yeh, H.-W. Liu, S.-R. Yang, Y.-T. Chen, K.-M. Chen and W.-S. Liao, "Effects of Fin Width on Device Performance and Reliability of Double-Gate n-Type FinFETs," *IEEE Transactions on Electron Devices*, vol. 60, no. 11, pp. 3639–3644, 2013.

210 S. E. Liu, J. S. Wang, Y. R. Lu, D. S. Huang, C. F. Huang, W. H. Hsieh, J. H. Lee, Y. S. Tsai, J. R. Shih, Y.-H. Lee and K. Wu, "Self-Heating Effect in FinFETs and Its Impact on Devices Reliability Characterization," in Waikoloa, HI, *IEEE International Reliability Physics Symposium*, 2014.

211 F. Stellari, K. A. Jenkins, A. J. Weger, B. Linder and P. Song, "Self-Heating Characterization of FinFET SOI Devices Using 2D Time Resolved Emission

Measurements," in Monterey, CA, *IEEE International Reliability Physics Symposium (IRPS)*, 2015.

212 N. Fuqua, *Reliability Engineering for Electronic Design*, Boca Raton, FL: CRC Press, 2020 (1987).

213 J. Qin, *A New Physics-of-Failure Based VLSI Circuit Reliability Simulation and Prediction Methodology (Thesis)*, Department of Mechanical Engineering, University of Maryland, 2007.

214 C. Liu, H.-C. Sagong, H. Kim, S. Choo, H. Lee, Y. Kim, H. Kim, B. Jo, M. Jin, J. Kim, S. Ha, S. Pae and J. Park, "Systematical Study of 14nm FinFET reliability: From Device Level Stress to Product HTOL," in Monterey, CA, *IEEE International Reliability Physics Symposium (IRPS)*, 2015.

215 E. Bender, J. Bernstein and A. Bensoussan, "Reliability Prediction of FinFET FPGAs by MTOL," *Microelectronics Reliability*, vol. 114, no. 10, 113809, 2020.

216 Xilinx, *Ultra96 Hardware User's Guide, Rev. 1, Version 0.9*, Avnet Inc., 2018. [Online]. Available: https://www.avnet.com/opasdata/d120001/medias/docus/187/Ultra96-HW-User-Guide-rev-1-0-V0_9_preliminary.pdf. [Accessed 2023]

217 Xilinx, *Xilinx Multi-Node Technology Leadership Continues with UltraScale+ Portfolio "3D on 3D" Solutions, Document WP472*, Avnet Inc., 2015. [Online]. Available: https://www.xilinx.com/products/silicon-devices/soc/zynq-ultrascale-mpsoc.html. [Accessed 2023]

218 C. Hu, S. Tam, F.-C. Hsu, P.-K. Ko, T.-Y. Chan and K. Terrill, "Hot-Electron-Induced MOSFET Degradation - Model, Monitor, and Improvement," *IEEE Transactions on Electron Devices*, vol. ED-32, no. 2, pp. 375–385, 1985.

219 A. Acovic, G. La Rosa and Y.-C. Sun, "A Review of Hot-Carrier Degradation Mechanisms in MOSFETs," *Microelectronics Reliability*, vol. 36, no. 7-8, pp. 845–869, 1996.

220 Environmental, Tenney, *TPS Junior Compact Temperature Test Chambers, Datasheet Thermal Product Solutions*, Tenney, 2022. [Online]. Available: https://www.tenney.com/sites/default/files/Tenney-Junior-Compact-Temperature_0.pdf. [Accessed 2023]

221 Xilinx, *Ultra96 USB-to-JTAG/UART Pod, v3*, Avnet Inc., 2018. [Online]. Available: https://www.avnet.com/opasdata/d120001/medias/docus/190/5362-PB-AES-ACC-U96-JTAG-V3b.pdf. [Accessed 2023]

222 Texas-Instruments, *TCA9548A Low-Voltage 8-Channel I2C Switch with Reset*, Dallas, TX: Texas Instrument, 2016.

223 M. Wang, Z. Liu, T. Yamashita, J. H. Stathis and C.-Y. Chen, "Separation of Interface States and Electron Trapping for Hot Carrier Degradation in Ultra-Scaled Replacement Metal Gate n-FinFET," in Monterey, CA, *IEEE International Reliability Physics Symposium (IRPS)*, 2015.

224 I. Messaris, N. Fasarakis, T. Karatsori, A. Tsormpatzoglou, G. Ghibaudo and C. A. Dimitriadis, "Hot Carrier Degradation Modeling of Short-Channel n-FinFETs," in

Columbus, OH, *73rd Annual Device Research Conference (DRC)*, pp. 183–184, 2015.

225 P. Magnone, F. Crupi, N. Wils, R. Jain, H. Tuinhout, P. Andricciola, G. Giusi and C. Fiegna, "Impact of Hot Carriers on nMOSFET Variability in 45- and 65-nm CMOS Technologies," *IEEE Transactions on Electron Devices*, vol. 58, no. 8, pp. 2347–2353, 2011.

226 X. Wang, P. P. Jain, D. Jiao and C. Kim, "Impact of Interconnect Length on BTI and HCI Induced Frequency Degradation," in Anaheim, CA, *IEEE International Reliability Physics Symposium (IRPS)*, 2012.

227 E. Bender and J. Bernstein, "Self-Heating Effects Measured in Fully Packaged FinFET Devices," in Niš, Serbia, *Proceeding of 2021 IEEE 32nd International Conference on Microelectronics (Miel)*, 2021.

228 Xilinx, *UltraScale Architecture System Monitor, UG580 (v1.10.1)*, Avnet Inc., 2021. [Online]. Available: https://docs.xilinx.com/v/u/en-US/ug580-ultrascale-sysmon. [Accessed 2023]

229 Rigol-Technologies, *DP800 Series Programmable Linear Power Supply - DSH04100-2021-06*, Rigol Technologies Inc., 2021. [Online]. Available: https://www.rigolna.com/products/dc-power-loads/dp800/. [Accessed 2023]

230 A. Djemouai, M. Sawan and M. Slamani, "New Frequency-Locked Loop Based on CMOS Frequency-to-Voltage Converter: Design and Implementation," *IEEE Transactions on Circuits and Systems—II: Analog and Digital Signal Processing*, vol. 48, no. 5, pp. 441–449, 2001.

231 H. Kufluoglu and M. Alam, "A Computational Model of NBTI and Hot Carrier Injection Time-Exponents for MOSFET Reliability," *Journal of Computational Electronics*, vol. 3, pp. 165–169, 2004.

232 Y. Huang, T. Yew, W. Wang, Y. Lee, J. Shih and K. Wu, "Re-investigating the Adequacy of Projecting Ring Oscillator Frequency Shift from Device Level Degradation," in Waikoloa, HI, *IEEE International Reliability Physics Symposium (IRPS)*, 2014.

233 Y. Huang, T. Yew, W. Wang, Y. Lee, J. Shih and K. Wu, "Re-investigation of Frequency Dependence of PBTI/TDDB and Its Impact on Fast Switching Logic Circuits," in Monterey, CA, *IEEE International Reliability Physics Symposium (IRPS)*, 2013.

234 K. Kim, W. Wang and K. Choi, "On-Chip Aging Sensor Circuits for Reliable Nanometer MOSFET Digital Circuits," *IEEE Transactions on Circuits and Systems II: Express Briefs*, vol. 57, no. 10, pp. 798–802, 2010.

235 E. Stott, Z. Guan, J. Levine, J. Wong and P. Cheung, "Variation and Reliability in FPGAs," *IEEE Design & Test*, vol. 30, no. 6, pp. 50–59, 2013.

236 E. Bender and J. Bernstein, "Product Failure Time Assessments Using Early Degradation Filtering," *Engineering Technololgy*, vol. 3, no. 5, 555625, 2021.

237 N. Drego, A. Chandrakasan, D. Boning and D. Shah, "Reduction of Variation-Induced Energy Overhead in Multi-Core Processors," *IEEE Transactions on*

Computer-Aided Design of Integrated Circuits and Systems, vol. 30, no. 6, pp. 891–904, 2011.

238 T. Efron and R. Tibshirani, *An Introduction to the Bootstrap*; 1st Edition, New York: Chapman and Hall/CRC, 1994.

239 W. Weibull, "A Statistical Distribution Function of Wide Applicability," *Journal of Applied Mechanics*, vol. 18, no. 3, pp. 293–297, 1951.

240 G. Klutke, P. Kiessler and M. Wortman, "A Critical Look at the Bathtub Curve," *IEEE Transactions on Reliability.*, vol. 52, no. 1, pp. 125–129, 2003.

241 R. Myers, K. Wong and H. Gordy, *Reliability Engineering for Electronic Systems*, New York, NY: John Wiley & Sons, 1964.

242 B. Stine, D. Boning and J. Chung, "Analysis and Decomposition of Spatial Variation in Integrated Circuit Processes and Devices," *IEEE Transactions on Semiconductor Manufacturing*, vol. 10, no. 1, pp. 24–41, 1997.

243 T. Siddiqua, S. Gurumurthi and M. Stan, "Modeling and Analyzing NBTI in the Presence of PV," in Santa Clara, CA, *12th International Symposium on Quality Electronic Design (ISQED)*, 2011.

244 B. Li, H. Masanori and S. Ulf, "From Process Variations to Reliability: A Survey of Timing of Digital Circuits in the Nanometer Era," *IPSJ Transactions on System LSI Design Methodology*, vol. 11, pp. 29–45, 2018.

245 S. Bhardwaj, W. Wang, R. Vattikonda, Y. Cao and S. Vrudhula, "Predictive Modeling of the NBTI Effect for Reliable Design," in San Jose, CA, *Proceedings of the 2006 IEEE Custom Integrated Circuits Conference*, 2006.

246 X. Yang, E. Weglarz and K. Saluja, "On NBTI Degradation Process in Digital Logic Circuits," in Bangalore, India, *VLSID '07: Proceedings of the 20th International Conference on VLSI Design held jointly with 6th International Conference: Embedded Systems*, 2007.

247 J. Suehle, "Ultra Thin Gate Oxide Reliability: Physical Models, Statistics, and Characterization," *IEEE Transactions on Electron Devices*, vol. 49, no. 6, pp. 958–971, 2002.

248 K. Cheung, "Soft Breakdown in Thin Gate Oxide - A Measurement Artifact," in Dallas, TX, *41st Annual International Reliability Physics Symposium (IRPS)*, 2003.

249 C. Henderson, *Course on Semiconductor Reliability: Time Dependent Dielectric Breakdown*, Semitracks Inc., 2002. [Online]. Available: https://www.semitracks.com/courses/reliability/semiconductor-reliability.php. [Accessed 2023]

250 N. Ravindra and J. Zhao, "Fowler-Nordheim Tunneling in Thin SiO_2 Films," *Smart Materials and Structures*, vol. 1, no. 3, pp. 197–201, 1992.

251 R. Degraeve, G. Groeseneken, R. Bellens, J. Ogier, M. Depas, P. Roussel and H. Maes, "New Insights in the Relation Between Electron Trap Generation and the Statistical Properties of Oxide Breakdown," *IEEE Transactions on Electron Devices*, vol. 45, no. 4, pp. 904–911, 1998.

252 H. Luo, F. En, X. Kong and X. Zhang, "The Different Gate Oxide Degradation Mechanism Under Constant Voltage/Current Stress and Ramp Voltage Stress," in

Lake Tahoe, CA, *IEEE International Integrated Reliability Workshop Final Report (Cat. No.00TH8515)*, pp. 141–143, 2000.

253 T. Pompl and M. Röhner, "Voltage Acceleration of Time-Dependent Breakdown of Ultra-Thin Gate Dielectric," *Microelectronics Reliability*, vol. 46, no. 2-3, pp. 1835–1841, 2005.

254 E. Wu, E. Nowak, A. Vayshenker, W. Lay and D. Harmon, "CMOS Scaling Beyond the 100-nm Node with Silicon-Dioxide Based Gate Dielectrics," *IBM Journal of Research and Development*, vol. 46, no. 2, pp. 287–298, 2002.

255 X. Li, J. Qin, B. Huang, X. Zhang and J. Bernstein, "A New SPICE Reliability Simulation Method for Deep Submicrometer CMOS VLSI Circuits," *Transactions on Device and Materials Reliability*, vol. 6, no. 2, pp. 247–257, 2006.

256 U. Costa, V. Freire, L. Malacarne, R. Mendes, S. Picoli Jr., E. de Vasconcelos and E. da Silva Jr., "An Improved Description of the Dielectric Breakdown in Oxides Based on a Generalized Weibull Distribution," *Physica A: Statistical Mechanics and its Applications*, vol. 361, no. 1, pp. 209–215, 2006.

257 A. Ghetti, "Gate Oxide Reliability: Physical and Computational Models," in Springer Series in Materials Science, vol 72, Berlin, Heidelberg, Dabrowski, J. and Weber, E. R. (eds), *Predictive Simulation of Semiconductor Processing* - Springer, p. 201+, 2004.

258 E. Wu and W. Abadeer, "Challenges for Accurate Reliability Projections in the Ultra Thin Oxide Regime," in San Diego, CA, *International Reliability Physics Symposium (IRPS)*, pp. 57–65, 1999.

259 G. Groeseneken, R. Degraeve, T. Nigam, G. Van den Bosch and H. Maes, "Hot Carrier Degradation and Time-Dependent Dielectric Breakdown in Oxides," *Microelectronic Engineering*, vol. 49, no. 1-2, pp. 27–40, 1999.

260 H. Maes, G. Groeseneken, R. Degraeve, J. Blauwe and G. den Bosch, "Assessment of Oxide Reliability and Hot Carrier Degradation in CMOS Technology," *Microelectronic Engineering*, vol. 40, no. 3-4, pp. 147–166, 1998.

261 H.-S. Kim, *Patent No. US 6873932B1 - Method and Apparatus for Predicting Semiconductor Device Lifetime*, United States Patent and Trademark Office, 2005. [Online]. Available: https://patentimages.storage.googleapis.com/7b/5f/b3/7a0df92897ecac/US6873932.pdf. [Accessed 2023]

262 J. Segura and F. C. Hawkins, *CMOS Electronics: How It Works, How It Fails*, Piscataway, NJ: Wiley IEEE Press, 2004.

263 D. Vasileska, "Semiconductor Device and Process Simulation," in Tempe, AZ, *Arizona State University Course EEE 533* - Spring, 2001.

264 K. Eriguchi and K. Ono, "Impacts of Plasma Process-Induced Damage on MOSFET Parameter Variability and Reliability," *Microelectronics Reliability (SI: Proceedings of ESREF 2015)*, vol. 55, no. 9-10, pp. 1269–2172, 2015.

265 M. Pagey, *Hot-Carrier Reliability Simulation in Aggressively Scaled MOS Transistors (Thesis)*, Nashville (Tennessee): Faculty of the Graduate School of Vanderbilt University, 2003.

266 A. Haggag and W. McMahon, "High Performance Chip Reliability from Short-Time-Tests," in Orlando, FL, *39th International Reliability Physics Symposium (IRPS)*, 2001.

267 J. Stathis and S. Zafar, "The Negative Bias Temperature Instability in MOS Devices: A Review," *Microelectronics Reliability*, vol. 46, no. 2-4, pp. 270–286, 2006.

268 S. Ogawa and N. Shiono, "Generalized Diffusion-Reaction Model for the Low-Field Charge-Buildup Instability at the Si- SiO_2 Interface," *Physical Review. B, Condensed matter*, vol. 51, no. 7, pp. 4218–4230, 1995.

269 S. Rashkeev, D. Fleetwood, R. Schrimpf and S. Pantelides, "Proton-Induced Defect Generation at the Si-SiO$_2$/Interface," *IEEE Transactions on Nuclear Science*, vol. 48, no. 6, pp. 2086–2092, 2001.

270 B. Deal Sklar, "Characteristics of the Surface-State Charge of Thermally Oxidized Silicon," *Journal of Electrochemical Society*, vol. 114, no. 3, pp. 266–274, 1967.

271 G. Chen, M. Li, C. Ang, J. Zheng and D. Kwong, "Dynamic NBTI of p-MOS Transistors and its Impact on MOSFET Scaling," *IEEE Electron Device Letters*, vol. 23, no. 12, pp. 734–736, 2002.

272 S. Zafar et al., "Evaluation of NBTI in HfO2 Gate-Dielectric Stacks with Tungsten Gates," *IEEE Electron Device Letters*, vol. 25, no. 3, pp. 153–155, 2004.

273 H. Aono et al., "Modeling of NBTI Degradation and its Impact on Electric Field Dependence of the Lifetime," in Phoenix, AZ, *International Reliability Physics symposium (IRPS)*, pp. 23–27, 2004.

274 G. Haller, M. Knoll, D. Braunig, F. Wulf and W. Fahrner, "Bias Temperature Stress on Metal-Oxide-Semiconductor Structures as Compared to Ionizing Irradiation and Tunnel Injection," *Journal of Applied Physics*, vol. 56, no. 6, pp. 1844–1850, 1984.

275 P. Chaparala, J. Shibley and P. Lim, "Threshold Voltage Drift in PMOSFETS Due to NBTI and HCI," in Lake Tahoe, CA, *IEEE International Integrated Reliability Workshop Final Report (Cat. No.00TH8515)*, pp. 95–97, 2000.

276 S. Mahapatra, P. Kumar and M. Alam, "Investigation and Modeling of Interface and Bulk Trap Generation During Negative Bias Temperature Instability," *IEEE Transaction on Electron Devices*, vol. 51, no. 9, pp. 1371–1379, 2004.

277 F. Jensen, *Electronic Component Reliability*, New York, NY: John Wiley & Sons, 1995.

278 M. Ohring, *Reliability and Failure of Electronic Materials and Devices*, Amsterdam, Boston: Academic Press, 2011.

279 E. Ogawa, A. Bierwag, K.-D. Lee, H. Matsuhashi, P. Justinson, et al., "Direct Observation of a Critical Length Effect in Dual-Damascene Cu/Oxide Interconnects," *Applied Physics Letters*, vol. 78, no. 18, pp. 2652–2645, 2001.

280 D. Ney, X. Federspiel, V. Girault, O. Thomas and P. Gergaud, "Stress-Induced Electromigration Backflow Effect in Copper Interconnects," *Transactions on Devices and Materials Reliability*, vol. 6, no. 2, pp. 175–180, 2006.

281 L. Doyen, E. Petitprez, P. Waltz, X. Federspiel, L. Arnaud and Y. Wouters, "Extensive Analysis of Resistance Evolution Due to Electromigration Induced Degradation," *Journal of Applied Physics*, vol. 104, no. 12, 123521, 2008.

282 T. May and M. Woods, "Alpha-Particle-Induced Soft Errors in Dynamic Memories," *IEEE Transactions on Electron Devices*, vol. 26, no. 1, pp. 2–9, 1979.

283 T. Yoo, S. H. Lee, K. L. Suk, E. K. Kim, W. K. Choi, D.-W. Kim and D. W. Kim, "Advanced Chip Last Process Integration for Fan Out WLP," in San Diego, CA, *IEEE 72nd Electronic Components and Technology Conference (ECTC)*, pp. 1371–1375, 2022.

284 JEDEC, *JESD94- Application Specific Qualification Using Knowledge Based Test Methodology*, Arlington, VA: JEDEC Solid State Technology Association, 2004.

285 JEDEC, *JESD47D, Stress-Test-Driven Qualification of Integrated Circuits*, Arlington, VA: JEDEC Solid State Technology Association, 2004.

286 JEDEC, *JEP148, Reliability Qualification of Semiconductor Devices Based on Physics of Failure Risk and Opportunity Assessment*, Arlington, VA: JEDEC Solid State Technology Association, 2004.

287 S. Ganesan, M. Pecht and S. Sharon Ling, "Use of High Temperature Operating Life Data to Mitigate Risks in Long Duration Space Applications," *Microelectronics Reliability*, vol. 46, no. 2-4, pp. 360–366, 2006.

288 F. McCluskey, E. Hakim, J. Fink, A. Fowler and M. Pecht, "Reliability Assessment of Electronic Components Exposed to Long-Term Non-Operating Conditions," *IEEE Trans. on Components, Packaging and Manufacturing Technology-Part A*, vol. 21, no. 2, pp. 352–359, 1998.

289 Intel, *25-GS3000, Intel Product Qualification Specification*," Intel® Centrino® Advanced-N 6235, Santa Clara, CA, 2000.

290 R. Blish and N. Durrant, *Semiconductor Device Reliability Failure Models, Technology Transfer #00053955AXFR*, Austin, TX: International SEMATECH, http://www.sematech.org, 2000.

291 M. Pecht and A. Dasgupta, "Physics of Failure: An Approach to Reliable Product Development," in Lake Tahoe, CA, *EEE 1995 International Integrated Reliability Workshop. Final Report*, pp. 1–4, 1995.

292 A. Cory, "Improved Reliability Prediction Through Reduced –Stress Temperature Cycling," in San Jose, CA, *IEEE 38th Annual International Reliability Physics Symposium*, pp. 231–236, 2000.

293 R. Van Gestel, K. De Zeeuw, L. Van Gemert and E. Bagerman, "Comparison of Delamination Effects Between Temperature Cycling Test and Highly Accelerated Stress Test in Plastic Packaged Devices," in San Diego, CA, *International Reliability Physics Symposium (IRPS)*, pp. 177–181, 1992.

294 D. Danielson, G. Marcyk, E. Babb and S. Kudva, "HAST Applications: Acceleration Factors and Results for VLSI Components," in Phoenix, AZ, *International Reliability Physics Symposium (IRPS)*, pp. 114–121, 1989.

295 J. Scalise, "Plastic Encapsulated Microcircuits (PEM) Qualification Testing," in Orlando, FL, *IEEE ECTC Proceedings 46th Electronic Components and Technology Conference*, pp. 392–397, 1996.

296 T. Parker and C. Webb, "A Study of Failures Identified During Board Level Environmental Stress Testing," *IEEE Transactions on Components, Hybrids, and Manufacturing Technology*, vol. 15, no. 6, pp. 1086–1092, 1992.

297 J. Carulli and T. Anderson, "The Impact of Multiple Failure Modes on Estimating Product Field Reliability," *IEEE Design and Test of Computers*, vol. 23, no. 2, pp. 118–126, 2006.

298 R. Merrett, J. Bryant and R. Studd, "An Appraisal of High Temperature Humidity Stress Tests for Assessing Plastic Encapsulated Semiconductor Components," in Phoenix, AZ, *21st International Reliability Physics Symposium, (IRPS)*, pp. 73–82, 1983.

299 C. Shirley and R. Blish, "Thin-Film Cracking and Wire Ball Shear in Plastic Dips Due to Temperature Cycle and Thermal Shock," in San Diego, CA, *25th International Reliability Physics Symposium*, pp. 238–249, 1987.

300 R. Blish and P. Vaney, "Failure Rate Model for Thin Film Cracking in Plastic ICs," in Las Vegas, NV, *29th Annual Proceedings International Reliability Physics Symposium (IRPS)*, pp. 22–29, 1991.

301 R. Zelenka, "A reliability model for interlayer dielectric cracking during temperature cycling," in Las Vegas, NV, *29th Annual Proceedings International Reliability Physics Symposium (IRPS)*, 1991.

302 S. Omi, K. Fujita, T. Tsuda and T. Maeda, "Causes of Cracks in SMD and Type Specific Remedies," *IEEE Transactions on Components, Hybrids, and Manufacturing Technology*, vol. 14, no. 4, pp. 818–823, 1991.

303 C. Hong, "Thin Film Cracking/Delamination Evaluation Using Assembly Test Chip," in Lake Tahoe, CA, *International Report on Wafer Level Reliability Workshop*, pp. 163–166, 1992.

304 K. X. Hu, C.-P. Yeh, B. Doot, A. Skipor and K. Wyatt, "Die Cracking in Flip-Chip-on-Board Assembly," in Las Vegas, NV, *Proceedings. 45th Electronic Components and Technology Conference*, pp. 293–299, 1995.

305 W. Wu, M. Held, P. Jacob, P. Scacco and A. Birolini, "Thermal stress related packaging failure in power IGBT modules," in Yokohama, Japan, *Proceedings of International Symposium on Power Semiconductor Devices and IC's: ISPSD '95*, pp. 330–334, 1995.

306 K.-Y. Chou, M.-J. Chen, C.-C. Lin, Y.-S. Su, C.-S. Hou and T.-C. Ong, "Die Cracking Evaluation and Improvement in ULSI Plastic Package," in Kobe, Japan, *ICMTS 2001. Proceedings of the 2001 International Conference on Microelectronic Test Structures, (Cat. No.01CH37153)*, pp. 239–244, 2001.

307 H. Nguyen, C. Salm, J. Vroemen, J. Voets, B. Krabbenborg, et al., "Test Chip for Detecting Thin Film Cracking Induced by Fast Temperature Cycling and

Electromigration in Multilevel Interconnect Systems," in Singapore, *Proceedings of 9th IPFA*, pp. 135–139, 2002.

308 M.-Y. Tsai, C. Hsu and C. Wang, "Investigation of Thermomechanical Behaviors of Flip Chip BGA Packages During Manufacturing Process and Thermal Cycling," *IEEE Transactions on Components and Packaging Technologies*, vol. 27, no. 3, pp. 568–576, 2004.

309 L. Annaniah, M. Devarajan and T. K. San, "An Investigation on Die Crack Detection Using Temperature Sensitive Parameter for High Speed LED Mass Production," *Results in Physics*, vol. 7, pp. 3882–3891, 2017.

310 K. Van Doorselaer and K. De Zeeuw, "Relation Between Delamination and Temperature Cycling Induced Failures in Plastic Packaged Devices," *IEEE Transactions on Components, Hybrids and Manufacturing Technology*, vol. 13, no. 4, pp. 879–882, 1990.

311 J. Emerson, J. Sweet and D. Peterson, "Evaluating Plastic Assembly Processes for High Reliability Applications Using HAST and Assembly Test Chips," in San Jose, CA, *International Reliability Physics symposium (IRPS)*, pp. 191–195, 1994.

312 M. Pecht, L. Nguyen and E. Hakim, *Plastic Encapsulated Microelectronics*, New York, NY: John Wiley & Sons Inc., 1995.

313 M. Amagi, "Polyimide Fatigue Induced Chip Surface Damage in DRAM's Lead-On-Chip (LOC) Packages," in Las Vegas, NV, *IEEE International Reliability Physics Symposium (IRPS)*, pp. 97–106, 1995.

314 N. Tanaka, M. Kitano, T. Kumazawa and A. Nishimura, "Evaluating IC-package Interface Delamination by Considering Moisture-Induced Molding-Compound Swelling," *IEEE Transactions on Components and Packaging Technologies*, vol. 22, no. 3, pp. 426–432, 1999.

315 T. Aihara, S. Ito, H. Sasajima and K. Oota, "Development of Reliability and Moldability on Fine Pitch Ball Grid Array by Optimizing Materials," *Journal of Electronic Packaging*, vol. 123, no. 1, p. 88, 2001.

316 I. Harvey, D. Turner, J. Ortowski and C. Herbert, "Optimization Case Study of CSP Temperature Cycle and Board Bending Reliability, RE1-3," in San Diego, CA, *IPC SMEMA Council Electronics Assembly Process Exhibition and Conference (APEX)*, 2001.

317 C. Chung, J. Fun, M. Huang and F. Tsai, "Study on Failure Mechanism of PCT Reliability for BT Substrate Based CSP (Chip Scale Packages)," *Transactions of the American Society of Mechanical Engineers (ASME)*, vol. 124, no. 4, pp. 334–339, 2002.

318 T. Lin, Z. Xiong, Y. Yao, L. Tok, Z. Yu, B. Njoman, K. Chua and Y. Ma, "Failure Analysis of Full Delamination on the Stacked Die Leaded Packages," *Journal of Electronic Packaging*, vol. 125, no. 3, pp. 392–399, 2003.

319 T. Saitoh, H. Matsuyama and M. Toya, "Delamination and Encapsulant Resin Cracking in LSI Plastic Packages Subjected to Temperature Cyclic Loading," *Journal of Electronic Packaging*, vol. 125, no. 3, pp. 420–425, 2003.

320 Z. Wei, L. H. Yam and L. Cheng, "Detection of Internal Delamination in Multi-Layer Composites Using Wavelet Packets Combined With Modal Parameter Analysis," *Composite Structures*, vol. 64, no. 3-4, pp. 377–387, 2004.

321 W.-S. Kwon, H.-J. Kim, K. Paik, S.-Y. Jang and S.-M. Hong, "Mechanical Reliability and Bump Degradation of ACF Flip Chip Packages Using BCB Bumping Dielectrics Under Temperature Cycling," Transactions of the ASME," *Journal of Electronic Packaging*, vol. 126, no. 2, pp. 202–207, 2004.

322 P.-H. Tsao, C. Huang, A. Lin, M.-J. Lii, D. Perng and N.-S. Tsai, "Cavity-Down Thermal-Enhanced Package Reliability Evaluation for Low-k Dielectric/Cu Interconnects," in Las Vegas, NV, *2004 Proceedings. 54th Electronic Components and Technology Conference (IEEE Cat. No.04CH37546)*, pp. 1191–1193, 2004.

323 W.-S. Kwon, M.-J. Yim, K.-W. Paik, S.-J. Ham and S.-B. Lee, "Thermal Cycling Reliability and Delamination of Anisotropic Conductive Adhesives Flip Chip on Organic Substrates With Emphasis on the Thermal Deformation," *Journal of Electronic Packaging*, vol. 127, no. 2, pp. 86–90, 2005.

324 T. Braun, K. Becker, M. Koch, V. Bader, R. Aschenbrenner and H. Reichl, "High-Temperature Reliability of Flip Chip Assemblies," *Microelectronics Reliability*, vol. 46, no. 1, pp. 144–154, 2006.

325 E. Prack and X. Fan, "Root Cause Mechanism for Delamination/Cracking in Stacked Die Chip Scale Packages," in Tokyo, Japan, *IEEE International Symposium on Semiconductor Manufacturing*, pp. 219–222, 2006.

326 H. Liu, X. Pang and S. Xu, "Failure Mechanism Study for Low-k Device Bond Pad Crack Post Temperature Cycle," in Singapore, *21st Electronics Packaging Technology Conference (EPTC)*, pp. 208–212, 2019.

327 B. Boettge, F. Naumann, S. Behrendt, M. Scheibel, S. Kaessner, S. Klengel, M. Petzold, K. Nickel, G. Hejtmann, A.-Z. Miric and R. Eisele, "Material Characterization of Advanced Cement-Based Encapsulation Systems for Efficient Power Electronics With Increased Power Density," in San Diego, CA, *IEEE 68th Electronic Components and Technology Conference*, pp. 1258–1269, 2018.

328 K. Kho, G. You, G. Tan, C.-T. Hsu and H. Guan, "Detecting Wire Bond Inter Layer Dielectric Crack by Dark Field Imaging," in Singapore, *IEEE 23rd Electronics Packaging Technology Conference (EPTC)*, pp. 104–107, 2021.

329 M. Amagai, H. Seno and K. Ebe, "Cracking Failures in Lead-on-Chip Packages Induced by Chip Backside Contamination," *IEEE Transactions on Components, Packaging and Manufacturing Technology - Part B*, vol. 18, no. 1, pp. 119–126, 1995.

330 R. Dias, A. Lucero, S. Niemeyer and S. Myers, "Failure Mechanisms in C4 Organic Packages," in Santa Clara, CA, *Intel Q&R Conference*, pp. 57–68, 1997.

331 E.-C. Ahn, T.-J. Cho, J.-B. Shim, H.-J. Moon, et al., "Reliability of Flip Chip BGA Package on Organic Substrate," in Las Vegas, NV, *50th Electronic Components and Technology Conference (Cat. No.00CH37070)*, pp. 1215–1220, 2000.

332 Y. Lin, W. Liu, Y. Guo and F. Shi, "Reliability Issues of Low-Cost Overmolded Flip Chip Packages," *IEEE Transaction On Advanced Packaging*, vol. 28, no. 1, pp. 79–88, 2005.

333 L. Mercado, H. Wieser and T. Hauck, "Multichip Package Delamination and Die Fracture Analysis," *IEEE Transactions on Advanced Packaging*, vol. 26, no. 2, pp. 152–159, 2003.

334 P. Tu, Y. Chan and J. Lai, "Effect of Intermetallic Compounds on the Thermal Fatigue of Surface Mount Solder Joints," *IEEE Transactions on Components, Packaging, and Manufacturing Technology - Part B*, vol. 20, no. 1, pp. 87–93, 1997.

335 R. Ghaffarian, "Accelerated Thermal Cycling and Failure Mechanisms for BGA and CSP Assemblies," *Transactions Of the ASME Journal of Electronic Packaging*, vol. 122, no. 4, pp. 335–340, 2002.

336 R. Pucha, K. Tunga, J. Pyland and S. Sitaraman, "Accelerated Thermal Cycling Guidelines for Electronic Packages in Military Avionics Thermal Environment," *Transactions of the ASME Journal of Electronic Packaging*, vol. 126, no. 2, pp. 256–264, 2004.

337 J. Suhling, H. Gale, R. Johnson, et al., "Thermal Cycling Reliability of Lead-Free Solders for Automotive Applications," in Las Vegas, NV, *The 9th Intersociety Conference on Thermal and Thermomechanical Phenomena In Electronic Systems (IEEE Cat. No.04CH37543)*, pp. 350–357, 2004.

338 G. Wang, S. Groothuis and P. Ho, "Packaging Effect on Reliability for Cu/Low-k Structure," in Phoenix, AZ, *IEEE 42nd Annual International Reliability Physics Symposium*, pp. 557–562, 2004.

339 J. Lau and W. Dauksher, "Reliability of a 1657 CCGA Package with $^{95.5}Sn^{3.9}Ag^{0.6}Cu$ Lead-free Solder Paste on PCBs," *Transactions of the ASME Journal of Electronic Packaging*, vol. 127, no. 2, pp. 96–105, 2005.

340 C. Birzer, S. Stoeckl, G. Schuetz and M. Fink, "Reliability Investigations of Leadless QFN Packages until End-of-life with Application-Specific Board-Level Stress Tests," in San Diego, CA, *IEEE ECTC 6th Electronic Components and Technology Conference*, pp. 594–600, 2006.

341 J. Davis, M. Bozack and J. Evans, "Effect of (Au, Ni)Sn_4 Evolution on Sn-^{37}Pb/ENIG Solder Joint Reliability Under Isothermal and Temperature-Cycled Conditions," *IEEE Transactions on Components and Packaging Technologies*, vol. 30, no. 1, pp. 32–41, 2007.

342 C.-C. Lee, T. Tran, Y. Yuan, et al., "Challenges in Temperature Cycling Test for Electronic Packages Containing Low-k/Cu Silicon," in Sparks, NV, *Proceedings 57th Electronic Components and Technology Conference*, pp. 1186–1192, 2007.

343 R. Ghaffarian, "Reliability of Package on Package (PoP) Assembly Under Thermal Cycles," in Las Vegas, NV, *IEEE Intersociety Conference on Thermal and Thermomechanical Phenomena in Electronic Systems (ITherm)*, pp. 472–476, 2019.

344 J. Uebbing, "Mechanisms of Temperature Cycle Failure In Encapsulated Optoelectronic Devices," in Las Vegas, NV, *19th International Reliability Physics Symposium, (IRPS)*, pp. 149–156, 1981.

345 T. Hund and P. Plunkett, "Improving Thermosonic Gold Ball Bond Reliability," *IEEE Transactions on Components, Hybrids, and Manufacturing Technology*, vol. CHMT-8, no. 4, pp. 446–456, 1985.

346 J. Park, B.-S. Kim, H.-J. Cha, Y.-B. Jo, S.-C. Shin, et al., "Interfacial Degradation Mechanism of Au-Al Bonding in Quad Flat Package," in Phoenix, AZ, *IEEE 42nd Annual International Reliability Physics Symposium*, pp. 569–570, 2004.

347 J. Xu, Y. Guo, Y. Su, R. Tang and X. Long, "Growth Kinetics of Intermetallic Compound in Solder Joints During Thermal Cycling: aA Review," in Singapore, *IEEE 23rd Electronics Packaging Technology Conference (EPTC)*, pp. 464–468, 2021.

348 W. Tang, X. Long and F. Yang, "Tensile Deformation and Mirostructures of Sn–$^{3.0}$Ag–$^{0.5}$Cu Solder Joints: Effect of Annealing Temperature," *Microelectronics Reliability*, vol. 104, 113555, 2020.

349 K. Striny and A. Schelling, "Reliability Evaluation of Aluminum-Metallized MOS Dynamic RAM's in Plastic Packages in High Humidity and Temperature Environments," *IEEE Transactions on Components, Hybrids, and Manufacturing Technology*, vol. CHMT-4, no. 4, pp. 476–481, 1981.

350 J. Emerson, D. Peterson and J. Sweet, "HAST Evaluation of Organic Liquid IC Encapsulant Using Sandia's Assembly Test Chips," in San Diego, CA, *International Reliability Physics Symposium (IRPS)*, pp. 951–956, 1992.

351 T. Tran, L. Yong, B. Williams, S. Chen and A. Chen, "Fine Pitch Probing and Wirebonding and Reliability of Aluminum Capped Copper Bond Pads," in Las Vegas, NV, *50th Electronic Components and Technology Conference (Cat. No.00CH37070)*, pp. 1674–1680, 2000.

352 S. Wagner, K. Hoeppner, M. Toepper, O. Wittler and K. Lang, "A Critical Review of Corrosion Phenomena in Microelectronic Systems," in Nuremberg, Germany, *PCIM Europe; International Conference for Power Electronics, Intelligent Motion, Renewable Energy and Energy Management*, pp. 1–7, 2014.

353 J. Wu, C. Lee, P. Zheng, et al., "Electromigration Reliability of SnAgxCuy Flip Chip Interconnects," in Las Vegas, NV, *54th Electronic Components and Technology Conference (IEEE Cat. No.04CH37546)*, pp. 961–967, 2004.

354 H. Balkan, "Flip Chip Electromigration: Impact of Test Conditions in Product Life Predictions," in Las Vegas, NV, *54th Electronic Components and Technology Conference (IEEE Cat. No.04CH37546)*, pp. 983–987, 2004.

355 T. Shao, I. H. Chen and C. Chen, "Electromigration Failure Mechanism of Sn$_{96.5}$Ag$_{3.5}$ Flip Chip Solder Bumps," in Las Vegas, NV, *Proceeding of 54th Electronic Components and Technology Conference (IEEE Cat. No.04CH37546)*, pp. 979–982, 2004.

356 C. Basaran, H. Ye, D. Hopkins, D. Frear and J. Lin, "Failure Modes of Flip Chip Solder Joints Under High Electric Current Density," *Transaction of the ASME Journal of Electronic Packaging*, vol. 127, no. 2, pp. 157–163, 2005.

357 M. Ding, G. Wang, B. Chao, P. Ho, P. Su, T. Uehling and D. Wontor, "A Study of Electromigration Failure in Pb-free Solder Joints," in San Jose, CA, *43th International Reliability Physics Symposium (IRPS)*, pp. 518–523, 2005.

358 M. Tajedini et al., "Electromigration Effect on the Pd Coated Cu Wirebond," in San Diego, CA, *IEEE 71st Electronic Components and Technology Conference (ECTC)*, 2021.

359 R. Ghaffarian, "Impact of CSP Assembly Underfill on Reliability," in San Diego, CA, *IPC SMEMA Council Electronics Assembly Process Exhibition and Conference (APEX), Session P-AD2/3*, pp. 1–7, 2001.

360 M. Khan, H. Fatemi, J. Romero and E. Delenia, "Effect of High Thermal Stability Mold Material on the Gold-Aluminum Bond Reliability in Epoxy Encapsulated VLSI Devices," in Anaheim, CA, *International Reliability Physics Symposium (IRPS)*, pp. 40–49, 1998.

361 P. Syndergaard and J. Young, "A Model of Temperature Cycling Performance in Plastic Encapsulated Packages," in Lake Tahoe, CA, *IEEE International Integrated Reliability Workshop (IRWS), CA Final Report*, pp. 37–41, 1994.

362 R. Raghunathan and S. Sitaraman, "Qualification Guidelines for Automotive Packaging Devices," in Las Vegas, NV, *THERM'2000. The Seventh Intersociety Conference on Thermal and Thermomechanical Phenomena in Electronic Systems (Cat. No.00CH37069)*, pp. 385–389, 2000.

363 L. Li, J. Xie, M. Ahmad and M. Brillhart, "Environmental Effects on Dielectric Films in Plastic Encapsulated Silicon Devices," in Sparks, NV, *IEEE ECTC, 57th Electronic Components and Technology Conference, 2007*, pp. 755–760, 2007.

364 R. Blish II, "Temperature Cycling and Thermal Shock Failure Rate Modeling," in Denver, CO, *35th International Reliability Physics Symposium (IRPS)*, pp. 110–115, 1997.

365 P. C. Paris, M. P. Gomez and W. E. Anderson, "A Rational Analytical Theory of Fatigue," in vol 13, Seattle, WA, *University of Washington - The Trend in Engineering*, pp. 9–14, 1961.

366 C. Shirley and C. Hong, "Optimal Acceleration of Cyclic THB Tests for Plastic-Packaged Devices," in Las Vegas, NV, *29th International Reliability Physics Symposium (IRPS)*, pp. 12–21, 1991.

367 W.-S. Lei and A. Kumar, "Delamination and Reliability Issues in Packaged Devices, Chapter 7," in Sunnyvale, CA, *Adhesion in Microelectronics - Applied Materials, Inc.*, pp. 267–312, 2014.

368 T. Ferguson and J. Qu, "Effect of Moisture on the Interfacial Adhesion of the Underfill/Soldermask Interface," *ASME Journal of Electronic Packaging*, vol. 124, no. 2, pp. 106–110, 2002.

369 K. Chen, D. Jiang, N. Kao and J. Lai, "Effects of Underfill Materials on the Reliability of Low-K Flip Chip Packaging," *Microelectronics Reliability*, vol. 46, no. 1, pp. 155–163, 2006.

370 G. Wang, P. Ho and S. Groothuis, "Chip-Packaging Interaction: A Critical Concern for Cu/Low k Packaging," *Microelectronics Reliability*, vol. 45, no. 7-8, pp. 1079–1093, 2004.

371 C. Zhai, U. Ozkan, A. Dubey, R. B. Sidharth II and R. Master, "Investigation of Cu/Low-k Film Delamination in Flip Chip Packages," in San Diego, CA, *IEEE 72nd Electronic Components and Technology Conference (ECTC)*, pp. 709–717, 2006.

372 S. Ray, S. Kiyono, K. Waite and L. Nicholls, "Qualification of Low-k 90 nm Technology Die with Pb-Free Bumps on a Build-up Laminate Package (PBGA) with Pb-Free Assembly Processes," in San Diego, CA, *72nd Electronic Components and Technology Conference (ECTC)*, pp. 139–144, 2006.

373 Y. Lin, C. Peng and K. Chiang, "Parametric Design and Reliability Analysis of Wire Interconnect Technology Wafer Level Packaging," *ASME Journal of Electronic Packaging*, vol. 124, no. 3, pp. 234–239, 2002.

374 X. Liu, V. Sooklal, M. Verges and M. Larson, "Experimental Study and Life Prediction on High Cycle Vibration Fatigue in BGA Packages," *Microelectronics Reliability*, vol. 46, pp. 1128–1138, 2006.

375 D. Noctor, F. Bader, A. Viera, P. Boysan, S. Golwalkar and R. Foehringer, "Attachment Reliability Evaluation and Failure Analysis of Thin Small Outline Packages (TSOP's) with Alloy 42 Lead Frames," *IEEE Transactions on Components, Hybrids and Manufacturing Technology*, vol. 16, no. 8, pp. 961–971, 1993.

376 Y. Jeon, Y. Lee and Y. Choi, "Thin Electroless Cu/OSP on Electroless Ni as a Novel Surface Finish for Flip Chip Solder Joints," in San Diego, CA, *IEEE 72nd Electronic Components and Technology Conference (ECTC)*, pp. 119–124, 2006.

377 M. Amagai, "The Effect of Polymer Die Attach Material on Solder Joint Reliability," in Paris, France, *Workshop on Mechanical Reliability of Polymer Materials and Plastic Packages of IC Devices, EEP-Vol 25, ASME*, pp. 223–230, 1998.

378 H.-J. Kim, J. Lee, K.-W. Paik, et al., "Effects of Cu/Al Intermetallic Compound (IMC) on Copper Wire and Aluminum Pad Bondability," *IEEE Transactions on Components and Packaging Technologies*, vol. 26, no. 2, pp. 367–374, 2003.

379 T. Uno and K. Tatsumi, "Thermal Reliability of Gold-Aluminum Bonds Encapsulated in Bi-Phenyl Epoxy Resin," *Microelectronics Reliability*, vol. 40, no. 1, pp. 145–153, 2002.

380 T. Heleine, R. Robert, M. Murcko and S.-C. Wang, "A Wire Bond Reliability Model," in Las Vegas, NV, *29th Annual Proceedings International Reliability Physics Symposium (IRPS)*, pp. 378–381, 1991.

381 K. Norris and A. Landzberg, "Reliability of Controlled Collapse Interconnections," *IBM Journal R & D*, vol. 13, no. 3, pp. 266–271, 1969.

382 J. Lau and Y.-H. Pao, *Solder Joint Reliability of BGA, CSP, Flip Chip and Fine Pitch SMT Assemblies*, New York, NY: McGraw Hill, 1997.

383 B. Chen, X. Shi, G. Li, K. Ang and J. Pickering, "Rapid Temperature Cycling Methodology for Reliability Assessment of Solder Interconnection in Tape Ball Grid Array Assembly," *Transactions of the ASME Journal of Electronic Packaging*, vol. 127, no. 4, pp. 466–473, 2005.

384 N. Strifas, C. Vaughan and M. Ruzzene, "Accelerated Reliability-Thermal and Mechanical Fatigue Solder Joints Methodologies," in Dallas (TX), *IEEE 40th Annual International Reliability Physics Symposium*, pp. 144–147, 2002.

385 J. Wu, P. Zheng, C. Lee, S. Hung and J. Lee, "A Study in Flip Chip UBM/Bump Reliability With Effects of SnPb Solder Composition," *Microelectronics Reliability*, vol. 46, no. 1, pp. 41–52, 2006.

386 W. Yiping, Z. Jinsong, W. Fengshun, A. Bing, W. Boyi and W. Lei, "Effects of UBM Thickness on Electromigration in Pb-free Solder Joints," in Las Vegas, NV, *Proceeding of 54th Electronic Components and Technology Conference (IEEE Cat. No.04CH37546)*, 2004.

387 W. McGarvey, "Autoclave vs. 85°C/85% R. H. Testing - A Comparison," in San Diego, CA, *17th International Reliability Physics Symposium (IRPS)*, pp. 136–142, 1979.

388 G. Di Giacomo, *Reliability of Electronic Packages and Semiconductor Devices*, New York, NY: McGraw Hill, The University of Maryland, 1997.

389 J. E. T. Osenbach, N. Chand, R. Comizzoli and H. Krautter, "Temperature-Humidity-Bias Behavior and Acceleration Factors for Nonhermetic Uncooled InP-Based Lasers," *Journal of Light Wave Technology*, vol. 15, no. 5, pp. 861–873, 1997.

390 D. Ryu and S. Chang, "Novel Concepts for Reliability Technology," *Microelectronics Reliability*, vol. 45, no. 3-4, pp. 611–622, 2006.

391 R. Howard, "Packaging Reliability-How to Define and Measure It," *IEEE Transaction On Components, Hybrids, and Manufacturing Technology*, vol. CHMT-5, no. 4, pp. 454–462, 1982.

392 O. Hallberg and D. Peck, "Recent Humidity Accelerations, A Base for Testing Standards," *Quality and Reliability Engineering International*, vol. 7, no. 3, pp. 169–180, 1991.

393 Intel, *25-GS0003, Procedure for Failure Mechanism Based Certifications of New Package Technologies*, Santa Clara, CA: Intel® Centrino®, 2000.

394 C. Dunn and J. McPherson, "Temperature Cycling Acceleration Factors for Aluminum Metallization Failure in VLSI Applications," in New Orleans, LA, *28th International Reliability Physics Symposium (IRPS)*, 1990.

395 M. Cooper, "Investigation of Arrhenius Acceleration Factor for Integrated Circuit Early Life Failure Region with Several Failure Mechanisms," *IEEE Transactions on Components and Packaging Technologies*, vol. 28, no. 3, pp. 561–563, 2004.

396 J. Jones and J. Hayes, "Estimation of System Reliability Using a "Non-Constant Failure Rate" Model," *IEEE Transactions on Reliability*, vol. 50, no. 3, pp. 286–288, 2001.

397 E. Moura, "A Method to Estimate the Acceleration Factor for Subassemblies," *IEEE Transactions on Reliability*, vol. 41, no. 3, pp. 396–399, 1992.

398 G. Cassanelli, G. Mura, F. Cesaretti, M. Vanzi and F. Fantini, "Reliability Predictions in Electronic Industrial Applications," *Microelectronics Reliability*, vol. 45, no. 9-11, pp. 1321–1326, 2005.

399 S. Moreau, T. Lequeu and R. Jerisian, "Comparative Study of Thermal Cycling and Thermal Shock Tests on Electronic Component Reliability," *Microelectronics Reliability*, vol. 44, no. 9-11, pp. 1343–1347, 2004.

400 J.-H. Koh and T.-G. Kim, "Reliability of Pb (Mg, Nb) O_3-Pb (Zr, Ti) O_3 Multilayer Ceramic Piezoelectric Actuators by Weibull Method," *Microelectronics Reliability*, vol. 46, no. 1, pp. 183–188, 2006.

401 L. Yang, J. Bernstein and J. Walls, "Physics based Predictive Reliability Study for Electronic Packages," in San Jose, CA, *40th IMAPS Symposium*, 2007.

402 R. Kotlowicz, "Comparative Compliance of Representative Lead Designs for Surface Mounted Components," *Transactions on Components, Hybrids, and Manufacturing Technology*, vol. 12, no. 4, pp. 431–448, 1989.

403 C. Li et al., "Damage Integral Methodology for Thermal and Mechanical Fatigue of Solder Joints," in New York, NY, Lau, J. H. (ed.), *Solder Joint Reliability, Theory and Applications* - Springer Science + Business Media, LLC, pp. 261–288, 1991.

404 P. Hall, "Creep and Stress Relaxation in Solder Joints," in New York, NY, Lau, J. H. (ed.), *Solder Joint Reliability, Theory and Applications* - Springer Science + Business Media, LLC, pp. 306–332, 1991.

405 T. Ju and Y. Lee, "Effects of Ceramic Ball Grid Array Package's Manufacturing Variations on Solder Joint Reliability," in vol 2256, Denver, CO, *Proceedings of the International Conference on Multichip Modules*, pp. 514–519, 1994.

406 T. Scharr, "Fatigue Properties of Base Copper Materials Used in the Fabrication of TAB Leads," in Atlanta, GA, *41st Electronic Components & Technology Conference*, pp. 848–853, 1991.

407 K. Dittmer, M. Poech, F. Wulff and M. Krumm, "Failure Analysis of Aluminum Wire Bonds in High Power Igbt Modules," in vol 390, *MRS Online Proceedings Library (OPL): Symposium S – Electronic Packaging Materials Science VIII* - Cambridge University Press, p. 251, 1995.

408 S. Peddada and I. R. Blish, "Modeling of Down Bond Failure Rates in 196 Lead MM Packages," in Santa Clara, CA, *Intel Q&R Conference*, pp. 61–71, 1993.

409 J. Hagge, "Mechanical Considerations for Reliable Interfaces in Next Generation Electronics Packaging," in Dayton, OH, *IEEE National Aerospace and Electronics Conference - NEPCON*, pp. 2021–2026, 1989.

410 Y. Chen and N. Mencinger, *"Intel Courses: Reliability Statistics,"* Semitracks Inc., 2000. [Online]. Available: https://www.semitracks.com/courses/reliability/ semiconductor-reliability.php. [Accessed 2023].

411 P. Lall, "Tutorial: Temperature as an Input to Microelectronics-Reliability Models," *IEEE Transactions On Reliability*, vol. 45, no. 1, pp. 3–9, 1996.

412 M. Cooper, "Investigation of Arrhenius Acceleration Factor for Integrated Circuit Early Life Failure Region with Several Failure Mechanisms," *IEEE Transactions on Components and Packaging Technologies*, vol. 28, no. 3, pp. 561–563, 2005.

Index

Reliability Prediction for Microelectronics, First Edition. Joseph B. Bernstein, Alain A. Bensoussan, and Emmanuel Bender.
© 2024 John Wiley & Sons Ltd. Published 2024 by John Wiley & Sons Ltd.